An Episodic History of Mathematics

Mathematical Culture Through Problem Solving

Library of Congress Control Number: 2010921168

Print ISBN: 978-0-88385-766-3

Electronic ISBN: 978-1-61444-605-7

Printed in the United States of America

Current Printing (last digit):
10 9 8 7 6 5 4 3 2

An Episodic History of Mathematics

Mathematical Culture Through Problem Solving

Steven G. Krantz
Washington University, St. Louis

Published and distributed by
The Mathematical Association of America

MAA TEXTBOOKS

Calculus Deconstructed: A Second Course in First-Year Calculus, Zbigniew H. Nitecki
Combinatorics: A Guided Tour, David R. Mazur
Combinatorics: A Problem Oriented Approach, Daniel A. Marcus
Complex Numbers and Geometry, Liang-shin Hahn
A Course in Mathematical Modeling, Douglas Mooney and Randall Swift
Cryptological Mathematics, Robert Edward Lewand
Differential Geometry and its Applications, John Oprea
Elementary Cryptanalysis, Abraham Sinkov
Elementary Mathematical Models, Dan Kalman
An Episodic History of Mathematics: Mathematical Culture Through Problem Solving, Steven G. Krantz
Essentials of Mathematics, Margie Hale
Field Theory and its Classical Problems, Charles Hadlock
Fourier Series, Rajendra Bhatia
Game Theory and Strategy, Philip D. Straffin
Geometry Revisited, H. S. M. Coxeter and S. L. Greitzer
Graph Theory: A Problem Oriented Approach, Daniel Marcus
Knot Theory, Charles Livingston
Lie Groups: A Problem-Oriented Introduction via Matrix Groups, Harriet Pollatsek
Mathematical Connections: A Companion for Teachers and Others, Al Cuoco
Mathematical Interest Theory, Second Edition, Leslie Jane Federer Vaaler and James W. Daniel
Mathematical Modeling in the Environment, Charles Hadlock
Mathematics for Business Decisions Part 1: Probability and Simulation (electronic textbook), Richard B. Thompson and Christopher G. Lamoureux
Mathematics for Business Decisions Part 2: Calculus and Optimization (electronic textbook), Richard B. Thompson and Christopher G. Lamoureux
The Mathematics of Choice, Ivan Niven
The Mathematics of Games and Gambling, Edward Packel
Math Through the Ages, William Berlinghoff and Fernando Gouvea
Noncommutative Rings, I. N. Herstein
Non-Euclidean Geometry, H. S. M. Coxeter
Number Theory Through Inquiry, David C. Marshall, Edward Odell, and Michael Starbird
A Primer of Real Functions, Ralph P. Boas
A Radical Approach to Lebesgue's Theory of Integration, David M. Bressoud
A Radical Approach to Real Analysis, 2nd edition, David M. Bressoud
Real Infinite Series, Daniel D. Bonar and Michael Khoury, Jr.
Topology Now!, Robert Messer and Philip Straffin
Understanding our Quantitative World, Janet Andersen and Todd Swanson

MAA Service Center
P.O. Box 91112
Washington, DC 20090-1112
1-800-331-1MAA FAX: 1-301-206-9789

To Marvin J. Greenberg,
an inspiring teacher.

Contents

Preface

Together with philosophy, mathematics is the oldest academic discipline known to mankind. Today mathematics is a huge and complex enterprise, far beyond the ken of any individual. Those of us who elect to study the subject can only choose a piece of it, and in the end must specialize drastically in order to make any contribution to the evolution of ideas.

An important development of twenty-first century life is that mathematical and analytical thinking have permeated all aspects of our world. We all need to understand the spread of diseases, the likelihood that we will contract SARS or hepatitis. We all must deal with financial matters. Finally, we all must deal with computers and databases and the Internet. Mathematics is an integral part of the theory and the operating systems that make all these computer systems work. Theoretical mathematics is used to design automobile bodies, to plan reconstructive surgery procedures, and to analyze prison riots. The modern citizen who is unaware of mathematical thought is lacking a large part of the equipment of life.

Thus it is worthwhile to have a book that will introduce the student to some of the genesis of mathematical ideas. While we cannot get into the nuts and bolts of Andrew Wiles's solution of Fermat's Last Theorem, we can describe some of the stream of thought that created the problem and led to its solution. While we cannot describe all the sophisticated mathematics that goes into the theory behind black holes and modern cosmology, we can indicate some of Bernhard Riemann's ideas about the geometry of space. While we cannot describe in specific detail the mathematical research that professors at the University of Paris are performing today, we can indicate the development of ideas that has led to that work.

Mathematical history is exciting and rewarding, and it is a significant slice of the intellectual pie. A good education consists of learning different methods of discourse, and certainly mathematics is one of the most well-developed and important modes of discourse that we have.

The purpose of this book, then, is to acquaint the student with mathematical language and mathematical life by means of a number of historically important mathematical vignettes. And the book will also serve to help the prospective teacher to become inured in some of the important ideas of mathematics—both classical and modern.

The focus in this text is on *doing*—getting involved with the mathematics and solving problems. This book is unabashedly mathematical: The history is primarily a device for feeding the reader some doses of mathematical meat. In the course of reading this book, the neophyte will become *involved* with mathematics by working on the same problems that Zeno and Pythagoras and Descartes and Fermat and Riemann worked on. This is a book to be read with pencil and paper in hand, and a calculator or computer close by. The student will want to experiment, to try things, to become a part of the mathematical process.

This history is also an opportunity to have some fun. Most of the mathematicians treated here were complex individuals who led colorful lives. They are interesting to us as people as well as scientists. There are wonderful stories and anecdotes to relate about Pythagoras and Galois and Cantor and Poincaré, and we do not hesitate to indulge ourselves in a little whimsy and gossip. This device helps to bring them to life, and thus to stimulate reader interest.

It should be clearly understood that this is in no sense a thoroughgoing history of mathematics, in the sense of the wonderful treatises of Boyer/Merzbach [BOM] or Katz [KAT] or Smith [SMI]. It is instead a collection of snapshots of aspects of the world of mathematics, together with some cultural information to put the mathematics into perspective. The reader will pick up history on the fly, while actually *doing mathematics*—developing mathematical ideas, working out problems, formulating questions.

And we are not shy about the things we ask the reader to do. This book will be accessible to students with a wide variety of backgrounds and interests and goals. Future mathematicians, future teachers, and future thinkers in all walks of life will benefit from studying this text. The book will give the student some exposure to calculus, number theory, mathematical induction, cardinal numbers, cartesian geometry, transcendental numbers, complex numbers, Riemannian geometry, homotopy theory, mechanics, cryptography, and several other exciting parts of the mathematical enterprise. Because it is our intention to introduce the student to what mathematicians think and what mathematicians value, we actually *prove* a number of important facts: **(i)** the existence of irrational numbers, **(ii)** the existence of transcendental numbers, **(iii)** Fermat's little theorem, **(iv)** the completeness of the real number system, **(v)** the fundamental theorem of algebra, and **(vi)** Dirichlet's theorem. The reader of this text will come away with a hands-on feeling for what mathematics is about and what mathematicians do.

Certainly one of the several audiences that I have in mind for this book is the undergraduate math major engaged in a capstone experience for his/her education. It is typical that such a student will look back on his course experience as a disjointed agglomeration of ideas that do not speak to each other very well. The purpose of the capstone is to draw things together in a natural way and to show the student our subject as an organic whole. This text is intended to be a catalyst to this process. It shows mathematics as a living, growing organism. A typical topic in this book draws on many different types of mathematics, and many different mathematical methods. Every chapter has exercises marked **Project**, and these are open-ended, thought-provoking problems that could lead to a class discussion or a term paper or even an honors project. Each of the history sections has references for further reading on the lives of the great mathematicians. Finally, virtually every section of the book ends with a little material called *Further Reading*; these squibs contain references to *The American Mathematical Monthly*, *The College Mathematics Journal*, and other periodicals of this type in which the student can do independent reading on the topic at hand and extend his/her horizons in a scholarly fashion.

This book is intended to be pithy and brisk. Chapters are short, and it will be easy for the student to browse around the book and select topics of interest to dip into. Each chapter will have an exercise set, and the text itself will be peppered with items labeled "For You to

Try". This device gives the student the opportunity to test his/her understanding of a new idea at the moment of impact. It will be both rewarding and reassuring. And it should keep interest piqued.

Richard Bonacci first enlisted me to write this book, and I thank him for his ideas and his encouragement. Klaus Peters provided many helpful comments and criticisms. Don Albers signed the book to the MAA, and helped me bring the project to fruition. Certainly the reviewers that he engaged in the writing process provided copious and detailed advice that have turned this into a more accurate and useful teaching tool. Zaven Karian served as a special editor for this project, and helped me to craft it into a useful tool. I am grateful to all.

The instructor teaching from this book will find grist for a number of interesting mathematical projects. Term papers, and even honors projects, will be a natural outgrowth of this text. The book can be used for a course in mathematical culture (for non-majors), for a course in the history of mathematics, for a course of mathematics for teacher preparation, as a guidebook to a student engaged in a capstone project, or for a course in problem-solving. We hope that it will help to bridge the huge and demoralizing gap between the technical world and the humanistic world. For certainly the most important thing that we do in our society is to communicate. My wish is to communicate mathematics.

SGK
St. Louis, MO

1

The Ancient Greeks and the Foundations of Mathematics

1.1 Pythagoras

1.1.1 Introduction to Pythagorean Ideas

Pythagoras (569–500 B.C.E.) was both a person and the namesake for a society (i.e., the *Pythagoreans*). He was also a political figure and a mystic. He was special in his time because, among other features, he involved women as equals in his activities. One critic characterized the man as "one tenth of him genius, nine-tenths sheer fudge." Pythagoras died, according to legend, in the flames of his own school—fired by political and religious bigots who stirred up the masses to protest against the enlightenment which Pythagoras sought to bring them. Classical references on the life and work of Pythagoras include [LAE] and [POR]. Modern sources are [DEV] and [OME].

As with many figures from ancient times, there is little specific that we know about Pythagoras's life. We know a bit about his ideas and his school, and we sketch some of these here.

The Pythagorean society was intensely mathematical in nature, but it was also quasi-religious. Among its tenets (according to [RUS]) were:

- To abstain from beans.
- Not to pick up what has fallen.
- Not to touch a white cock.
- Not to break bread.
- Not to step over a crossbar.
- Not to stir the fire with iron.
- Not to eat from a whole loaf.
- Not to pluck a garland.
- Not to sit on a quart measure.
- Not to eat the heart.
- Not to walk on highways.
- Not to let swallows share one's roof.

- When the pot is taken off the fire, not to leave the mark of it in the ashes, but to stir them together.
- Not to look in a mirror beside a light.
- When you rise from the bedclothes, roll them together and smooth out the impress of the body.

The Pythagoreans embodied a passionate spirit that is remarkable to our eyes:

> Bless us, divine Number, thou who generatest gods and men.

and

> Number rules the universe.

The Pythagoreans are remembered for two monumental contributions to mathematics. The first of these was to establish the importance of, and the necessity for, *proofs* in mathematics: that mathematical statements, especially geometric statements, must be established by way of rigorous proof. Prior to Pythagoras, the ideas of geometry were generally rules of thumb that were derived empirically, merely from observation and (occasionally) measurement. Pythagoras also introduced the idea that a great body of mathematics (such as geometry) could be derived from a small number of postulates.[1] The second great contribution was the discovery of, and proof of, the fact that not all numbers are commensurate. More precisely, the Greeks prior to Pythagoras believed with a profound and deeply held passion that everything was built on the whole numbers. Fractions arise in a concrete manner: as ratios of the sides of triangles with sides of whole-number length (and are thus *commensurable*—this antiquated terminology has today been replaced by the word "rational")—see Figure 1.1.

Pythagoras proved the result that we now call *the Pythagorean theorem*, but we do not know what technique he used to prove this seminal result. We shall discuss a variety of approaches in what follows. The result says that the legs a, b and hypotenuse c of a right triangle (Figure 1.2) are related by the formula

$$a^2 + b^2 = c^2 . \qquad (\star)$$

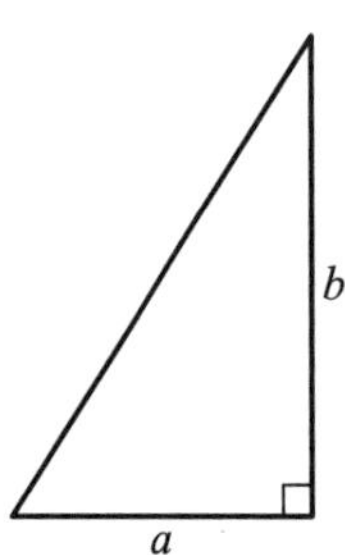

Figure 1.1. A geometric representation of the fraction $\frac{b}{a}$.

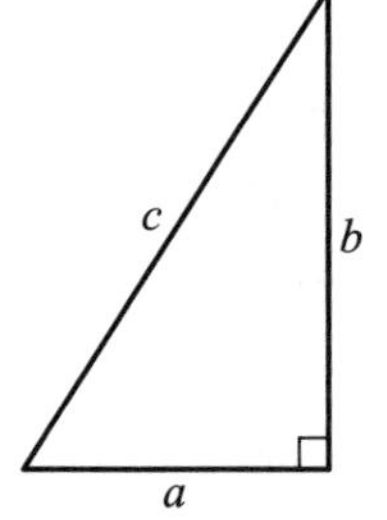

Figure 1.2. The right triangle in the Pythagorean theorem.

This theorem has perhaps more proofs than any other result in mathematics—over fifty altogether. And in fact it is one of the most ancient mathematical results. There is evidence that the Babylonians and the Chinese knew this theorem before Pythagoras.

[1]This notion was brought to full flower by Euclid, as we discuss below.

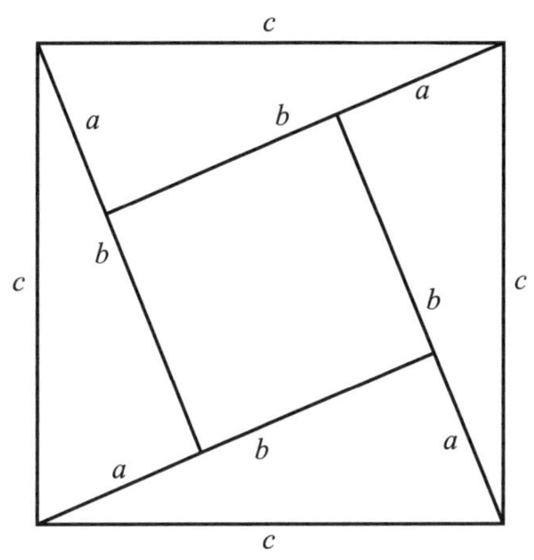

Figure 1.3. A visual demonstration of the Pythagorean Theorem

Quite surprisingly, one modern proof of the Pythagorean theorem was devised by President James Garfield (who almost chose a career of teaching mathematics rather than going into politics). We now provide one of the simplest and most classical arguments. Refer to Figure 1.3.

Proof of the Pythagorean Theorem. Observe that we have four right triangles and a square packed into a larger square. Each triangle has legs a and b, and we take it that $b > a$. Of course, on the one hand, the area of the larger square is c^2. On the other hand, the area of the larger square is the sum of the areas of its component pieces.

Thus we calculate that

$$\begin{aligned} c^2 &= (\text{area of large square}) \\ &= (\text{area of triangle}) + (\text{area of triangle}) \\ &\quad + (\text{area of triangle}) + (\text{area of triangle}) + (\text{area of small square}) \\ &= \frac{1}{2} \cdot ab + \frac{1}{2} \cdot ab + \frac{1}{2} \cdot ab + \frac{1}{2} \cdot ab + (b-a)^2 \\ &= 2ab + [a^2 - 2ab + b^2] \\ &= a^2 + b^2. \end{aligned}$$

This calculation proves the Pythagorean theorem. □

For You to Try If $c = 10$ and $a = 6$ then can you determine what b must be in the Pythagorean theorem?

Other proofs of the Pythagorean theorem will be explored in the exercises, as well as later on in the text.

Now Pythagoras noticed that, if $a = 1$ and $b = 1$, then $c^2 = 2$. He wondered whether there was a rational number c that satisfied this last identity. His stunning conclusion was this:

Theorem. *There is no rational number c such that $c^2 = 2$.*

Proof. Suppose that the conclusion is false. Then there *is* a rational number $c = \alpha/\beta$, expressed in lowest terms (i.e., α and β are integers that have no integer factors in common) such that $c^2 = 2$. This translates to

$$\frac{\alpha^2}{\beta^2} = 2 \qquad \text{or} \qquad \alpha^2 = 2\beta^2.$$

We conclude that the right-hand side is even, hence so is the left-hand side. Therefore $\alpha = 2m$ for some integer m.

But then

$$(2m)^2 = 2\beta^2 \qquad \text{or} \qquad 2m^2 = \beta^2.$$

So we see that the left-hand side is even, so β is even.

But now both α and β are even—the two numbers have a common factor of 2. This statement contradicts the hypothesis that α and β have no common integer factors. Thus it cannot be that c is a rational number. Instead, c must be irrational. □

For You to Try Use the argument just presented to show that 7 does not have a rational square root.

For You to Try Why does the argument just presented break down when 2 is replaced by 4 (of course 4 *does* have a rational square root)?

The result of the last theorem had a profound effect on the thinking of the times (about 500 B.C.E.). For it established irrefutably that there were new numbers besides the rationals to which everyone had been wedded. And these numbers were inescapable: they arose in such simple contexts as the calculation of the diagonal of a square. Because of this result of Pythagoras, the entire Greek approach to the number concept had to be rethought. In fact it can be argued that the Greek approach to mathematics became, as a consequence, based in geometry rather than the theory of numbers

1.1.2 Pythagorean Triples

It is natural to ask which triples of integers (a, b, c) satisfy $a^2 + b^2 = c^2$. Such a trio of numbers is called a *Pythagorean triple*.

The most famous and standard Pythagorean triple is $(3, 4, 5)$. But there are many others, including $(5, 12, 13)$, $(7, 24, 25)$, $(20, 21, 29)$, and $(8, 15, 17)$. What would be a complete list of all Pythagorean triples? Are there only finitely many of them, or is there an infinite list?

It has been known since the time of Euclid that there are infinitely many Pythagorean triples, and there is a formula that generates all of them.[2] We may derive it as follows. First, we may as well suppose that a and b are relatively prime—they have no factors in common. We call this a *reduced triple*. Therefore a and b are not both even, so one of them is odd. Say that b is odd.

Now certainly $(a + b)^2 = a^2 + b^2 + 2ab > a^2 + b^2 = c^2$. From this we conclude that $c < a + b$. So let us write $c = (a + b) - \gamma$ for some positive integer γ. Plugging this

[2] It may be noted, however, that the ancients did not have adequate notation to write down formulas as such.

expression into the Pythagorean formula ($\star$) yields

$$a^2 + b^2 = (a + b - \gamma)^2$$

or

$$a^2 + b^2 = a^2 + b^2 + \gamma^2 + 2ab - 2a\gamma - 2b\gamma\,.$$

Cancelling, we find that

$$\gamma^2 = 2a\gamma + 2b\gamma - 2ab\,. \tag{$\dagger$}$$

The right-hand side is even (every term has a factor of 2), so we conclude that γ is even. Let us write $\gamma = 2m$, for m a positive integer.

Substituting this last expression into ($\dagger$) yields

$$4m^2 = 4am + 4bm - 2ab$$

or

$$ab = 2am + 2bm - 2m^2\,.$$

The right-hand side is even, so we conclude that ab is even. Since we have already noted that b is odd, we can only conclude that a is even. Now equation ($\star$) tells us

$$c^2 = a^2 + b^2\,.$$

Since the sum of an odd number and an even number is an odd number, we see that c^2 is odd. Hence c is odd.

Thus the numbers in a reduced Pythagorean triple are never all even and never all odd. In fact two of them are odd and one is even. It is convenient to write $b = s-t$ and $c = s+t$ for some integers s and t (one of them even and one of them odd). Then ($\star$) tells us that

$$a^2 + (s-t)^2 = (s+t)^2\,.$$

Multiplying things out gives

$$a^2 + (s^2 - 2st + t^2) = (s^2 + 2st + t^2)\,.$$

Cancelling like terms and regrouping gives

$$a^2 = 4st\,.$$

We already know that a is even, so this is expected.

Since st must be a perfect square (because 4 is a perfect square and a^2 is a perfect square), it is now useful to write $s = u^2$, $t = v^2$. Therefore

$$a^2 = 4u^2v^2$$

and hence

$$a = 2uv\,.$$

In conclusion, we have learned that a reduced Pythagorean triple must take the form

$$(2uv, u^2 - v^2, u^2 + v^2)\,, \tag{$\ddagger$}$$

with $u > v$. Conversely, any triple of the form $(2uv, u^2 - v^2, u^2 + v^2)$ is most certainly a Pythagorean triple. This may be verified directly:

$$\begin{aligned}[2uv]^2 + [u^2 - v^2]^2 &= [4u^2v^2] + [u^4 - 2u^2v^2 + v^4] \\ &= u^4 + 2u^2v^2 + v^4 \\ &= [u^2 + v^2]^2 .\end{aligned}$$

Take a moment to think about what we have discovered. Every Pythagorean triple must have the form (‡). That is to say, $a = 2uv$, $b = u^2 - v^2$, and $c = u^2 + v^2$. Here u and v are any integers of our choosing.

As examples:

- If we take $u = 2$ and $v = 1$ then we obtain $a = 2 \cdot 2 \cdot 1 = 4$, $b = 2^2 - 1^2 = 3$, and $c = 2^2 + 1^2 = 5$. Of course (4, 3, 5) is a familiar Pythagorean triple. We certainly know that $4^2 + 3^2 = 5^2$.
- If we take $u = 3$ and $v = 2$ then we obtain $a = 2 \cdot 3 \cdot 2 = 12$, $b = 3^2 - 2^2 = 5$, and $c = 3^2 + 2^2 = 13$. Indeed $(12, 5, 13)$ is a Pythagorean triple. We may calculate that $12^2 + 5^2 = 13^2$.
- If we take $u = 5$ and $v = 3$ then we obtain $a = 2 \cdot 5 \cdot 3 = 30$, $b = 5^2 - 3^2 = 16$, and $c = 5^2 + 3^2 = 34$. You may check that (30, 16, 34) is a Pythagorean triple, for $30^2 + 16^2 = 34^2$.

For You to Try Find all Pythagorean triples in which one of the terms is 5.

For You to Try Find all Pythagorean triples in which all three terms are less than 30.

Further Reading

David E. Kullman, What's harmonic about the harmonic series?, *The College Mathematics Journal* 32(2001), 201–203.

Jean-P. Quadrat, Jean B. Lasserre, and Jean-B. Hiriart-Urruty, Pythagoras' theorem for areas, *The American Mathematical Monthly* 108(2001), 549–551.

1.2 Euclid

1.2.1 Introduction to Euclid

Certainly one of the towering figures in the mathematics of the ancient world was Euclid of Alexandria (325 B.C.E.–265 B.C.E.). Although Euclid is not known so much (as were Archimedes and Pythagoras) for his original and profound insights,[3] he has had an incisive effect on human thought. After all, Euclid wrote a treatise (consisting of thirteen Books)—now known as Euclid's *Elements*—which has been continuously in print for over 2000 years and has been through myriad editions. It is still studied in detail today, and continues

[3]To be fair, the Euclidean algorithm—which is fundamental to number theory—is named after Euclid. We shall say more about this idea later in the book. And the original proof that there are infinitely many primes, also discussed later, is attributed to Euclid.

to have a substantial influence over the way that we think about mathematics. Good sources for background on Euclid are [DUA], [FRA].

Not a great deal is known about Euclid's life, although it is fairly certain that he had a school in Alexandria. In fact "Euclid" was quite a common name in his day, and various accounts of Euclid the mathematician's life confuse him with other Euclids (one a prominent philosopher). One appreciation of Euclid comes from Proclus (411 C.E.–485 C.E.), one of the last of the ancient Greek philosophers:

> Not much younger than these [pupils of Plato] is Euclid, who put together the *Elements*, arranging in order many of Eudoxus's theorems, perfecting many of Theaetus's, and also bringing to irrefutable demonstration the things which had been only loosely proved by his predecessors. This man lived in the time of the first Ptolemy; for Archimedes, who followed closely upon the first Ptolemy makes mention of Euclid, and further they say that Ptolemy once asked him if there were a shortened way to study geometry than the *Elements*, to which he replied that "there is no royal road to geometry." He is therefore younger than Plato's circle, but older than Eratosthenes and Archimedes; for these were contemporaries, as Eratosthenes somewhere says. In his aim he was a Platonist, being in sympathy with this philosophy, whence he made the end of the whole *Elements* the construction of the so-called Platonic figures.

As often happens with scientists and artists and scholars of immense accomplishment, there is disagreement, and some debate, over exactly who or what Euclid actually was. The three schools of thought are these:

- Euclid was an historical character—a single individual—who in fact wrote the *Elements* and the other scholarly works that are commonly attributed to him.
- Euclid was the leader of a team of mathematicians working in Alexandria. They all contributed to the creation of the complete works that we now attribute to Euclid. They even continued to write and disseminate books under Euclid's name after his death.
- Euclid was not an historical character at all. In fact "Euclid" was a *nom de plume*—an allonym if you will—adopted by a group of mathematicians working in Alexandria. They took their inspiration from Euclid of Megara (who *was* in fact an historical figure), a prominent philosopher who lived about 100 years before Euclid the mathematician is thought to have lived.

Most scholars today subscribe to the first theory—that Euclid was certainly a unique person who created the *Elements*. But we acknowledge that there is at least anecdotal evidence for the other two scenarios. Certainly Euclid had a vigorous school of mathematics in Alexandria, and there is little doubt that his students participated in his projects.

It is thought that Euclid must have studied in Plato's Academy in Athens, for it is unlikely that there would have been another place where he could have learned the geometry of Eudoxus and Theaetus on which the *Elements* are based.

Another famous story and quotation about Euclid is this. A certain pupil of Euclid, at his school in Alexandria, came to Euclid after learning just the first proposition in the geometry of the *Elements*. He wanted to know what he would gain by putting in all this study, doing all the necessary work, and learning the theorems of geometry. At this, Euclid called over his slave and said, "Give him threepence since he must needs make gain by what he learns."

What is important about Euclid's *Elements* is the paradigm it provides for the way that mathematics should be studied and recorded. Euclid begins with several definitions of terminology and ideas for geometry, and then he records five important postulates (or axioms) of geometry. A version of these postulates is as follows:

P1 Through any pair of distinct points there passes a line.

P2 For each segment $\overline{AB}$ and each segment $\overline{CD}$ there is a unique point E on $\overleftrightarrow{AB}$ such that B is between A and E and $\overline{CD} \cong \overline{BE}$ (Figure 1.4(a)).

P3 For each point C and each point A distinct from C there exists a circle with center C and radius CA (Figure 1.4(b)).

P4 All right angles are congruent.

These are the standard four axioms which give our Euclidean conception of geometry. The fifth axiom, a topic of intense study for two thousand years, is the so-called parallel postulate (in Playfair's formulation):

P5 For each line ℓ and each point P that does not lie on ℓ there is a unique line m through P such that m is parallel to ℓ (Figure 1.4(c)).

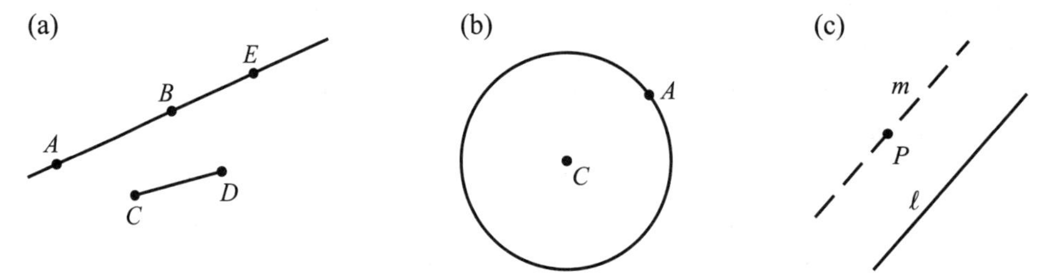

Figure 1.4. Euclid's second axiom.

Of course, prior to this enunciation of his celebrated five axioms, Euclid had defined point, line, "between", circle, and the other terms that he uses. Although Euclid borrowed freely from mathematicians both earlier and contemporaneous with himself, it is generally believed that the famous "Parallel Postulate", that is Postulate **P5**, is of Euclid's own creation.

It should be stressed that the *Elements* are not simply about geometry. In fact Books VII–IX deal with number theory. It is here that Euclid proves his famous result that there are infinitely many primes (treated elsewhere in this book) and also his celebrated "Euclidean algorithm" for long division. Book X deals with irrational numbers, and books XI–XIII treat three-dimensional geometry. In short, Euclid's *Elements* are an exhaustive treatment of virtually all the mathematics that was known at the time. And it is presented in a strictly rigorous and axiomatic manner that has set the tone for the way that mathematics is presented and studied today. Euclid's *Elements* is perhaps most notable for the clarity with which theorems are formulated and proved. The standard of rigor that Euclid set was to be a model for the inventors of calculus nearly 2000 years later.

Noted algebraist B. L. van der Waerden (1903–1996) assesses the impact of Euclid's *Elements* in this way:

> Almost from the time of its writing and lasting almost to the present, the *Elements* has exerted a continuous and major influence on human affairs. It was the primary source of geometric reasoning, theorems, and methods at least until the advent of non-Euclidean geometry in the 19th century. It is sometimes said that, next to the Bible, the *Elements* may be the most translated, published, and studied of all the books produced in the Western world.

Indeed, there have been more than 1000 editions of Euclid's *Elements*. It is arguable that Euclid was and still is the most important and most influential mathematics teacher of all time. It may be added that a number of other books by Euclid survive until now. These include *Data* (which studies geometric properties of figures), *On Divisions* (which studies the division of geometric regions into subregions having areas of a given ratio), *Optics* (which is the first Greek work on perspective), and *Phaenomena* (which is an elementary introduction to mathematical astronomy). Several other books of Euclid—including *Surface Loci*, *Porisms*, *Conics*, *Book of Fallacies*, and *Elements of Music*—have all been lost.

1.2.2 The Ideas of Euclid

Now that we have set the stage for who Euclid was and what he accomplished, we give an indication of the kind of mathematics for which he is remembered. We discuss the infinitude of primes and the Euclidean algorithm elsewhere in the book (Chapters 11, 20). Here we concentrate on Euclidean geometry.

We shall begin by stating some simple results from planar geometry and proving them in the style of Euclid. For the student with little background in proofs, this will open up a whole world of rigorous reasoning and geometrical analysis. Let us stress that, in the present text, we are only scratching the surface.

In the ensuing discussion we shall use the fundamental notion of *congruence*. In particular, two triangles are congruent if their corresponding sides and angles are equal. See Figure 1.5. There are a variety of ways to check that two triangles are congruent:[4]

- If the two sets of sides may be put in one-to-one correspondence so that corresponding pairs are equal, then the two triangles are congruent. We call this device "side-side-side" or SSS. See Figure 1.6.

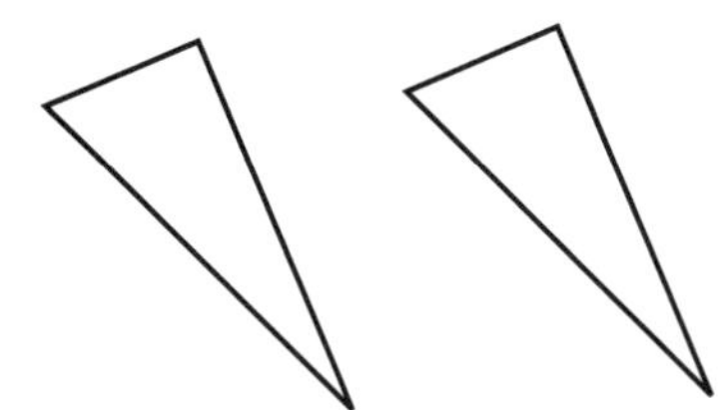

Figure 1.5. Two congruent triangles.

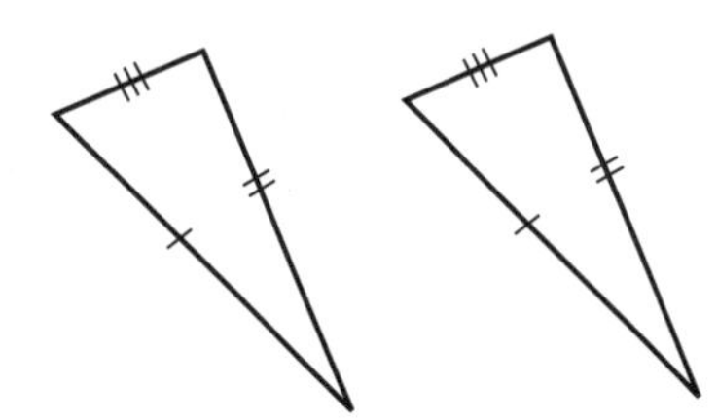

Figure 1.6. The side-side-side axiom.

[4] In this discussion we use corresponding markings to indicate sides or angles that are equal. Thus if two sides are each marked with a single hash mark, then they are equal in length. If two angles are marked with double hash marks, then they are equal in size.

- If just one side and its two adjacent angles correspond in each of the two triangles, so that the two pairs of angles are equal and each of the corresponding sides is equal, then the two triangles are congruent. We call this device "angle-side-angle" or ASA. See Figure 1.7.
- If two sides and the included angle correspond in each of the two triangles, so that the two pairs of sides are equal, and the included angles are equal, then the two triangles are congruent. We call this device "side-angle-side" or SAS. See Figure 1.8.

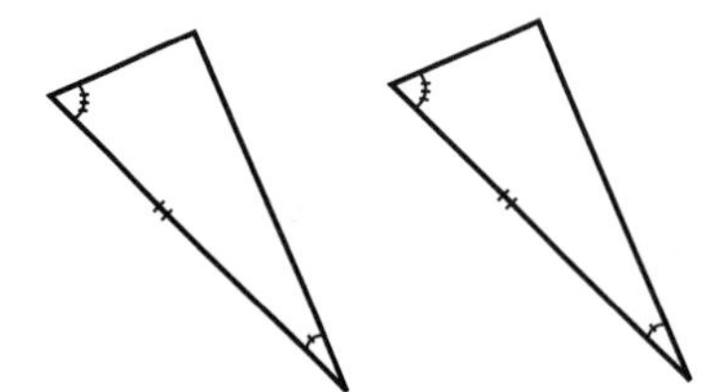

Figure 1.7. The angle-side-angle axiom.

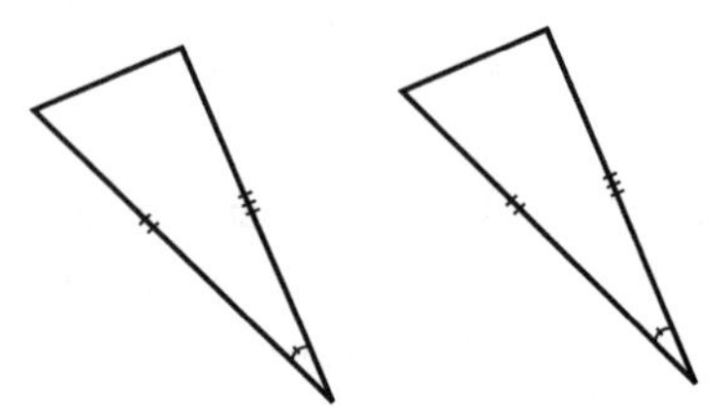

Figure 1.8. The side-angle-side axiom.

We shall take these three paradigms for congruence as intuitively obvious (Euclid actually provides arguments for these three paradigms). You may find it useful to discuss them in class.

Theorem 1.1. *Let $\triangle ABC$ be an isosceles triangle with equal sides $\overline{AB}$ and $\overline{AC}$. See Figure 1.9. Then the angles $\angle B$ and $\angle C$ are equal.*

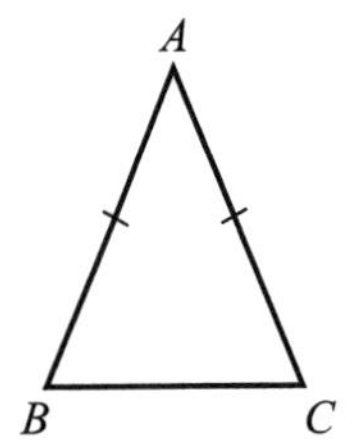

Figure 1.9. An isosceles triangle.

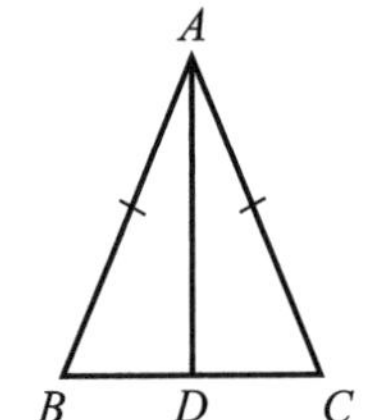

Figure 1.10. The median.

Proof. Draw the *median* from the vertex A to the opposite side BC (here the definition of the median is that it bisects the opposite side). See Figure 1.10. Thus we have created two subtriangles $\triangle ABD$ and $\triangle ACD$. Notice that these two smaller triangles have all corresponding sides equal (Figure 1.11): side $\overline{AB}$ in the first triangle equals side $\overline{AC}$ in the second triangle; side $\overline{AD}$ in the first triangle equals side $\overline{AD}$ in the second triangle; and side $\overline{BD}$ in the first triangle equals side $\overline{DC}$ in the second triangle (because the median bisects

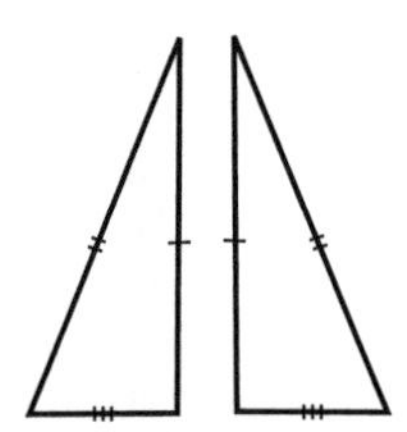

Figure 1.11. Analysis of an isosceles triangle.

side $\overline{BC}$). As a result (by SSS), the two subtriangles are *congruent*. All the corresponding artifacts of the two triangles are the same. We may conclude, therefore, that $\angle B = \angle C$. □

Corollary 1.1. *Let $\triangle ABC$ be an isosceles triangle as in the preceding theorem (Figure 1.9). Then the median from A to the opposite side BC is also perpendicular to BC.*

Proof. We have already observed that the triangles $\triangle ABD$ and $\triangle ACD$ are congruent. In particular, the angles $\angle ADB$ and $\angle ADC$ are equal. But those two angles also must sum up to 180° or π radians. The only possible conclusion is that each angle is 90° or a right angle. □

A basic fact, which is equivalent to the Parallel Postulate **P5**, is as follows.

Theorem 1.2. *Let ℓ and m be parallel lines, as in Figure 1.12. Let p be a transverse line which intersects both ℓ and m. Then the alternating angles α and β (as shown in the figure) are equal.*

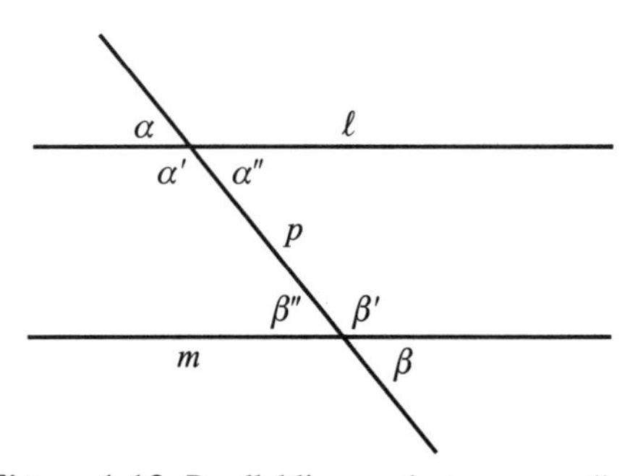

Figure 1.12. Parallel lines and a transverse line.

The proof is intricate, and would take us far afield. We shall omit it (but see [GRE]). An immediate consequence of Theorem 1.2 is this simple corollary:

Corollary 1.2. *Let lines ℓ and m be parallel lines as in the theorem, and let p be a transversal. Then the alternating angles α' and β' are equal. Also α'' and β'' are equal.*

Proof. Notice that

$$\alpha + \alpha' = 180^\circ = \beta + \beta'.$$

Since $\alpha = \beta$, we may conclude that $\alpha' = \beta'$.

The proof that $\alpha'' = \beta''$ follows similar lines, and we leave it for you to discuss in class. □

Now we turn to some consequences of this seminal idea.

Theorem 1.3. *Let $\triangle ABC$ be any triangle. Then the sum of the three angles in this triangle is equal to a halfline (i.e., to 180°).*

Proof. Examine Figure 1.13. Observe that $\angle\beta = \angle\beta'$ and $\angle\gamma = \angle\gamma'$. It follows that

$$\text{sum of angles in triangle} = \alpha + \beta + \gamma = \alpha + \beta' + \gamma' = \text{a line} = 180^\circ.$$

That is what was to be proved. □

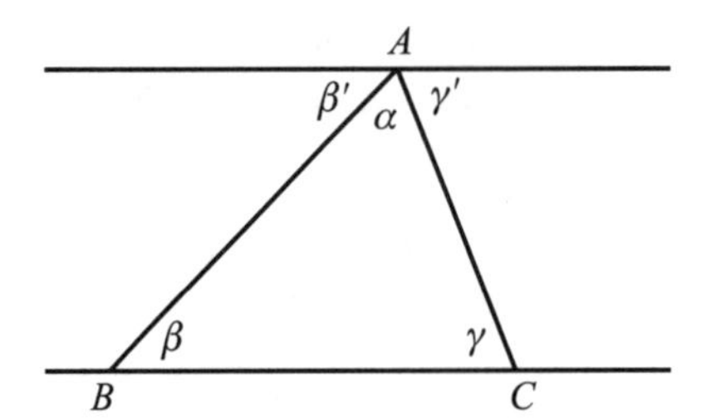

Figure 1.13. The sum of the angles in a triangle.

A companion result to the last theorem is this:

Corollary 1.3. *Let $\triangle ABC$ be any triangle and let τ be an exterior angle (see Figure 1.14). Then τ equals the sum of the other two interior angles α and β.*

We have defined the necessary terminology in context. The exterior angle τ is determined by the two sides $\overline{AC}$ and $\overline{BC}$ of the triangle—but is *outside* the triangle. This exterior angle is adjacent to an interior angle γ, as the figure shows. The assertion is that τ is equal to the sum of the *other two angles* α and β.

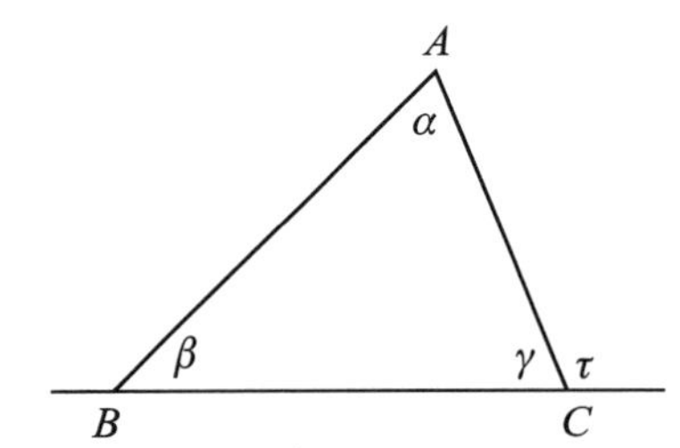

Figure 1.14. An exterior angle of a triangle.

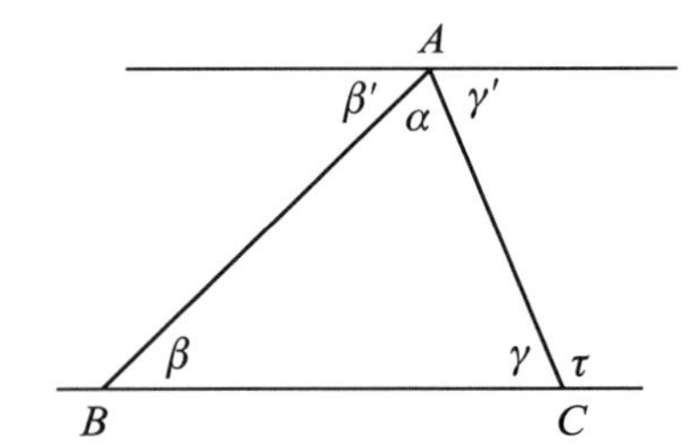

Figure 1.15. More on the exterior angle.

Proof of Corollary 1.3. According to Figure 1.15, the angle τ is certainly equal to $\alpha + \beta'$. Also $\beta = \beta'$ and $\gamma = \gamma'$. Thus

$$\begin{aligned} 180^\circ &= \gamma' + \alpha + \beta' \\ &= \gamma' + \tau. \end{aligned}$$

It follows that

$$\begin{aligned} \tau &= 180^\circ - \gamma' \\ &= 180^\circ - \gamma \\ &= \alpha + \beta. \end{aligned}$$

That is the desired result. □

Further Reading

Jerzy Czyz and William Self, The rationals are countable: Euclid's proof, *The College Mathematics Journal* 34(2003), 367–369.

D. H. Lehmer, Euclid's algorithm for large numbers, *The American Mathematical Monthly* 45(1938), 227–233.

1.3 Archimedes

1.3.1 The Genius of Archimedes

Archimedes (287 B.C.E.–212 B.C.E.) was born in Syracuse, Sicily. His father was Phidias, the astronomer. Archimedes developed into one of the most gifted, powerful, and creative mathematicians who ever lived. You can read about the life of Archimedes in [AAB] and [DIJ].

One of Archimedes's achievements was to develop methods for calculating areas and volumes of various geometric figures. We shall imitate one of Archimedes's techniques—the method of exhaustion that he learned from Eudoxus (408 B.C.E.–355 B.C.E.)—to approximate the area inside a circle to any desired degree of accuracy. This gives us a method for in turn approximating the value of π. It can be said that Archimedes turned the method of exhaustion to a fine art, and that some of his calculations were tantamount to the foundations of integral calculus (which was actually not fully developed until nearly 2000 years later).

Archimedes grew up in privileged circumstances. He was closely associated with, and perhaps even related to, Hieron King of Syracuse; he was also friends with Gelon, son of Hieron. He studied in Alexandria and developed there a relationship with Conon of Samos; Conon was someone whom Archimedes admired as a mathematician and cherished as a friend.

When Archimedes returned from his studies to his native city he devoted himself to pure mathematical research. During his lifetime, he was regularly called upon to develop instruments of war in the service of his country. And he was no doubt better known to the populace at large, and also appreciated more by the powers that be, for that work than for his pure mathematics. Among his other creations, Archimedes is said to have created (using his understanding of leverage) a device that would lift enemy ships out of the water and overturn them. Another of his creations was a burning mirror that would set enemy ships afire. Archimedes himself set no value on these contrivances, and declined even to leave any written record of them.

Perhaps the most famous story about Archimedes concerns a crown that was specially made for his friend King Hieron. It was alleged to be manufactured of pure gold, yet Hieron suspected that it was actually part silver. Archimedes puzzled over the proper method to determine whether this was true (without modifying or destroying the crown!). Then, one day, as Archimedes was stepping into his bath, he observed the water running over and had an inspiration. He determined that the excess of bulk that would be created by the introduction of alloy into the crown could be measured by putting the crown and equal weights of gold and silver separately into a vessel of water—and then noting the difference of overflow. If the crown were pure gold then it would create the same amount of overflow as the equal weight of gold. If not, then there was alloy present.

Archimedes is said to have been so overjoyed with his new insight that he sprang from his bath—stark naked—and ran home down the middle of the street shouting "*Eureka! Eureka!*", which means "I have found it! I have found it!" To this day, in memory of Archimedes, people cry Eureka! to celebrate a satisfying discovery.

Another oft-told story of Archimedes concerns his having said to Hieron, "Give me a place to stand and I will move the earth." What Archimedes meant by this bold assertion is

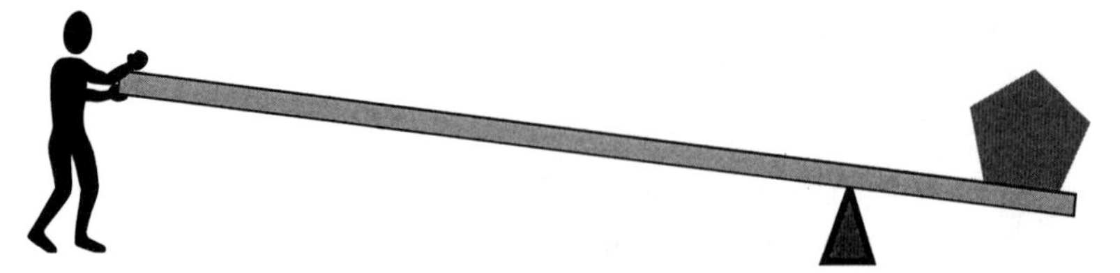

Figure 1.16. The power of a lever.

illustrated in Figure 1.16. Archimedes was one of the first to study and appreciate the power of levers. He realized that a person of modest strength could move a very great weight with the assistance of the leverage afforded by a very long arm. Not fully understanding this principle, Hieron demanded of Archimedes that he give an illustration of his ideas. And thus Archimedes made his dramatic claim. As a practical illustration of the idea, Archimedes arranged a lever system so that Hieron himself could move a large and fully laden ship.

One of Archimedes's inventions that lives on today is a water screw that he devised in Egypt for the purpose of irrigating crops. The same mechanism is used now in electric water pumps as well as hand-powered pumps in third world countries.

Archimedes died during the capture of Syracuse by the troops of Marcellus in 212 B.C.E. Even though Marcellus gave explicit instructions that neither Archimedes nor his house were to be harmed, a soldier became enraged when Archimedes would not divert his attention from his mathematics and obey an order. Archimedes is reported to have said sternly to the soldier, "Do not disturb my circles!" Thus Archimedes fell to the sword. Later in this book we tell the story of how Sophie Germain became enthralled by this story of Archimedes's demise, and was thus inspired to become one of the greatest female mathematicians who ever lived.

Next we turn our attention to Archimedes's study of the area of a circle.

1.3.2 Calculating the Area of a Circle in the Spirit of Archimedes

Begin by considering a regular hexagon with side length 1 (Figure 1.17). We divide the hexagon into triangles (Figure 1.18). Notice that each of the central angles of each of the triangles must have measure $360°/6 = 60°$. Since the sum of the angles in a triangle is $180°$, and since each of these triangles certainly has two equal sides and hence two equal angles, we may now conclude that all the angles in each triangle have measure $60°$. See Figure 1.19.

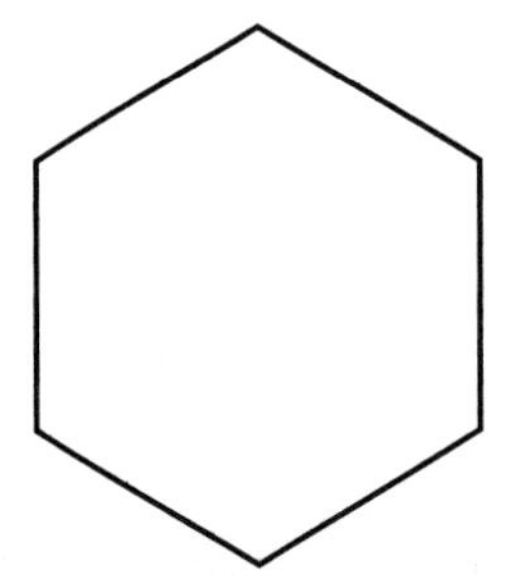

Figure 1.17. A regular hexagon.

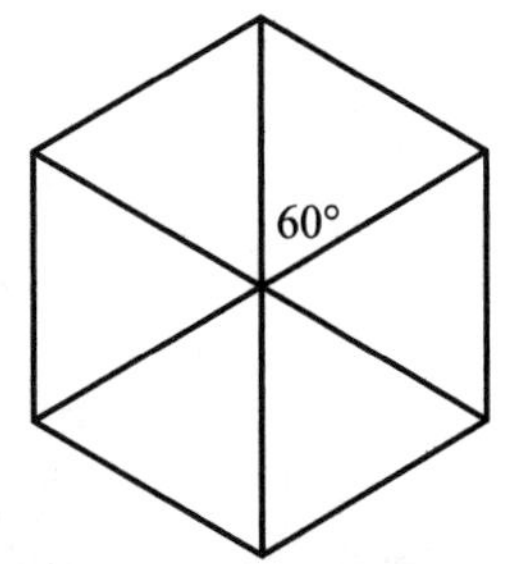

Figure 1.18. The hexagon divided into triangles.

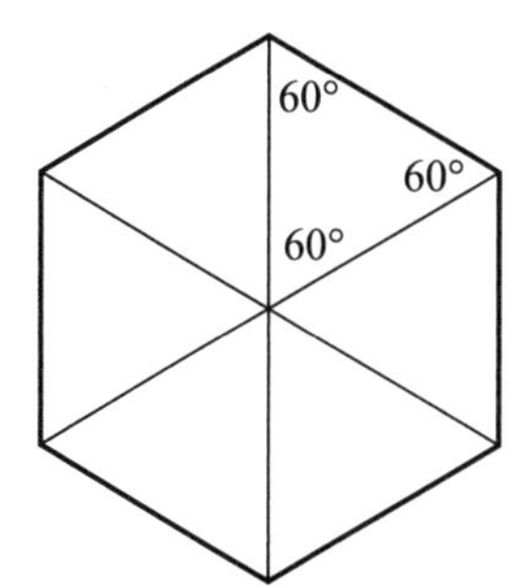

Figure 1.19. The angle measure in each triangle.

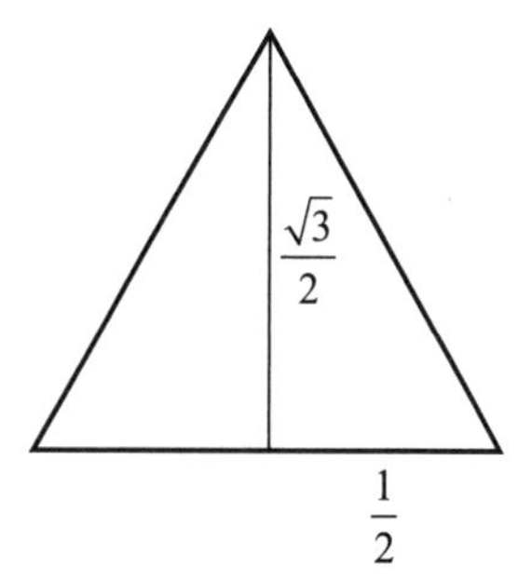

Figure 1.20. Division of the triangle.

But now we may use the Pythagorean theorem to analyze one of the triangles. We divide the triangle in two—Figure 1.20. Thus the triangle is the union of two right triangles (see Corollary 1.1). We know that the hypotenuse of one of these right triangles—which is the same as the side of one of the six equilateral triangles—is 1 and the base is $1/2$. Thus the Pythagorean theorem tells us that the height of the right triangle is $\sqrt{1^2 - (1/2)^2} = \sqrt{3}/2$. We may conclude then that the area of this right triangle, as shown in Figures 1.19 and 1.20, is

$$A(T) = \frac{1}{2} \cdot (\text{base}) \cdot (\text{height}) = \frac{1}{2} \cdot \frac{1}{2} \cdot \frac{\sqrt{3}}{2} = \frac{\sqrt{3}}{8}.$$

Therefore the area of the full equilateral triangle, with all sides equal to 1, is twice this or $\sqrt{3}/4$.

Now of course the full regular hexagon is made up of six of these equilateral triangles, so the area inside the hexagon is

$$A(H) = 6 \cdot \frac{\sqrt{3}}{4} = \frac{3\sqrt{3}}{2}.$$

We think of the area inside the regular hexagon as being a crude approximation to the area inside the circle: Figure 1.21. Thus the area inside the circle is very roughly the area inside the hexagon. Of course we know from other considerations that the area inside this circle is $\pi \cdot r^2 = \pi \cdot 1^2 = \pi$. Thus, putting our ideas together, we find that

$$\begin{aligned} \pi &= (\text{area inside unit circle}) \\ &\approx (\text{area inside regular hexagon}) \\ &= \frac{3\sqrt{3}}{2} \approx 2.598\ldots \end{aligned}$$

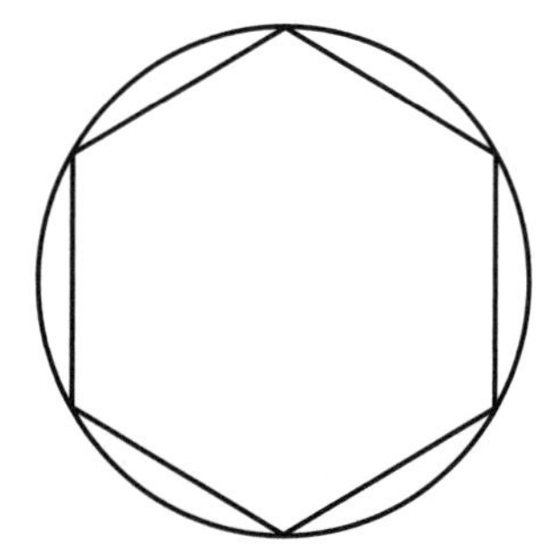

Figure 1.21. Area inside the hexagon approximates the area inside the circle.

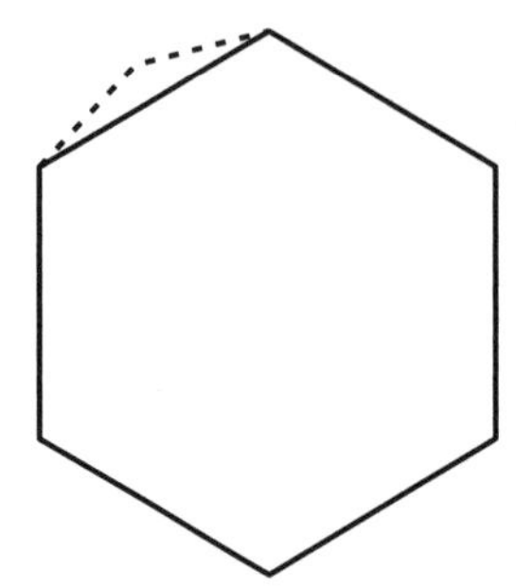

Figure 1.22. Increasing the number of sides in the inscribed polygon.

It is known that the true value of π is $3.14159265\ldots$. So our approximation is quite crude. The way to *improve* the approximation is to increase the number of sides in the approximating polygon. In fact what we shall do is *double* the number of sides to 12. Figure 1.22 shows how we turn one side into two sides; doing this six times creates a regular 12-sided polygon.

Notice that we create the regular 12-sided polygon (a dodecagon) by adding small triangles to each of the edges of the hexagon. Our job now is to calculate the area of the twelve-sided polygon. Thus we need to calculate the lengths of the edges. Examine a blown-up picture of the triangle that we have added (Figure 1.23). We use the Pythagorean theorem to calculate the length x of a side of the new dodecagon. It is

$$x = \sqrt{\left(\frac{1}{2}\right)^2 + \left(1 - \frac{\sqrt{3}}{2}\right)^2} = \sqrt{\frac{1}{4} + \left(1 - \sqrt{3} + \frac{3}{4}\right)} = \sqrt{2 - \sqrt{3}}.$$

Now let us focus attention on the dodecagon, divided into twelve isosceles triangles (Figure 1.24). We have just calculated that each side of the dodecagon has length $\sqrt{2-\sqrt{3}}$. If we can calculate the area of each of the congruent subtriangles, then we can obtain the area of the entire dodecagon (by multiplying by 12). Examine Figure 1.25. This is one of the 12 triangles that makes up the dodecagon. It has base $\sqrt{2-\sqrt{3}}$. Each of the two sides has length 1. Thus we may use the Pythagorean theorem to determine that the *height* of the

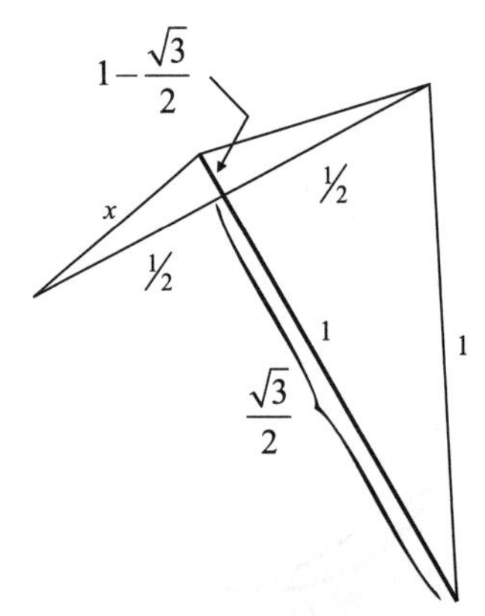

Figure 1.23. The triangle that we have added.

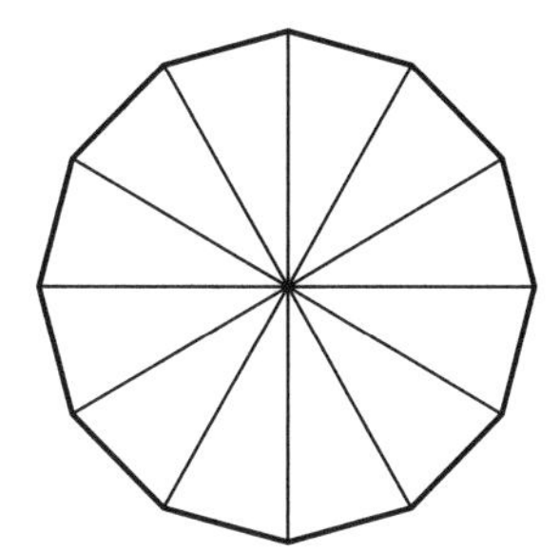

Figure 1.24. The dodecagon divided into triangles.

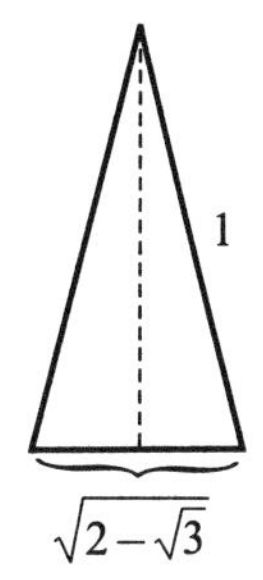

Figure 1.25. The area of a triangle making up the dodecagon.

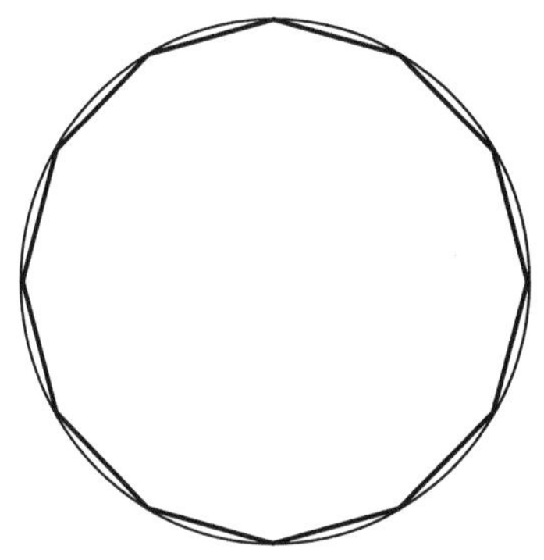

Figure 1.26. Approximation to the area inside a circle.

triangle is

$$h = \sqrt{1^2 - \left(\frac{\sqrt{2-\sqrt{3}}}{2}\right)^2} = \sqrt{1 - \frac{2-\sqrt{3}}{4}} = \sqrt{\frac{2+\sqrt{3}}{4}}\,.$$

We conclude that the area of the triangle is

$$A(T) = \frac{1}{2} \cdot (\text{base}) \cdot (\text{height}) = \frac{1}{2} \cdot \sqrt{2-\sqrt{3}} \cdot \sqrt{\frac{2+\sqrt{3}}{4}} = \frac{\sqrt{4-3}}{4} = \frac{1}{4}\,.$$

Hence the area of the dodecagon is

$$A(D) = 12 \cdot \frac{1}{4} = 3\,.$$

Examining Figure 1.26, and thinking of the area inside the dodecagon as an approximation to the area inside the unit circle, we find that

$$\pi = (\text{area inside unit circle}) \approx (\text{area inside regular dodecagon}) = 3\,.$$

This is obviously a better approximation to π than our first attempt. At least we now have the "3" right! Now let us do one more calculation in an attempt to improve the estimate. After that we will seek to find a pattern in these calculations.

Now we consider a regular 24-sided polygon (an icositetragon). As before, we construct this new polygon by erecting a small triangle over each side of the dodecagon.

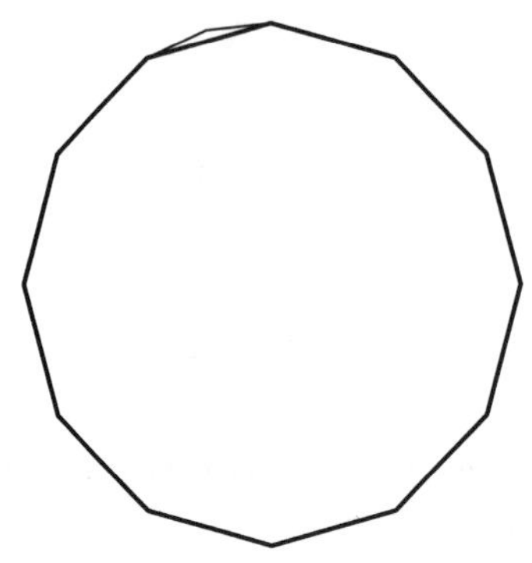

Figure 1.27. The icositetragon.

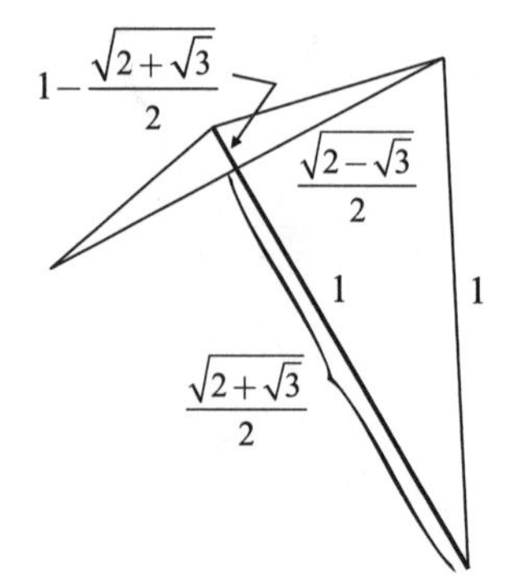

Figure 1.28. Adding a triangle.

See Figure 1.27. We examine a blowup (Figure 1.28) of one of these triangles, just as we did above for the dodecagon. We first solve the right triangle with base $\sqrt{2-\sqrt{3}}/2$ and hypotenuse 1—using the Pythagorean theorem, of course—to find that it has height $\sqrt{2+\sqrt{3}}/2$. Then we see that the smaller right triangle has base $1-\sqrt{2+\sqrt{3}}/2$ and height $\sqrt{2-\sqrt{3}}/2$. Thus, again by the Pythagorean theorem, the hypotenuse of the small right triangle is $\sqrt{2-\sqrt{2+\sqrt{3}}}$.

But the upshot is that the icositetragon is made up of isosceles triangles, as in Figure 1.29, having base $\sqrt{2-\sqrt{2+\sqrt{3}}}$ and side length 1. We may divide the triangle into two right triangles, as indicated in the figure, and then solve one of the right triangles using the Pythagorean theorem. The solution is that the height of this right triangle is $\sqrt{2+\sqrt{2+\sqrt{3}}}/2$. Altogether, then, the area of the triangle which is one twenty-fourth of the polygon is

$$
\begin{aligned}
A(T) &= \frac{1}{2}\cdot(\text{base})\cdot(\text{height}) \\
&= \frac{1}{2}\cdot\sqrt{2-\sqrt{2+\sqrt{3}}}\cdot\frac{\sqrt{2+\sqrt{2+\sqrt{3}}}}{2} \\
&= \frac{\sqrt{2-\sqrt{3}}}{4}.
\end{aligned}
$$

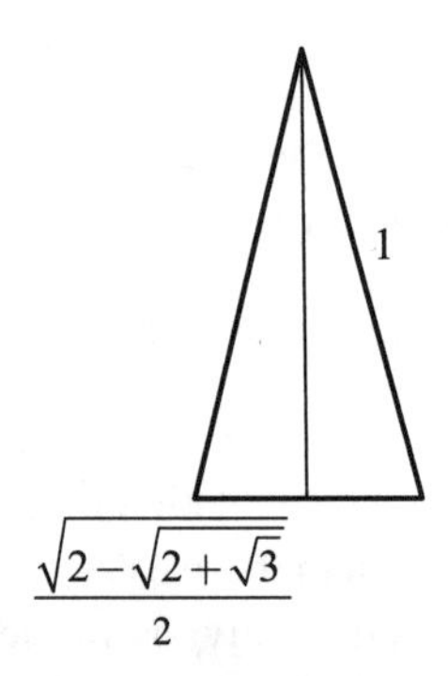

Figure 1.29. The icositetragon broken up into triangles.

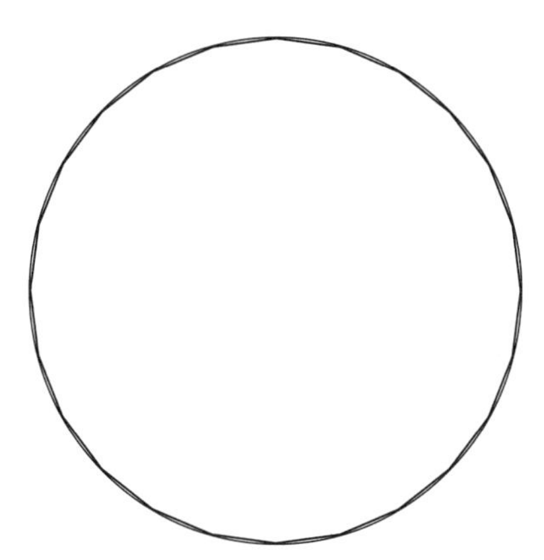

Figure 1.30. Approximation of the area inside a circle.

We conclude that the area of the 24-sided regular polygon is

$$A(P) = 24 \cdot \frac{\sqrt{2-\sqrt{3}}}{4} = 6\sqrt{2-\sqrt{3}}\,.$$

Examining Figure 1.30, and thinking of the area inside the icositetragon as an approximation to the area inside the unit circle, we find that

$$\pi = (\text{area inside unit circle}) \approx (\text{area inside regular 24-gon}) \approx 3.1058\,.$$

We see that, finally, we have an approximation to π that is accurate to one decimal place.

Of course the next step is to pass to a polygon of 48 sides. We shall not repeat all the steps of the calculation but just note the high points. First, we construct the regular 48-gon by placing small triangles along each of the edges of the dodecagon. See Figure 1.31. Now, once again, we must (blowing up the triangle construction) examine a figure like 1.32. The usual calculation shows that the side of the small added triangle has length $\sqrt{2-\sqrt{2+\sqrt{2+\sqrt{3}}}}$. Thus we end up examining a new isosceles triangle, which is $1/48$th of the 48-sided polygon. See Figure 1.33.

The usual calculations, just as we did for the polygons having 6 or 12 or 24 sides, show

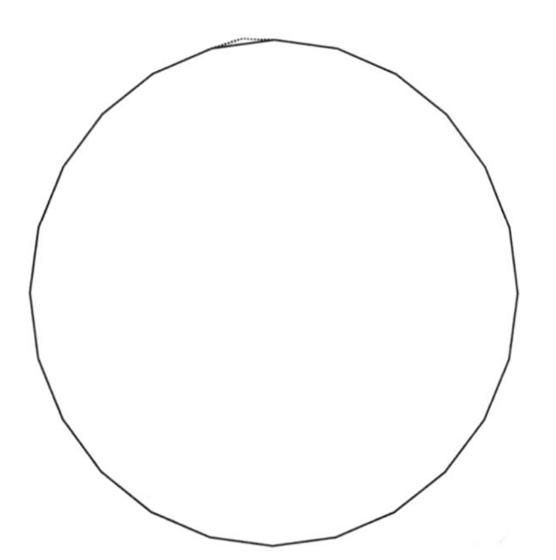

Figure 1.31. The regular 48-gon.

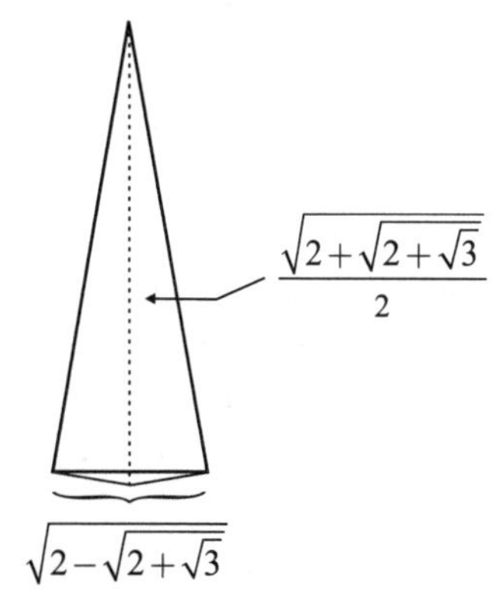

Figure 1.32. Adding a triangle.

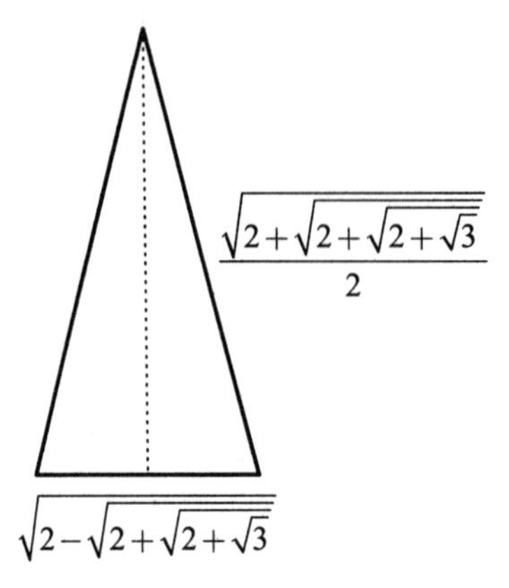

Figure 1.33. The area of a triangle inside the 48-gon.

that this new triangle has base $\sqrt{2-\sqrt{2+\sqrt{2+\sqrt{3}}}}$ and height $\sqrt{2+\sqrt{2+\sqrt{2+\sqrt{3}}}}/2$. Thus the area is

$$\begin{aligned}
A(T) &= \frac{1}{2}\cdot(\text{base})\cdot(\text{height}) \\
&= \frac{1}{2}\cdot\sqrt{2-\sqrt{2+\sqrt{2+\sqrt{3}}}}\cdot\sqrt{2+\sqrt{2+\sqrt{2+\sqrt{3}}}}/2 \\
&= \frac{\sqrt{2-\sqrt{2+\sqrt{3}}}}{4}.
\end{aligned}$$

The polygon comprises 48 such triangles, so the total area of the polygon is

$$A(P) = 48\cdot\frac{\sqrt{2-\sqrt{2+\sqrt{3}}}}{4} = 12\sqrt{2-\sqrt{2+\sqrt{3}}}.$$

Thinking of the area inside the 48-sided regular polygon as an approximation to the area inside the unit circle, we find that

$$\begin{aligned}
\pi &= (\text{area inside unit circle}) \\
&\approx (\text{area inside 48-sided regular polygon}) \\
&\approx 3.1326\,.
\end{aligned}$$

This is obviously a better approximation to π than our last three attempts. It is accurate to one decimal place, and the second decimal place is close to being right.

And now it is clear what the pattern is. The next step is to examine a regular polygon with 96 sides. The usual calculations will show that this polygon breaks up naturally into 96 isosceles triangles, and each of these triangles has area

$$A(T) = \frac{\sqrt{2-\sqrt{2+\sqrt{2+\sqrt{3}}}}}{4}.$$

Thus the area of the polygon is

$$A(P) = 96 \cdot \frac{\sqrt{2 - \sqrt{2 + \sqrt{2 + \sqrt{3}}}}}{4} = 24 \cdot \sqrt{2 - \sqrt{2 + \sqrt{2 + \sqrt{3}}}}.$$

We then see that

$$\begin{aligned} \pi &= (\text{area inside unit circle}) \\ &\approx (\text{area inside 96-sided regular polygon}) \\ &\approx 3.13935\,. \end{aligned}$$

This is certainly an improved approximation to the true value of π, which is $3.14159265\ldots$.

The next regular polygon in our study has 192 sides. It breaks up naturally into 192 isosceles triangles, each of which has area

$$A(T) = \frac{\sqrt{2 - \sqrt{2 + \sqrt{2 + \sqrt{2 + 3}}}}}{4}\,.$$

Thus the area of the regular 192-gon is

$$A(P) = 192 \cdot \frac{\sqrt{2 - \sqrt{2 + \sqrt{2 + \sqrt{2 + 3}}}}}{4} = 48 \cdot \sqrt{2 - \sqrt{2 + \sqrt{2 + \sqrt{2 + 3}}}}\,.$$

We then see that

$$\begin{aligned} \pi &= (\text{area inside unit circle}) \\ &\approx (\text{area inside 192-sided regular polygon}) \\ &\approx 3.14103\,. \end{aligned}$$

This new approximation of π is accurate to nearly three decimal places.

Archimedes himself considered regular polygons with nearly 500 sides. His method did not yield an approximation as accurate as ours. But, historically, it was one of the first estimations of the size of π.

For You to Try The method that we have used here to estimate the area inside a circle can also be used to estimate the area under a parabola (such as $y = -x^2 + 1$) and above the x-axis. It can also be used for 3-dimensional volumes, such as estimating the area inside a sphere. Discuss these new problems in class.

Further Reading

Frank Burk, Archimedes' quadrature and Simpson's rule, *The College Mathematics Journal* 18(1987), 222–223.

Walter Rudin, A generalization of a theorem of Archimedes, *The American Mathematical Monthly* 80(1973), 794–796.

Exercises

1. Verify that the number $\sqrt{17}$ is irrational.

2. The number $\alpha = \sqrt[5]{9}$ is that unique positive real number that satisfies $\alpha^5 = 9$. Verify that this α is irrational.

3. Let m be any positive whole number (i.e., a natural number). Show that $\sqrt{m}$ is either a positive whole number or is irrational. Discuss this problem in class.

4. Let m be any positive whole number (i.e., a natural number). Show that $\sqrt[3]{m}$ is either a positive whole number or is irrational. Discuss this problem in class.

5. Develop a new verification of the Pythagorean theorem using the diagram in Figure 1.34. Observe that the figure contains four right triangles and a square, but the configuration is different from that in Figure 1.3. Now we have a large square in a tilted position inside the main square. Using the labels provided in the figure, observe that the area of each right triangle is $ab/2$. And the area of the inside square is c^2. Finally, the area of the large, outside square is $(a + b)^2$. Put all this information together to derive Pythagoras's formula.

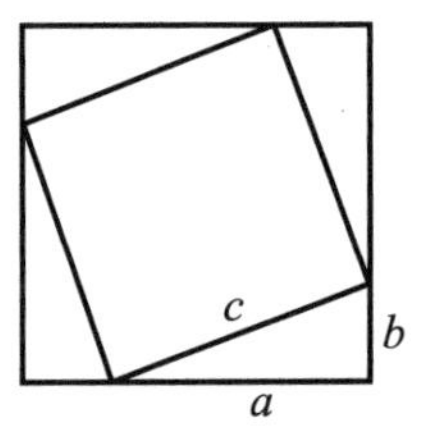

Figure 1.34. Verification of the Pythagorean theorem.

6. Explain the reasoning represented in Figure 1.35 to discover yet another proof of the Pythagorean theorem.

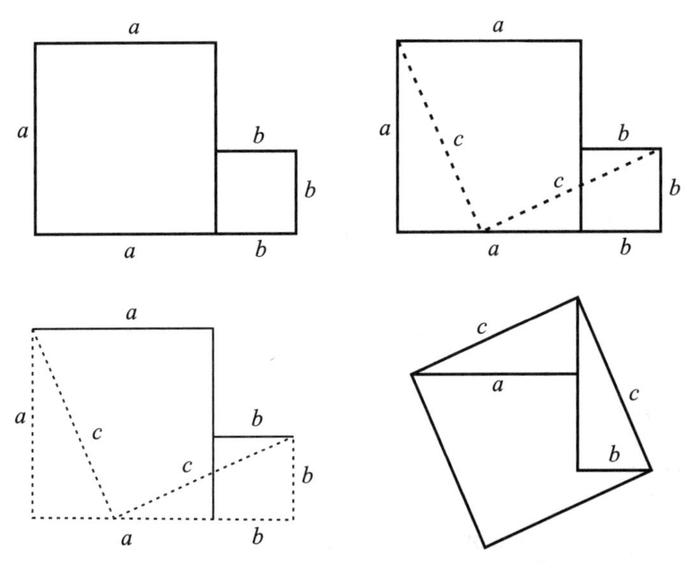

Figure 1.35. Another proof of the Pythagorean theorem.

7. Find all Pythagorean triples in which one of the three numbers is 7. Explain your answer.

8. Find all Pythagorean triples in which each of the three numbers is less than 35. Explain your answer.

9. The famous Waring problem (formulated in 1770) was to show that every positive integer can be written as the sum of *at most* four perfect squares. David Hilbert was the mathematician who finally solved this problem in 1909. So, for example,

$$11 = 3^2 + 1^2 + 1^2 + 1^2$$

and

$$87 = 2^2 + 3^2 + 5^2 + 7^2$$

and

$$31 = 5^2 + 2^2 + 1^2 + 1^2 .$$

Find the Waring/Hilbert decomposition of 101. Find the Waring/Hilbert decomposition of 1001. Write a computer program that will perform this job for you. Discuss this problem in class.

10. This is a good problem for class discussion. Refer to the Waring problem in Exercise 9. Formulate a version of the Waring problem for cubes instead of squares. How many cubes will it take to compose any positive integer? Write a computer program to test your hypothesis. Find a decomposition of 101 into cubes. Find a decomposition of 1001 into cubes.

11. We can locate any point in the plane with an ordered *pair* of real numbers. See Figure 1.36. Discuss this idea in class. Now use your understanding of the Pythagorean theorem to derive a formula for the distance in the plane between the points $(0, 0)$ and (a, b).

12. Refer to Exercise 11. Use the idea there to find a formula for the distance between two planar points (x, y) and (x', y').

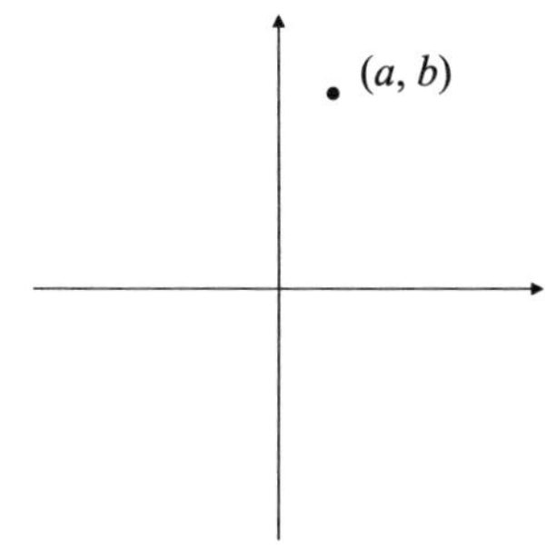

Figure 1.36. Using ordered pairs of numbers to plot points in the plane.

13. Refer to Exercise 12. If we can locate any point in the plane with an ordered *pair* of real numbers, then we can locate any point in 3-dimensional space with an ordered *triple* of numbers—see Figure 1.37. Discuss this idea in class. Now use your understanding of the Pythagorean theorem to derive a formula for the distance in 3-dimensional space between two points (x, y, z) and (x', y', z').

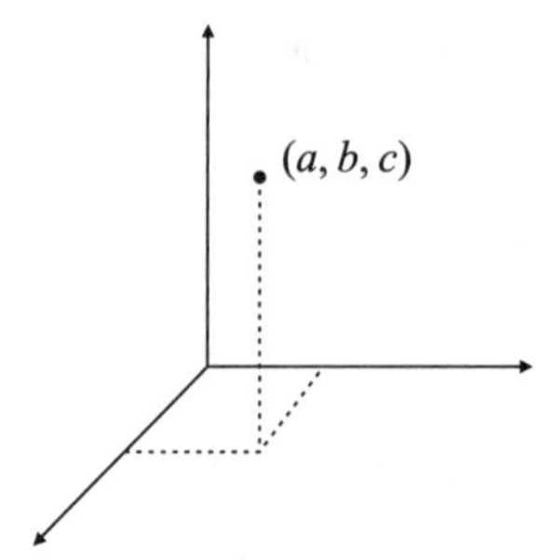

Figure 1.37. Using ordered triples of numbers to plot points in space.

14. **Project:** The *Law of Cosines* says that if $\triangle ABC$ is any triangle with vertices A, B, C, angles α, β, γ, and opposite edges a, b, c (See Figure 1.38) then

$$a^2 + b^2 - 2ab\cos\gamma = c^2 .$$

This is obviously a generalization of the Pythagorean theorem, for if $\gamma = \pi/2$ (or 90°), then the equation becomes $a^2 + b^2 = c^2$.

Find a synthetic proof of the Law of Cosines, similar in spirit to the proof of the Pythagorean theorem presented in the text.

Explain why there is not a similar equation involving sines. A full and detailed consideration of laws like this appears in [YEO], [HAR].

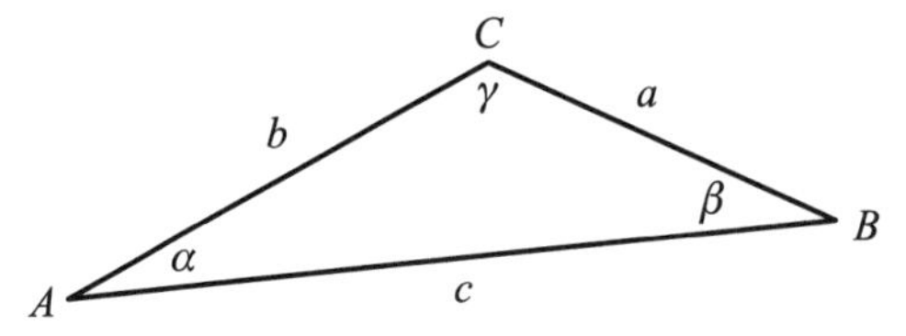

Figure 1.38. The Law of Cosines.

15. **Project:** Some of the more "obvious" results in Euclidean geometry are actually quite difficult to prove. As an example, consider this theorem:

Theorem. *Let $\triangle ABC$ be a triangle. Suppose that the angle bisector emanating from A has the same length as the angle bisector emanating from B. Then $\triangle ABC$ is an isosceles triangle with the angles at A and B equal.*

Attempt to construct a synthetic proof of this theorem. As a hint, note that the length of an angle bisector is equal to the product of its adjacent sides minus the product of the lengths of the two pieces into which it divides the opposite side.

As an alternative approach, picture your triangle in the Cartesian plane, introduce coordinates, and use methods of Cartesian geometry to prove the theorem.

The references [BIB], [HEM], [WEN] give thorough and classical treatments of plane geometry.

16. **Project:** Use the method of Archimedes to calculate the area inside an ellipse with minor axis $2a$ and major axis $2b$. The book [STE] contains delightful discussions of Archimedes's methodology.

2

Zeno's Paradox and the Concept of Limit

2.1 The Context of the Paradox

Ancient Greek mathematics—from about 500 B.C.E.to 100 C.E.—enjoyed many successes. The sieve of Eratosthenes, the discovery of infinitely many prime numbers, and the Pythagorean theorem are cornerstones of mathematics that live on today. We shall discuss all of these in the present book. But the mathematics of the Greeks was marked by one huge gap. They simply could not understand the concept of "limit". The popular formulations of the limit question were dubbed "Zeno's paradox" (named after the mathematician and Eleatic philosopher Zeno, 490 B.C.E.–435 B.C.E.), and these questions were hotly debated in the Greek schools and forums.

In fact Euclid's *Elements* (see [EUC]) contains over 40 different formulations of Zeno's paradox; we shall discuss several of them here. For this is what mathematicians do: When they cannot solve a problem, they re-state it and turn it around and try to find other ways to look at it. This is nothing to be ashamed of. As the great classic work on problem-solving—Pólya's *How to Solve It* [POL]—will tell you, one of the mathematician's most powerful tools is to restate a problem. We shall encounter this technique repeatedly in the present book.

But, unfortunately, this method of re-statement did not help the Greeks. Like all people in all civilizations, they had an interlocking system of beliefs to which their reasoning was wedded. And their scientific beliefs were intertwined with their religious beliefs. For example, Pythagoras was not simply a person. "Pythagoras" (or "the Pythagoreans") was the name for a society of people who developed the ideas to which we now attach his name. And that society was both a religious organization and a scientific laboratory.

One of the overriding Greek philosophical concerns was whether everything in the universe was "one", or whether the universe contained independent entities. The discussions of these matters were vigorous and subtle. Certainly Zeno's paradoxes, which live on today, were an outgrowth of the question of "oneness". We shall consider this matter in further detail in the considerations that follow. Suffice it to say for the moment that the issue of oneness had a powerful effect on the Greeks' ability to think about mathematical questions.

To put the matter bluntly, and religious beliefs aside, the Greeks were uncomfortable with division, they had rather limited mathematical notation, and they had a poor understanding of limits. It must be said that the Greeks made great strides with the tools that they had available, and it is arguable that Archimedes at least had a good intuitive grasp of the limit concept. Our knowledge has advanced a bit since that time. Today we have more experience and a broader perspective. Mathematics is now more advanced, and more carefully thought out. After we state Zeno's paradox (in several different formulations!), we shall be able to analyze it quickly and easily.

Further Reading

R. W. Hamming, Mathematics on a distant planet, *The American Mathematical Monthly* 105(1998), 640–650.

James C. Kirby, Bargaining theory, or Zeno's used cars, *The College Mathematics Journal* 27(1996), 285–286.

2.2 The Life of Zeno of Elea

Little is known of the life of Zeno of Elea. Our main source of information concerning this influential thinker is Plato's dialogue *Parmenides.* Although Plato gives a positive account of Zeno's teachings, he does not necessarily believe all the paradoxes that we usually attribute to Zeno. Further reading on the life and works of Zeno may be enjoyed in the sources [BAR] and [LAE].

The philosopher Diogenes Laertius also wrote of Zeno's life, but his reports are today deemed to be unreliable.

Zeno was certainly a philosopher, and was the son of Teleutagoras. He was a pupil and friend of the more senior philosopher Parmenides, and studied with him in Elea in southern Italy at the school which Parmenides had founded. This was one of the leading pre-Socratic schools of Greek philosophy, and was quite influential.

Parmenides's philosophy of "monism" claimed that the great diversity of objects and things that exist are merely a single external reality. This reality he called "Being". Parmenides asserted that "all is one" and that change or "non-Being" are impossible. The argument described here is often summarized with the phrase "the One vs. the Many". Zeno's thinking was strongly influenced by his teacher Parmenides. Zeno and Parmenides visited Athens together around the year 450 B.C.E. It is believed that Socrates met with the two men at that time. Zeno had already written a book before his trip to Athens, and this one book is really Zeno's claim to fame. As far as we know, Socrates was 20 years old, Zeno 40 years old, and Parmenides 65 years old at the time of the meeting. Zeno was something of the celebrity of the group—largely because of his book. Proclus describes the book in loving detail. It contains Zeno's 40 paradoxes concerning the continuum.

Of particular interest is the fact that Zeno argued for the One by trying to contradict the existence of the Many. By this means Zeno is credited with developing a method of indirect argument whose purpose is not victory but rather the discovery of truth. We now call this type of reasoning a *dialectic.*

As indicated, Zeno attempted to answer objections to Parmenides's theory of the existence of the One by showing that the hypothesis of the existence of the Many, both in time and in space, would lead to more serious inconsistencies.

What we today commonly call "Zeno's paradoxes" grew out of his wrestling with the "One vs. Many" dialectic. Thus Zeno's standard list of paradoxes certainly includes the tortoise and the hare and the man walking towards the wall, as described below. But it also includes more philosophical musings as we now relate:

(1) If the Existent is Many, it must be at once infinitely small and infinitely great—infinitely small, because its parts must be indivisible and therefore without magnitude; infinitely great, because, that any part having magnitude may be separate from any other part, the intervention of a third part having magnitude is necessary, and that this third part may be separate from the other two the intervention of other parts having magnitude is necessary, and so on *ad infinitum.*

(2) In like manner the Many must be numerically both finite and infinite—numerically finite, because there are as many things as there are, neither more nor less; numerically infinite, because, that any two things may be separate, the intervention of a third thing is necessary, and so on *ad infinitum.*

(3) If all that is in space, space itself must be in space, and so on *ad infinitum.*

(4) If a bushel of corn turned out upon the floor make a noise, each grain and each part of each grain must make a noise likewise; but, in fact, it is not so.

In fact even greater influence was had on the ancient Greeks by Zeno's paradox of predication. According to Plato, this conundrum ran as follows:

> If existences are many, they must be both like and unlike (unlike, inasmuch as they are not one and the same, and like, inasmuch as they agree in not being one and the same). But this is impossible; for unlike things cannot be like, nor like things unlike. Therefore existences are not many.

In the second decade of the fourth century B.C.E., the Greeks resumed the pursuit of truth in earnest. It was felt that Zeno's paradox of predication must be dealt with before there could be any discussion of the problem of knowledge and the problem of being could be resumed. Plato thus directs his serious students to the study of this question, and offers his own theory of the immanent[1] idea as a solution of the paradox.

Zeno took his teacher Parmenides's dictum "The Ent is, the Non-ent is not" and interpreted it anew.[2] To Zeno, this was a declaration of the Non-ent's absolute nullity. Thus Zeno developed the theory of the One as opposed to the theory of the Many. As a result of his efforts, the Eleaticism of Parmenides was forever silenced.

After meeting with Socrates in Athens, Zeno returned to the Italian town of Elea. Diogenes Laertius reports that Zeno died in a heroic attempt to remove a tyrant from the city. In fact Diogenes reports in great detail of the heroic deeds and the torture of Zeno at the hands of the tyrant. Diogenes also gives some material about Zeno's theory of cosmology.

Now let us look at the provenance of the paradoxes. They were well known in Plato's day, as they bore on Parmenides's rather prominent monistic theory of "Being". In other words, these paradoxes were offered as proof that everything was one, and could not be divided. Of them, Plato wrote

[1]Concerning the relationship of the world to the mind.
[2]Here "Ent" is an enunciation of the concept of oneness.

> ...a youthful effort, and it was stolen by someone, so that the author had no opportunity of considering whether to publish it or not. Its object was to defend the system of Parmenides by attacking the common conceptions of things.

In fact Plato claimed that Zeno's book was circulated without his knowledge. Proclus goes on to say

> ...Zeno elaborated forty different paradoxes following from the assumption of plurality and motion, all of them apparently based on the difficulties deriving from an analysis of the continuum.

The gist of Zeno's arguments, and we shall examine them in considerable detail below, is that if anything can be divided then it can be divided infinitely often. This leads to a variety of contradictions, especially because Zeno also believed that a thing which has no magnitude cannot exist.

In fact Simplicius (490 C.E.–560 C.E.) was the last head of Plato's academy, in the early sixth century. He explained Zeno's argument against the existence of any item of zero magnitude as follows:

> For if it is added to something else, it will not make it bigger, and if it is subtracted, it will not make it smaller. But if it does not make a thing bigger when added to it nor smaller when subtracted from it, then it appears obvious that what was added or subtracted was nothing.

It is a measure of how seriously Zeno's ideas were taken at the time that Aristotle, in his work *Physics*, gives four of Zeno's arguments: the Dichotomy, the Achilles, the Arrow, and the Stadium. For the Dichotomy, Aristotle describes Zeno's argument as follows:

> There is no motion because that which is moved must arrive at the middle of its course before it arrives at the end.

In greater detail: In order to traverse a line segment it is necessary to reach its midpoint. To do this one must reach the 1/4 point, to do this one must reach the 1/8 point and so on ad infinitum. Hence motion can never begin. The argument here is now answered by the well-known infinite sum

$$\frac{1}{2}+\frac{1}{4}+\frac{1}{8}+\cdots=1$$

On the one hand, Zeno can argue that the sum $1/2+1/4+1/8+\cdots$ never actually reaches 1, but more perplexing to the human mind is the attempts to sum $1/2+1/4+1/8+\cdots$ backwards. Before traversing a unit distance we must get to the middle, but before getting to the middle we must get 1/4 of the way, but before we get $1/4$ of the way we must reach $1/8$ of the way etc. See Figure 2.1. This argument makes us realize that we can never get

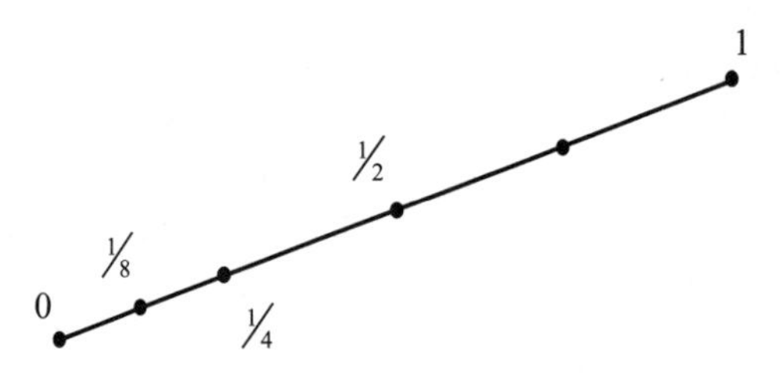

Figure 2.1. Zeno's paradox.

started since we are trying to build up this infinite sum from the "wrong" end. Indeed this is a clever argument which still puzzles us today. We shall spend considerable time in the present text analyzing this particular argument of Zeno.

The Arrow paradox is discussed by Aristotle as follows:

> If, says Zeno, everything is either at rest or moving when it occupies a space equal to itself, while the object moved is in the instant, the moving arrow is unmoved.

The argument rests on the fact that if in an indivisible instant of time the arrow moved, then indeed this instant of time would be divisible (for example in a smaller 'instant' of time the arrow would have moved half the distance). Aristotle argues against the paradox by claiming:

> ...for time is not composed of indivisible 'nows', no more than is any other magnitude.

It is easy to see, from what we have said, that Zeno's paradoxes have been important in the development of the notion of infinitesimals. In fact some modern writers believe that Zeno aimed his paradoxes against those who were introducing infinitesimals. Anaxagoras and the followers of Pythagoras—both of whom had a theory of incommensurables—are also thought by some to be the targets of Zeno's arguments.

The most famous of Zeno's paradoxes, and the one most frequently quoted and described, is undoubtedly Achilles and the hare (to be discussed in detail shortly). Aristotle, in his *Physics*, says:

> ...the slower when running will never be overtaken by the quicker; for that which is pursuing must first reach the point from which that which is fleeing started, so that the slower must necessarily always be some distance ahead.

Neither Plato nor Aristotle fully appreciated the significance of Zeno's arguments. In fact Aristotle called them "fallacies", without being able to refute them.

The celebrated twentieth-century philosopher Bertrand Russell (1872–1970) paid due homage to Zeno when he wrote:

> In this capricious world nothing is more capricious than posthumous fame. One of the most notable victims of posterity's lack of judgment is the Eleatic Zeno. Having invented four arguments all immeasurably subtle and profound, the grossness of subsequent philosophers pronounced him to be a mere ingenious juggler, and his arguments to be one and all sophisms. After two thousand years of continual refutation, these sophisms were reinstated, and made the foundation of a mathematical renaissance ....

There is no question that Zeno's ideas, and his cogent arguments, remained vital and influential even into modern times. Isaac Newton wrestled with the ideas when he was inventing his calculus (see [GLE]). It was not until A. Cauchy in the nineteenth century that a cogent manner was devised for dealing with many of the issues that Zeno raised (see our Chapter 12 below). It is well known that man wrestled with the idea of infinity for many hundreds of years; many nineteenth century mathematicians forbade any discussion or mention of the concept of infinity (see [KAP2]). And infinity is the obverse idea to infinitesimals. The histories of the two ideas are intimately bound up (see also [KAP1]).

As to Zeno's cosmology, it is by no means disjoint from his monistic ideas. Diogenes Laertius asserts that Zeno proposed a universe consisting of several worlds, composed of

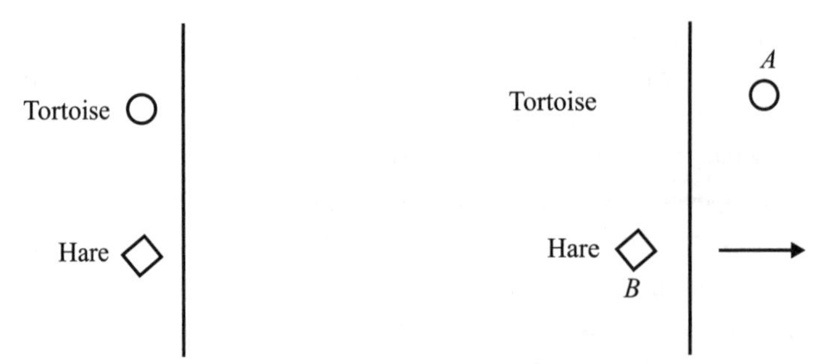

Figure 2.2. First formulation of Zeno's paradox.

Figure 2.3. The tortoise gets a head start.

"warm" and "cold", "dry" and "wet" but no void or empty space. It is not immediately clear that these contentions are consistent with the spirit of Zeno's paradoxes, but there is evidence that this type of belief was prevalent in the fifth century B.C.E., particularly associated with medical theory, and it may have been Zeno's version of a belief held by the Eleatic School.

Now let us turn our attention to the mathematical aspects of Zeno's ideas. We begin our studies by stating some versions of Zeno's paradox. Then we will analyze them, and compare them with our modern notion of limit that was developed by Cauchy and others in the nineteenth century. In the end, we will solve this 2000-year-old problem that so mightily baffled the Greeks.

Further Reading

Darrell P. Minor, Parrondo's paradox: Hope for losers!, *The College Mathematics Journal* 34(2003), 15–20

John Stillwell, The continuum problem, *The American Mathematical Monthly* 109(2002), 286–297.

2.3 Consideration of the Paradoxes

We consider several distinct formulations of the paradoxes. There is a common theme running through all of them.

Zeno's Paradox, First Formulation. A tortoise and a hare are in a race. See Figure 2.2. Now everyone knows that a hare can run faster than a tortoise (for specificity, let us say that the hare runs ten times as fast as the tortoise), so it is decided to give the tortoise a head start. Thus the tortoise is allowed to advance 10 feet before the hare begins—Figure 2.3. Hence the race starts with the tortoise at point A and the hare at point B.

Now first the hare must advance to the point A. But while she is doing that, the tortoise will have moved ahead a bit and he will be at a new point A' (Figure 2.4). The hare, in order to catch up, must move to point A'. Of course, while the hare is doing that, the tortoise will have moved ahead to some new point A''. Now the hare must catch up to point A''.

You can see the problem. Every time the hare tries to catch up with the tortoise, the tortoise will move ahead. The hare can never catch up. Thus the tortoise will win the race. □

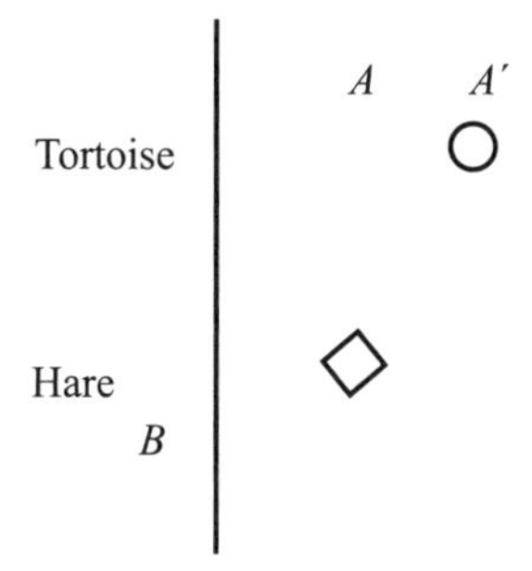

Figure 2.4. The tortoise advancing on the hare.

For You to Try Apply the analysis just given to two children who are each packing sand into a bucket. One child is twice as fast as the other: she packs two cups of sand per minute while the slower boy packs only one cup of sand per minute. But the slower child is allowed to begin with 3 cups of sand already in his bucket. Discuss how the bucket-packing will progress.

Zeno's Paradox, Second Formulation. A woman is walking towards a wall—Figure 2.5. But first she must walk halfway to the wall (Figure 2.6). And then she must walk half the remaining distance to the wall. See Figure 2.7. And so forth. In short, she will never actually *reach* the wall—because at each increment she has half the remaining distance to go. Figure 2.8 illustrates the incremental positions of the woman. □

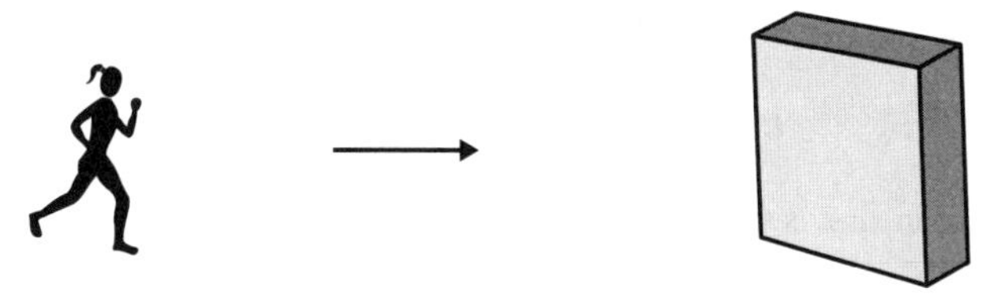

Figure 2.5. Second formulation of Zeno's paradox.

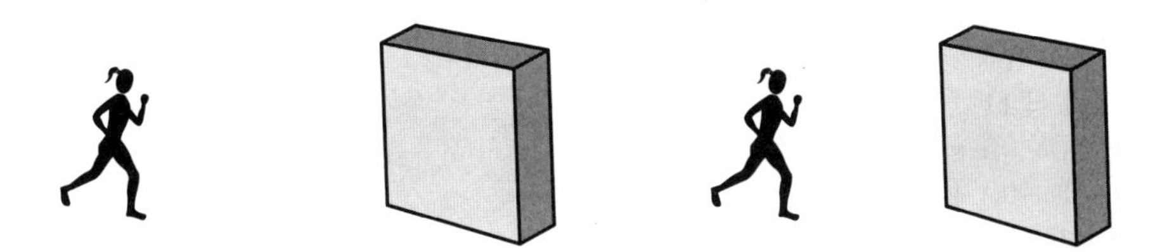

Figure 2.6. Halfway to the wall.

Figure 2.7. Advancing on the wall.

Zeno's Paradox, Third Formulation. Motion is impossible. For if an object moves in a straight line from 0 to A, then it first must reach $\frac{1}{2}A$. See Figure 2.9. But before it can reach $\frac{1}{2}A$ it must reach $\frac{1}{4}A$. *Ad infinitum.* Thus the motion can never begin. □

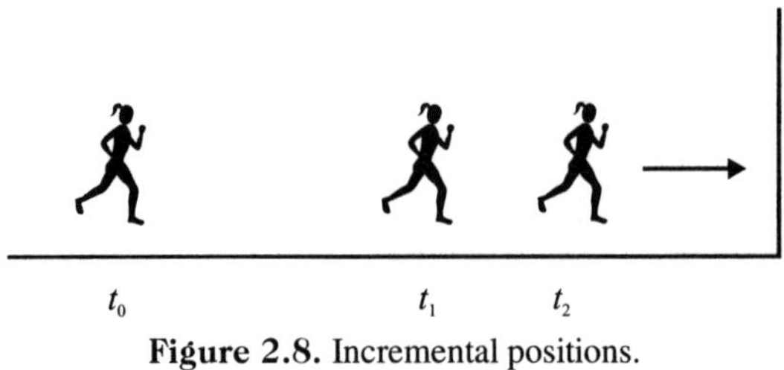

Figure 2.8. Incremental positions.

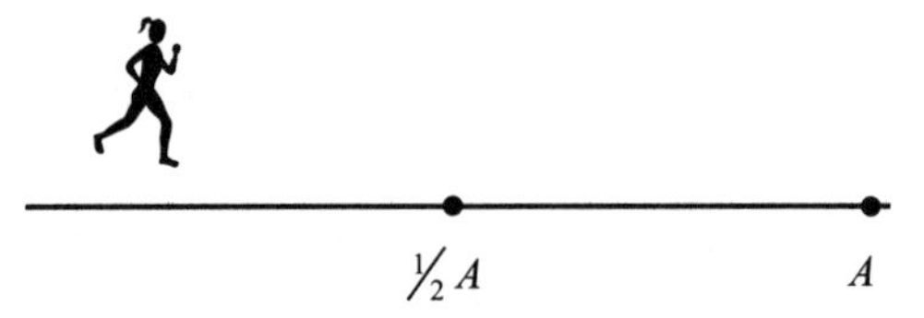

Figure 2.9. Third formulation of Zeno's paradox.

What is Really Going On?

Let us examine the first version of the paradox to see what is really going on. For specificity, let us suppose that the tortoise moves at the rate of 1 foot per second, and the hare moves at the rate of 10 feet per second. It takes the hare 1 second to catch up to the tortoise's head-start position (10 feet advanced) at A. During that 1 second, the tortoise has of course advanced 1 foot. It takes the hare 0.1 seconds to advance that additional foot. During that 0.1 seconds, the tortoise has advanced 0.1 of a foot. It takes the hare 0.01 seconds to catch up that much space. During that time, the tortoise advances another 0.01 feet. And so forth.

To summarize, if we add up all the units of distance that the tortoise will travel during this analysis, we obtain

$$D_T = 10 + 1 + 0.1 + 0.01 + \cdots .$$

A similar calculation shows that the hare travels

$$D_H = 10 + 1 + 0.1 + 0.01 + \cdots .$$

Now we see that our decimal notation comes to the rescue (and the Greeks *definitely* did not have decimal notation). The sum $D_T = D_H$ equals $11.111\ldots$ feet. To see this just sum up the terms:

$10 + 1 = 11$

$10 + 1 + 0.1 = 11.1$

$10 + 1 + 0.1 + 0.01 = 11.11$

and so forth.

Now take out your pencil and paper and divide 9 into 1 (or do it on your calculator if you must). You will obtain the answer $0.111\ldots$.[3] Thus we see that the total distance that the tortoise (or hare) travels during our analysis is $D = 11\frac{1}{9}$ feet. *What does this number mean?*

The point of the number D is that this is the place where the hare and the tortoise meet—they are in the same position. After that, the hare will pull ahead and win the race.

[3] In the next section we shall discuss infinite repeating decimal representations for rational numbers.

But we can say more. The total length of time that it takes the tortoise (or the hare) to get to position D is

$$\mathcal{T} = 1 + 0.1 + 0.01 + \cdots = 1.1111\cdots = 1\frac{1}{9} \text{ seconds.}$$

Our conclusion is that, after $1\frac{1}{9}$ seconds, the tortoise and the hare will have reached the same point. In the ensuing time, the hare will still be traveling ten times as fast as the tortoise, so of course it will pull ahead and win the race.

For You to Try Refer back to the preceding **For You to Try** unit. Assume that each child has a *very large* bucket. Do an analysis like the one that we did for the tortoise and the hare to determine when the faster girl will equal the slower boy in sand packing (and thereafter pass him).

Further Reading

Florian Cajori, History of Zeno's arguments on motion: phases in the development of the theory of limits, *The American Mathematical Monthly* 22 (1915), 179–186.

Leigh Atkinson, Where do functions come from?, *The College Mathematics Journal*, 33 (2002), 107–112.

2.4 Decimal Notation and Limits

In our analysis of Zeno's paradox, we came across an interesting idea: that of repeating decimal expansions. The specific one that came up in the last section was .11111 .... We were conveniently able to observe that this is just $1/9$. But what does (for instance) the decimal expansion 0.57123123123123... represent (*if anything*)? Let us do a little analysis.

Let $x = 0.57123123123123\ldots$. Now consider the number

$$1000x = 571.23123123123123\ldots.$$

We subtract these two numbers in the traditional manner:[4]

$$\begin{array}{rcl} 1000x & = & 571.23123123123123\ldots \\ x & = & 0.57123123123123\ldots \\ \hline 999x & = & 570.66 \end{array}$$

Notice how all the 123s cancel out! It is convenient to write the resulting equation as

$$999x = \frac{57066}{100}.$$

Then we find that

$$x = \frac{57066}{99900} = \frac{9511}{16650}.$$

We see that, with a bit of algebraic manipulation, we were able to express a repeating decimal as a rational fraction.

[4] The choice of $1000x$ rather than $100x$ or $10000x$ is motivated by the fact that it results in useful cancellations, as we shall see.

For You to Try Express the number $x = 43.75417171717\ldots$ as a rational fraction.

Rest assured that the ancient Greeks certainly considered the questions we are discussing here. But they were not equipped to come up with the answers that you have seen. They did not have the notation nor the concept of decimal number. But they certainly set in place the beginnings of the more complete understanding that we have today.

Further Reading

Saradakanta Ganguli, The Indian origin of the modern place-value arithmetical notation, *The American Mathematical Monthly* 39 (1932), 251–256.

David A. Smith, What's significant about a digit?, *The College Mathematics Journal* 20 (1989), 136–139.

2.5 Infinite Sums and Limits

The ideas we have considered so far actually beg a much more general question. When we studied Zeno's paradox, in the rendition with the tortoise and the hare, we considered the sum

$$10 + 1 + 0.1 + 0.01 + \cdots .$$

This might more conveniently be written as

$$10^1 + 10^0 + 10^{-1} + 10^{-2} + \cdots$$

or perhaps as

$$10 + \left(\frac{1}{10}\right)^0 + \left(\frac{1}{10}\right)^1 + \left(\frac{1}{10}\right)^2 + \cdots .$$

Observe that, after the first term, this is a sum of all the nonnegative powers of a fixed number, namely $1/10$. But that is an interesting notion, is it not? How can we sum all the powers of a fixed number? Let us pose the question a bit more abstractly.

Let the fixed number be $\sigma > 0$. Consider the sum

$$S = 1 + \sigma + \sigma^2 + \sigma^3 + \cdots .$$

We call this a *geometric series* in powers of σ. Our goal is to actually sum this series—to find an explicit formula for the infinite sum on the right.

In order to understand S, let us multiply both sides by σ. So

$$\sigma \cdot S = \sigma + \sigma^2 + \sigma^3 + \cdots .$$

Adding 1 to both sides yields

$$1 + \sigma \cdot S = 1 + \sigma + \sigma^2 + \sigma^3 + \cdots .$$

But now we recognize the right-hand side as S. So we can rewrite the last equation as

$$1 + \sigma \cdot S = S$$

or

$$1 = S \cdot (1 - \sigma).$$

Finally, we conclude that

$$S = \frac{1}{1 - \sigma}.$$

Put in other words, what we have learned is that

$$1 + \sigma + \sigma^2 + \sigma^3 + \cdots = \frac{1}{1 - \sigma}.$$

EXAMPLE 2.1. Calculate the sum

$$1 + \frac{1}{10} + \left(\frac{1}{10}\right)^2 + \left(\frac{1}{10}\right)^3 + \cdots.$$

Solution. We recognize this as the series that we encountered in our study of Zeno's paradox. But now we have a simple and direct way to analyze it. We see that this is a geometric series with $\sigma = 1/10$. Thus the sum is

$$S = \frac{1}{1 - 1/10} = \frac{10}{9}.$$

This answer is consistent with the one that we obtained when we discussed Zeno.

EXAMPLE 2.2. Calculate the sum

$$T = \frac{2}{3} + \left(\frac{2}{3}\right)^2 + \left(\frac{2}{3}\right)^3 + \cdots.$$

Solution. This is not precisely in our standard form for a geometric series. But we may write

$$T = \frac{2}{3} \cdot \left[1 + \frac{2}{3} + \left(\frac{2}{3}\right)^2 + \cdots\right] = \frac{2}{3} \cdot S,$$

where S is a standard geometric series in powers of $2/3$. Thus $S = 1/[1 - 2/3] = 3$ and hence

$$T = \frac{2}{3} \cdot 3 = 2.$$

For You to Try Find the sum of the series

$$4 + \frac{4 \cdot 5}{6} + \frac{4 \cdot 25}{36} + \frac{4 \cdot 125}{216} + \cdots.$$

Further Reading

Franklin Kemp, Infinite series flow chart for $\sum$, *The Two-Year College Mathematics Journal* 13(1982), 199.

I. M. Sheffer, Convergence of multiply-infinite series, *The American Mathematical Monthly* 52(1945), 365–376.

2.6 Finite Geometric Series

Thus far we have been examining the question of summing an infinite geometric series of the form

$$1 + \sigma + \sigma^2 + \sigma^3 + \cdots .$$

It seems reasonable to consider also the sums of finite geometric series such as

$$1 + 3 + 3^2 + \cdots + 3^{100} .$$

The idea is best understood by way of an example.

EXAMPLE 2.3. Find the sum of the series

$$1 + \left(\frac{1}{3}\right) + \left(\frac{1}{3}\right)^2 + \left(\frac{1}{3}\right)^3 + \cdots + \left(\frac{1}{3}\right)^{100} .$$

Solution. It would be quite tedious to actually *add up* this series—even with the aid of a calculator. Let us instead use some mathematical reasoning to tame the problem.

Our idea is to express this sum in terms of infinite geometric series. Namely, we may write

$$\begin{aligned}
&1 + \left(\frac{1}{3}\right) + \left(\frac{1}{3}\right)^2 + \left(\frac{1}{3}\right)^3 + \cdots + \left(\frac{1}{3}\right)^{100} \\
&\quad = \left[1 + \left(\frac{1}{3}\right) + \left(\frac{1}{3}\right)^2 + \left(\frac{1}{3}\right)^3 + \cdots\right] - \left[\left(\frac{1}{3}\right)^{101} + \left(\frac{1}{3}\right)^{102} + \left(\frac{1}{3}\right)^{103} + \cdots\right] \\
&\quad = \left[1 + \left(\frac{1}{3}\right) + \left(\frac{1}{3}\right)^2 + \left(\frac{1}{3}\right)^3 + \cdots\right] - \left(\frac{1}{3}\right)^{101} \cdot \left[1 + \left(\frac{1}{3}\right) + \left(\frac{1}{3}\right)^2 + \cdots\right].
\end{aligned}$$

Now we know that

$$1 + \left(\frac{1}{3}\right) + \left(\frac{1}{3}\right)^2 + \left(\frac{1}{3}\right)^3 + \cdots = \frac{1}{1 - (1/3)} = \frac{3}{2} .$$

In conclusion,

$$1 + \left(\frac{1}{3}\right) + \left(\frac{1}{3}\right)^2 + \left(\frac{1}{3}\right)^3 + \cdots + \left(\frac{1}{3}\right)^{100} = \frac{3}{2} - \left(\frac{1}{3}\right)^{101} \cdot \frac{3}{2} = \frac{3}{2} \cdot \left[1 - \left(\frac{1}{3}\right)^{101}\right] .$$

The method used in this last example is a cute trick, but not entirely satisfactory. For suppose we attempted to sum

$$1 + 3 + 3^2 + 3^3 + \cdots + 3^{100}$$

by the same method. It would fail, just because

$$1 + 3 + 3^2 + 3^3 + \cdots \qquad (*)$$

cannot be added. In other words, the sum $(*)$ increases without bound.[5] So it cannot be manipulated arithmetically as we did in the last example.

[5] A mathematician might say that the limit is $+\infty$.

Let us now develop a somewhat different technique. We will imitate the methodology of the last section. Let

$$S = 1 + \sigma + \sigma^2 + \sigma^3 + \cdots + \sigma^K .$$

Multiplying both sides by σ, we find that

$$\begin{aligned}\sigma \cdot S &= \sigma + \sigma^2 + \sigma^3 + \sigma^4 + \cdots + \sigma^{K+1} \\ &= \left[1 + \sigma + \sigma^2 + \sigma^3 + \sigma^4 + \cdots + \sigma^K\right] + (\sigma^{K+1} - 1) \\ &= S + (\sigma^{K+1} - 1) .\end{aligned}$$

Rearranging, we see that

$$S \cdot (\sigma - 1) = \sigma^{K+1} - 1$$

or

$$S = \frac{\sigma^{K+1} - 1}{\sigma - 1} . \qquad (\star)$$

Now let us do an example to illustrate the utility of this new formula.

EXAMPLE 2.4. Calculate the sum

$$S = 1 + 3 + 3^2 + 3^3 + \cdots + 3^{100} .$$

Solution. We apply formula ($\star$) with $\sigma = 3$ and $K = 100$. Thus

$$S = \frac{3^{101} - 1}{3 - 1} = \frac{1}{2} \cdot \left(3^{101} - 1\right) .$$

For You to Try Use your calculator to calculate the last sum, and compare your result with the answer that we obtained through mathematical reasoning.

EXAMPLE 2.5. Calculate the sum

$$T = \left(\frac{3}{4}\right)^{10} + \left(\frac{3}{4}\right)^{11} + \left(\frac{3}{4}\right)^{12} + \cdots + \left(\frac{3}{4}\right)^{30} .$$

Solution. We write

$$\begin{aligned}T &= \left[1 + \frac{3}{4} + \left(\frac{3}{4}\right)^2 + \cdots + \left(\frac{3}{4}\right)^{30}\right] - \left[1 + \frac{3}{4} + \left(\frac{3}{4}\right)^2 + \cdots + \left(\frac{3}{4}\right)^{9}\right] \\ &= \frac{(3/4)^{31} - 1}{(3/4) - 1} - \frac{(3/4)^{10} - 1}{(3/4) - 1} \\ &= 4 \cdot \left[\left(\frac{3}{4}\right)^{10} - \left(\frac{3}{4}\right)^{31}\right] .\end{aligned}$$

For You to Try Calculate the sum

$$W = \left(\frac{-5}{7}\right)^{12} + \left(\frac{-5}{7}\right)^{13} + \left(\frac{-5}{7}\right)^{14} + \cdots + \left(\frac{-5}{7}\right)^{45} .$$

For You to Try Calculate the sum

$$V = \left(\frac{6}{11}\right)^4 + \left(\frac{6}{11}\right)^6 + \left(\frac{6}{11}\right)^8 + \cdots + \left(\frac{6}{11}\right)^{30}.$$

Can you discern a pattern in your answers? Is it possible to look at a sum of the form

$$\alpha^j + \alpha^{j+1} + \cdots + \alpha^k$$

for $0 < j < k$ and just write down the answer?

Further Reading

Curtis Cooper, Geometric series and a probability problem, *The American Mathematical Monthly* 93(1986), 126–127.

Michael W. Ecker, A novel approach to geometric series, *The College Mathematics Journal* 29(1998), 419–420.

2.7 Some Useful Notation

This is a good opportunity to learn some useful and fun mathematical notation. The symbol

$$\sum_{j=1}^{N} a_j$$

is a shorthand for the sum

$$a_1 + a_2 + a + 3 + \cdots + a_N .$$

The symbol $\sum$ is the Greek letter *sigma* (the cognate of "S" in our alphabet), and stands for sum. The lower limit $j = 1$ tells where the sum, or series, begins. The upper limit "N" (or "$j = N$") tells where the sum (or series) stops.

EXAMPLE 2.6. Write out the sum

$$\sum_{j=1}^{8} [j^2 + j].$$

Solution. According to our rule, this is

$$\begin{aligned} &(1^2 + 1) + (2^2 + 2) + (3^2 + 3) + (4^2 + 4) + (5^2 + 5) \\ &\qquad + (6^2 + 6) + (7^2 + 7) + (8^2 + 8) \\ &= 2 + 6 + 12 + 20 + 30 + 42 + 56 + 72 \\ &= 240 . \end{aligned}$$

□

EXAMPLE 2.7. Write out the sum

$$\sum_{j=5}^{10} \frac{j}{j+1}.$$

Solution. Notice that we are stretching our new notation by beginning the sum at an index other than 1. It equals

$$\begin{aligned} &\frac{5}{5+1} + \frac{6}{6+1} + \frac{7}{7+1} + \frac{8}{8+1} + \frac{9}{9+1} + \frac{10}{10+1} \\ &= \frac{5}{6} + \frac{6}{7} + \frac{7}{8} + \frac{8}{9} + \frac{9}{10} + \frac{10}{11} \\ &\approx 5.2634\,. \end{aligned}$$

□

We can also use the summation notation to denote an infinite series. For example,

$$\begin{aligned} \sum_{j=0}^{\infty} \left(\frac{1}{2}\right)^j &= \left(\frac{1}{2}\right)^0 + \left(\frac{1}{2}\right)^1 + \left(\frac{1}{2}\right)^2 + \left(\frac{1}{2}\right)^3 + \cdots \\ &= 1 + \left(\frac{1}{2}\right)^1 + \left(\frac{1}{2}\right)^2 + \left(\frac{1}{2}\right)^3 + \cdots. \end{aligned}$$

And we know, from our earlier studies, that in fact this sum equals 2.

Further Reading

Donald E. Knuth, Two notes on notation, *The American Mathematical Monthly* 99(1992), 403–422.

David P. Kraines, Vivian Y. Kraines, and David A. Smith, Sum the alternating harmonic series, *The College Mathematics Journal* 20(1989), 433–435.

2.8 Concluding Remarks

Geometric series arose very naturally for us in our consideration of Zeno's paradox. In fact the Greeks were well aware of geometric series. They occur, in essence, in Euclid IX-35 [EUC], and also in Archimedes's quadrature of the parabola. Today, geometric series arise frequently in engineering analysis, in the study of the way that plants grow, and in many other applications of the mathematical sciences. They are a primary example of the mathematical modeling of nature. They also have considerable intrinsic interest—they are simply fascinating mathematical objects to study.

Exercises

1. Use geometric series to analyze the second version of Zeno's paradox.

2. Formulate a version of Zeno's paradox that involves division by 3 instead of division by 2. Discuss this question in class.

3. Calculate the sum

$$\left(\frac{4}{3}\right)^5 + \left(\frac{4}{3}\right)^6 + \left(\frac{4}{3}\right)^7 + \cdots + \left(\frac{4}{3}\right)^{50}.$$

4. Calculate the sum

$$\left(\frac{2}{7}\right)^3 + \left(\frac{2}{7}\right)^6 + \left(\frac{2}{7}\right)^9 + \cdots + \left(\frac{2}{7}\right)^{81}.$$

5. Calculate the sum

$$\sum_{j=0}^{\infty} \left(\frac{6}{13}\right)^j = 1 + \left(\frac{6}{13}\right) + \left(\frac{6}{13}\right)^2 + \cdots.$$

6. Calculate the sum

$$\sum_{j=3}^{\infty} \left(\frac{12}{17}\right)^j = \left(\frac{12}{17}\right)^3 + \left(\frac{12}{17}\right)^4 + \left(\frac{12}{17}\right)^5 + \cdots.$$

7. Calculate the sum

$$\left(\frac{17}{21}\right)^5 + \left(\frac{17}{21}\right)^{10} + \left(\frac{17}{21}\right)^{15} + \cdots.$$

8. A certain radioactive material has the property that half the substance present decays every three hours. If there are 10 grams present at 10:00A.M.on Monday, then how much material will remain at 10:00A.M.on Thursday of that same week? [*Hint:* You cannot solve this problem just using techniques of arithmetic. You must use the lessons of this chapter.]

9. A population of bacteria reproduces constantly. As a result, the total number of bacteria doubles every 6 hours. If there are 10,000 bacteria present at 9:00A.M.on Tuesday, then how many bacteria will be present at 9:00A.M.on Saturday of that same week? [*Hint:* You cannot solve this problem just using techniques of arithmetic. You must use the lessons of this chapter.]

10. It begins snowing some time before noon. At noon, a snow plow begins to clear the street. It clears two blocks in the first hour and one block in the second hour. When did it start snowing? [*Hint:* You will not be able to actually write down an equation or formula to solve this problem. But you can use the ideas from this chapter to set up an analysis of the problem. Use your computer or calculator to do some numerical approximations for the situation described. In other words, think of this as a problem of mathematical modeling. Use the calculating machinery to emulate the snowfall and come up with an approximate answer. Discuss this problem in class.]

11. A sponge absorbs water at a steady rate. As a result, the volume of the sponge increases by a factor of one tenth each hour. If the sponge begins at noon having volume 0.8 cubic feet, then what will be the volume of the sponge at the same time on the next day?

12. You deposit \$1000 in the bank on January 2, 2006. The bank pays 5% interest, compounded daily (this means that $1/365$ of the interest is paid each day, and the interest is added to the principal). How much money will be in your account on January 2, 2007? [*Hint:* Bear in mind that, when interest is calculated on the second day, there will be interest paid on the interest from the first day. And so forth. Thus the amount of increase in money is greater with each passing day. Discuss this problem in class.]

13. Project: Examine the infinite sum

$$\sum_{j=1}^{\infty} \frac{1}{j} = 1 + \frac{1}{2} + \frac{1}{3} + \cdots .$$

This series in fact diverges. By this we mean that the partial sums become larger and larger (without bound) as N gets large. Set

$$S_N \equiv \sum_{j=1}^{N} \frac{1}{j} .$$

Use your calculator or computer to calculate S_{100}, S_{1000}, $S_{1000000}$. You will find that the answers you obtain are indeed larger and larger, but they are not *very large.* In fact $S_{1000000}$ is only about 13. In order for S_N to exceed 100, N must be about 2.66×10^{43}. So this series diverges *very slowly.* Examine the partial sums S_2, S_4, S_8, and so forth to see whether you can discern a pattern that will provide convincing evidence that

$$\lim_{N \to \infty} S_N = +\infty .$$

You can read about series like this one in context in the text [KRA1].

14. Project: Let us use an unusual method to "calculate" the length of the diagonal of a square of side 1. Examine Figure 2.10.

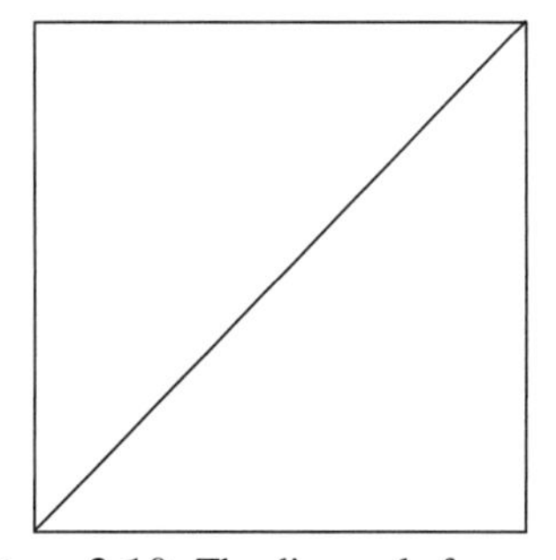

Figure 2.10. The diagonal of a square.

We will approximate the diagonal with a piecewise linear curve with four edges parallel to the sides of the square—see Figure 2.11. Each piece has length $1/2$, so the length of the approximating curve is $1/2 + 1/2 + 1/2 + 1/2 = 2$.

Now we obtain a finer approximation by using a piecewise linear curve with shorter components—each having length $1/4$. See Figure 2.12. The sum of the lengths of these eight pieces is 2. So we see that our second approximation also has value 2.

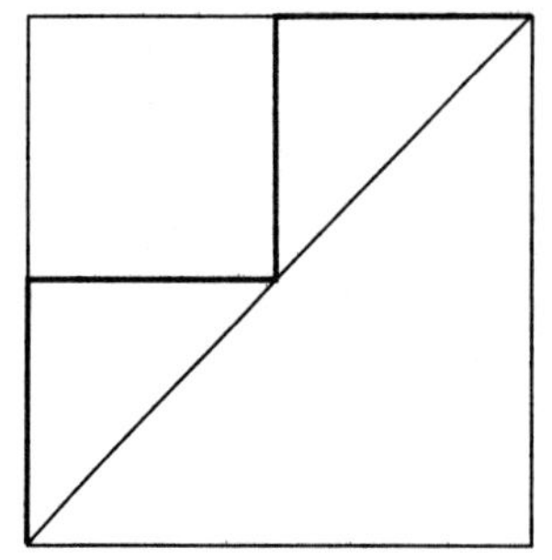

Figure 2.11. First approximation to the diagonal of a square.

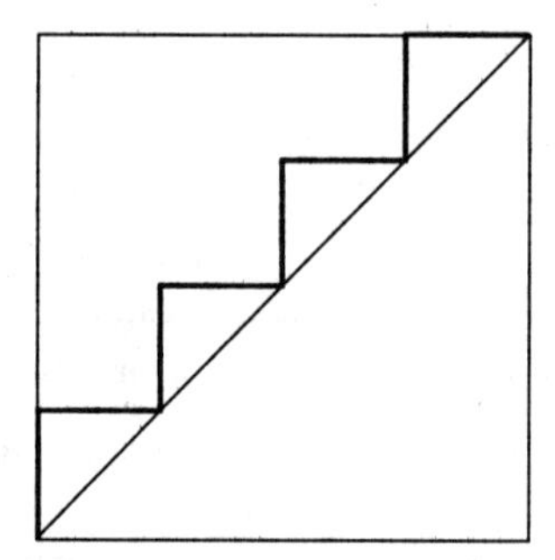

Figure 2.12. Second approximation to the diagonal of a square.

Continuing in this fashion, with piecewise linear curves having components of length $1/8$, $1/16$, etc., we find that each approximation has length 2. Since the approximating curves clearly tend to the diagonal of the square (just look at the picture!), we conclude that the diagonal of the square has length 2.

What—if anything—is wrong with this reasoning? How can you make peace with the Pythagorean theorem? See [DUN], [KEN] [CGK] for more conundra in calculus.

15. Project: Imitate the argument in Exercise 14 to calculate the circumference of a circle. Again, what is the flaw in the reasoning? The reference [STE] will give you further ideas in this topic.

16. Project: If an arbitrary hexagon is inscribed in any conic section, and opposite pairs of sides are extended until they meet, then the three intersection points will lie on a straight line. This famous theorem of projective geometry is due to Pascal. Draw some pictures to convince yourself that it is true. Find a justification for the assertion. See [CAS] for further discussion of projective geometry.

3

The Mystical Mathematics of Hypatia

3.1 Introduction to Hypatia

One of the great minds of the ancient world was Hypatia of Alexandria (370 C.E.–430 C.E.). Daughter of the astronomer and mathematician Theon, she flourished during the reign of the Emperor Arcadius. Precious little is known of Hypatia's life and work, but further reading may be found in [DZI] and [HEA].

Historians believe that Theon tried to raise the "perfect human being" in his daughter Hypatia. He nearly succeeded, in that Hypatia was a physical beauty and had dazzling intellect. She had remarkable grace and was an accomplished athlete. She was a dedicated scholar possessing a towering intellect.

Hypatia soon outstripped her father and her teachers and became the leading intellectual light of Alexandria. She was a powerful teacher, and communicated strong edicts to her pupils. Among these were:

> All formal dogmatic religions are fallacious and must never be accepted by self-respecting persons as final.
>
> Reserve your right to think, for even to think wrongly is better than not to think at all.

The classical Hypatia.

A modern Hypatia.

> Neo-Platonism is a progressive philosophy, and does not expect to state final conditions to men whose minds are finite. Life is an unfoldment, and the further we travel the more truth we can comprehend. To understand the things that are at our door is the best preparation for understanding those that lie beyond.
>
> Fables should be taught as fables, myths as myths, and miracles as poetic fantasies. To teach superstitions as truths is a most terrible thing. The child mind accepts and believes them, and only through great pain and perhaps tragedy can he be in after years relieved of them. In fact men will fight for a superstition quite as quickly as for a living truth—often more so, since a superstition is so intangible you cannot get at it to refute it, but truth is a point of view, and so is changeable.

The writings of Hypatia have all been lost. What we know of her thoughts comes from citations and quotations in the work of others.

Hypatia was a pagan thinker at the time when the Roman world was converted to Christianity. Thus, in spite of her many virtues, she made enemies. Chief among these was Cyril, the Bishop of Alexandria. According to legend, he inflamed a mob of Christians against her. They set upon her as she was leaving her Thursday lecture, and she was dragged to a church where it was planned that she would be forced to recant her beliefs. But the mob grew out of control. Her clothes were rent from her body, she was beaten mercilessly, and then she was dismembered. The skin was flayed from her body with oyster shells. Her remains were then burned. The book [DZI] considers a variety of accounts of Hypatia and her demise. It is difficult to tell which are apocryphal.

Hypatia is remembered today for her work on Apollonius's theory of conics, and for her commentary on Diophantus. All of these theories survive to the present time, and are still studied intensely. She also did work, alongside her father, on editing Euclid's *Elements*. The surviving presentation of Euclid's classic work bears Hypatia's mark.

Certainly Hypatia was one of the great thinkers of all time, and it is appropriate for us to pay her due homage. But we have no detailed knowledge of her work—certainly no firsthand knowledge. So what we can do is to study conic sections with Hypatia in mind, knowing that she certainly left her mark on this subject. We will give some of the classical ideas, as Hypatia herself would have conceived them, and also some of the modern ideas—based on the analytic geometry of René Descartes (see Chapter 6).

It was Apollonius, Hypatia's inspiration, who first understood that all of the conic sections can be realized as slices of a fixed cone. He also gave the names to the conic sections that we use today. Examine Figure 3.1. It shows a cone with two nappes (branches). We

Figure 3.1. A cone with two nappes.

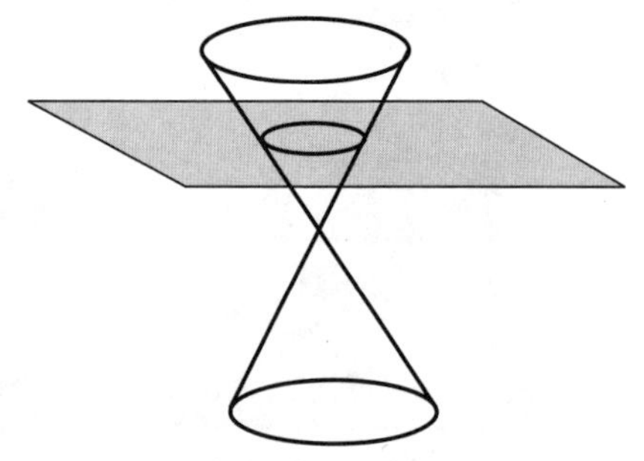

Figure 3.2. A circle.

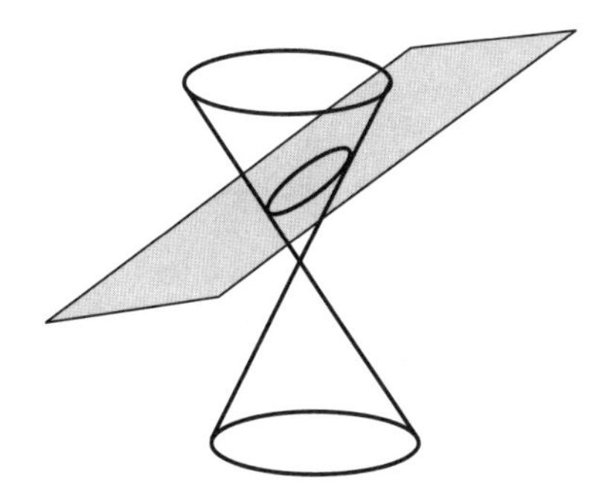

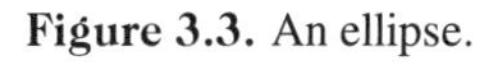

Figure 3.3. An ellipse.

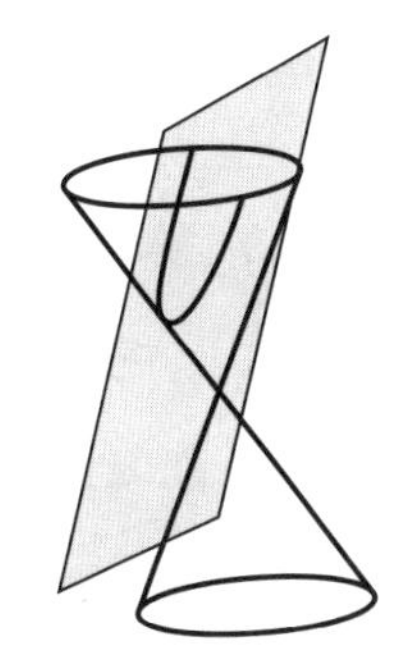

Figure 3.4. A parabola.

slice this cone with a plane. Depending on the way that the plane intersects the cone, the result will give different types of curves. Figure 3.2 shows a circle. Figure 3.3 shows an ellipse. Figure 3.4 exhibits a parabola. And Figure 3.5 gives us a hyperbola. Figure 3.6 shows each of these curves on a planar set of axes.

Of course it is intuitively clear how one can examine the intersection of the plane and the cone in Figures 3.2–3.5 to see where the circle, ellipse, parabola, and hyperbola in Figure 3.6 come from. But it would be advantageous, and certainly aesthetically pleasing, to have a synthetic definition of each of these figures that makes sense *in the context of the plane*. This we shall now discuss.

The Circle: A circle with center P and radius r is just the set of all points in the plane that have distance r from the point P. Examine Figure 3.7. It clearly exhibits this geometric definition. And you can see that we have made this definition *without any reference to the cone*. The cone is of course interesting for historical reasons: it is the genesis of these figures, and suggests that they are related. But each can be studied intrinsically, and for its own merits.

The Ellipse: Fix two points F_1 and F_2 in the plane. Fix a positive number a such that $2a$ is greater than the distance from F_1 to F_2. Consider the locus of points P in the plane with

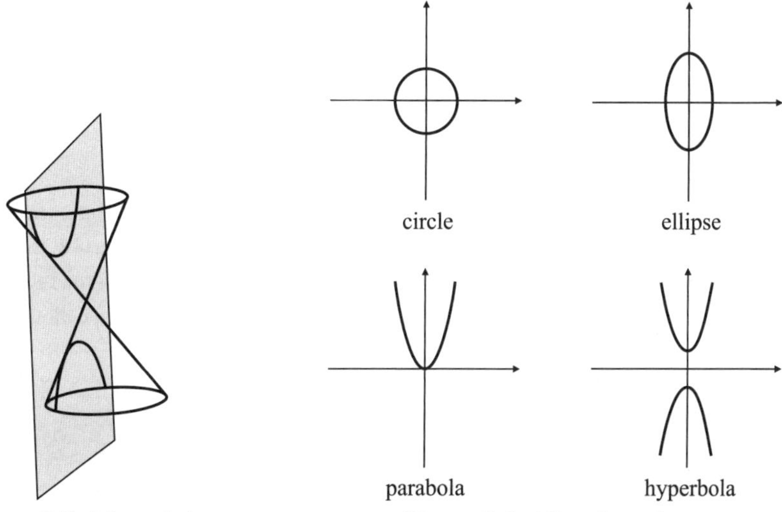

Figure 3.5. A hyperbola.

Figure 3.6. All conic sections.

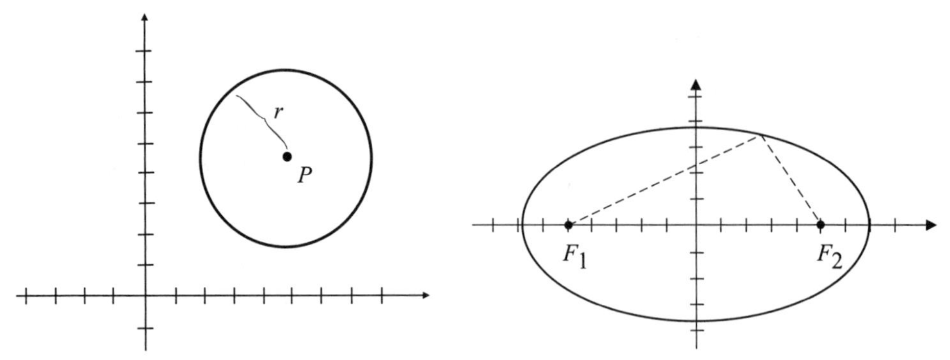

Figure 3.7. A circle with center P and radius r.

Figure 3.8. The locus of an ellipse.

the property that the distance of P to F_1 plus the distance of P to F_2 is equal to $2a$. This locus is called an *ellipse*. Refer to Figure 3.8.

The two points F_1, F_2 are called the *foci* of the ellipse and the midpoint of the segment $\overline{F_1 F_2}$ is called the *center* of the ellipse. The chord of the ellipse passing through the two foci is called the *major axis* of the ellipse. The perpendicular chord, passing through the center of the ellipse, is called the *minor axis* of the ellipse. See Figure 3.9.

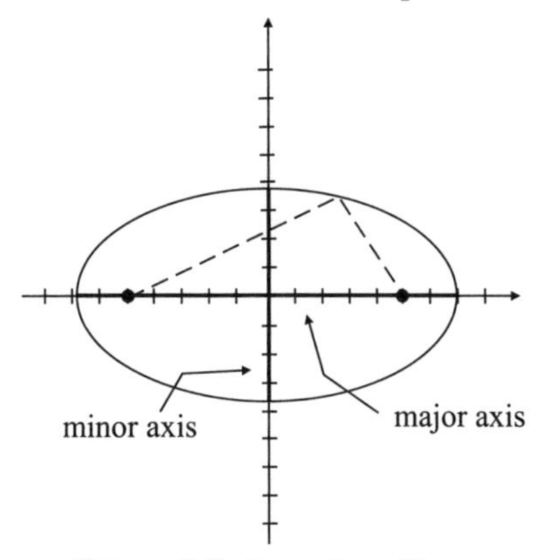

Figure 3.9. Axes of an ellipse.

For You to Try What happens to the ellipse as the two foci tend towards each other? As they coalesce into a single point? Does another conic section result?

The Parabola: Fix a point P in the plane and a line ℓ that does not pass through P. The set of points that are equidistant from P and ℓ is a parabola. See Figure 3.10. The point that is on the perpendicular segment from P to ℓ and halfway between the two is called the *vertex* of the parabola. The point P is called the *focus*, and the line ℓ is called the *directrix*.

For You to Try Let $P = (2, 0)$ and let ℓ be the line $\{(x, y) : x = -2\}$. Sketch the resulting parabola. Where will the vertex lie?

The Hyperbola: Fix two distinct points F_1, F_2 in the plane. Fix a positive number a that is less than half the distance of F_1 to F_2. Consider the locus of points P with the property

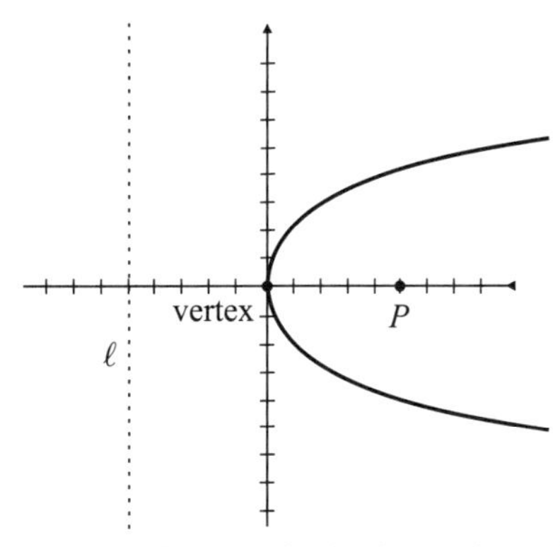

Figure 3.10. Synthetic description of a parabola.

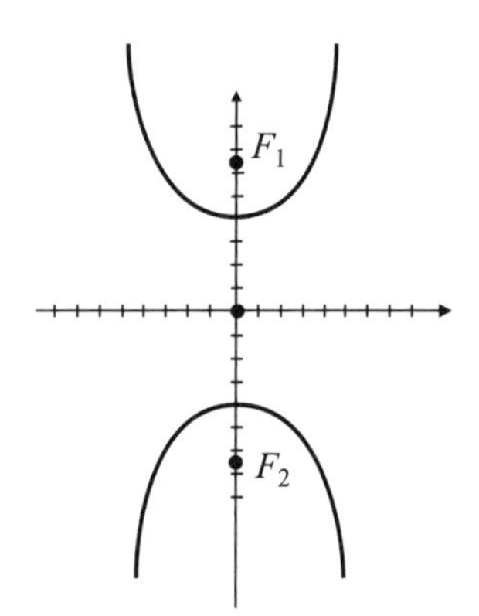

Figure 3.11. Foci, center, and vertices of a hyperbola.

that the *difference* of the distances $|P - F_1|$ and $|P - F_2|$ equals $2a$. This is a hyperbola. The points F_1, F_2 are called *foci* of the hyperbola. The midpoint of the two foci is called the *center* of the hyperbola. The line through the two foci intersects the hyperbola in two points called the *vertices* of the hyperbola. All of these attributes are exhibited in Figure 3.11.

For You to Try Let $F_1 = (-2, 0)$ and $F_2 = (2, 0)$. Let $a = 1$. Discuss the resulting hyperbola. Does it open up-down or left-right? Can you sketch the graph?

Further Reading

Debra Charpentier, Women mathematicians, *The Two-Year College Mathematics Journal* 8(1977), 73–79.

Michael A. B. Deakin, Hypatia and her mathematics, *The American Mathematical Monthly* 101(1994), 234–243.

3.2 What is a Conic Section?

Now we shall attempt to unify the preceding discussion. What do the circle, the ellipse, the parabola, and the hyperbola have in common? What are their common features?

One of the beauties of Descartes's conception of geometry is that it allows us to think of conic sections in terms of equations.

As an example, consider the parabola. Let us suppose that the directrix is the line $y = a > 0$ and the focus is the origin $O = (0, 0)$. The parabola defined by these two pieces of data is the set of points which are equidistant from the focus and the directrix. Let (x, y) be such a point. Then the distance of (x, y) to O is $\sqrt{x^2 + y^2}$. The distance of (x, y) to the directrix is $|y - a|$—see Figure 3.12. So the equation is then

$$\sqrt{x^2 + y^2} = |y - a| .$$

Squaring both sides gives

$$x^2 + y^2 = y^2 - 2ay + a^2$$

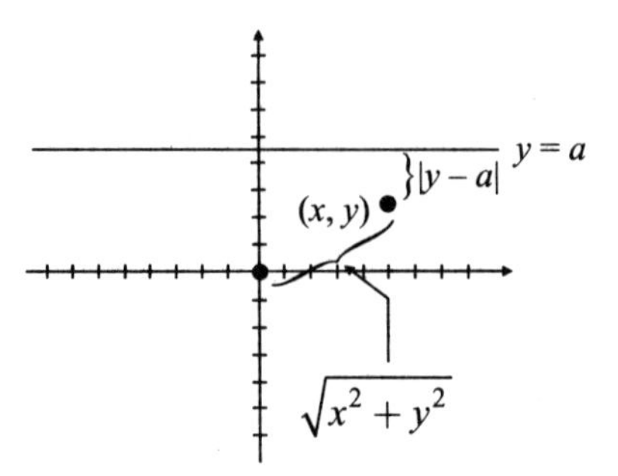

Figure 3.12. Distance to the directrix.

or

$$y = -\frac{1}{2a}x^2 + \frac{a}{2}.$$

See Figure 3.13.

A characteristic of the equation of a parabola is that one variable (in this case x) is squared and the other (in this case y) is not. Because of the positioning of the directrix and focus, a parabola such as we have been discussing must open either up or down. See Figure 3.13.

If instead we were to set up the geometry so that the directrix is $x = a > 0$ and the focus is the origin, then the equation would be

$$x = -\frac{1}{2a}y^2 + \frac{a}{2}.$$

Again, we see that one variable (in this case y) is squared and the other (in this case x) is not. Because of the positioning of the directrix and focus, a parabola such as we have been discussing must open either left or right. See Figure 3.14.

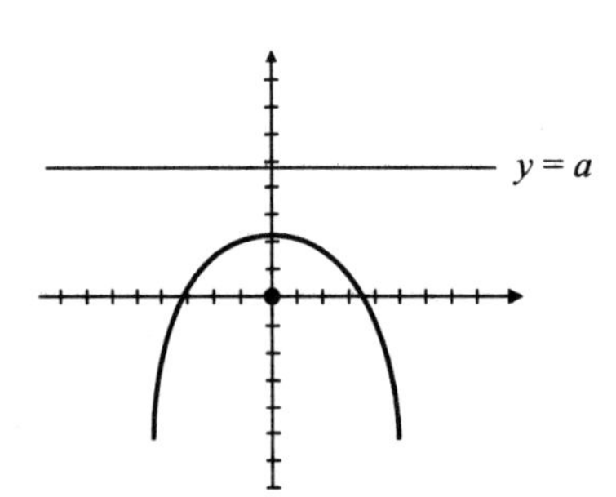

Figure 3.13. Downward opening parabola.

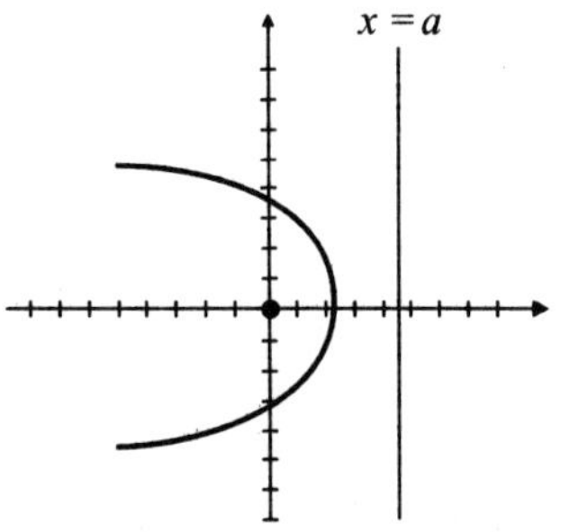

x = a

Figure 3.14. Left- or right-opening parabola.

More generally, the equation of an up-down opening parabola will have the form

$$y - b = c(x - a)^2.$$

Such a parabola will have vertex at the point (a, b) and will open up if $c > 0$ and down if $c < 0$. See Figure 3.15. The equation of a left-right opening parabola will have the form

$$x - a = c(y - b)^2.$$

Such a parabola will have vertex at the point (a, b) and will open to the right if $c > 0$ and to the left if $c < 0$. See Figure 3.16.

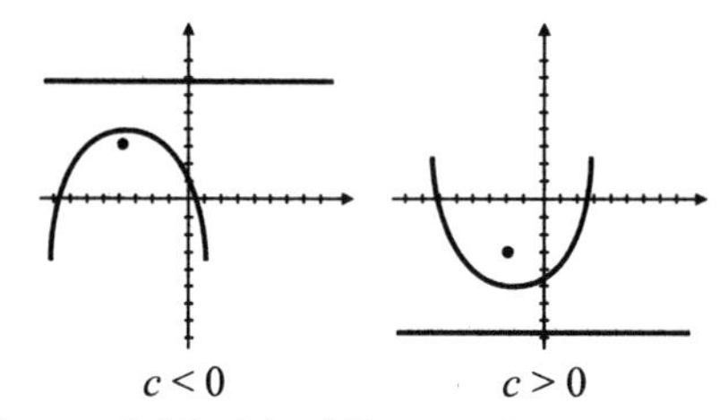

Figure 3.15. The difference between upward and downward opening parabolas.

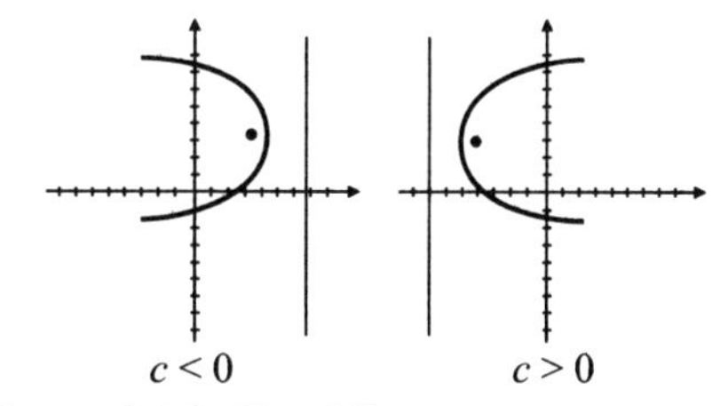

Figure 3.16. The difference between leftward and rightward opening parabolas.

For You to Try Discuss the parabola $y^2 - 4x - 2y = 10$. Does it open up-down or left-right? How can you tell? Can you sketch the graph?

An analysis similar to the one just given for the parabola, but a bit more complicated, yields that the equation of an ellipse will have the form

$$\frac{(x-c_1)^2}{a^2} + \frac{(y-c_2)^2}{b^2} = 1\,.$$

The center of this ellipse is the point (c_1, c_2). If we put in $y = c_2$ and solve for x we find that $x = c_1 \pm a$. Thus the left and right extreme points of the ellipse are $(c_1 - a, c_2)$ and $(c_1 + a, c_2)$. If instead we put $x = c_1$ and solve for y then we find that $y = c_2 \pm b$. Thus the upper and lower extreme points of the ellipse are $(c_1, c_2 - b)$ and $(c_1, c_2 + b)$. Refer to Figure 3.17 for a picture of this ellipse.

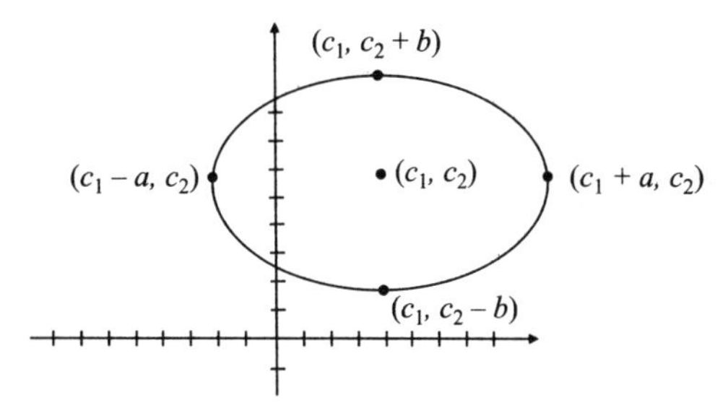

Figure 3.17. The ellipse.

For You to Try Discuss the ellipse $4x^2 + 8y^2 + 16x + 32y = 16$. Which direction is the major axis (the long direction) of the ellipse? Which direction is the minor axis (the short direction) of the ellipse? Can you sketch it?

Yet another analysis of the same type—and we shall omit the details—shows that the equation of a hyperbola has the form

$$\frac{(x-c_1)^2}{a^2} - \frac{(y-c_2)^2}{b^2} = \pm 1\,. \qquad (**)$$

If the right-hand side of $(**)$ is $+1$, then the hyperbola opens left-right. In fact take $y = c_2$; you can then solve for x and find that $x = c_1 \pm a$. So the vertices of the hyperbola are at $(c_1 - a, c_2)$ and $(c_1 + a, c_2)$. See Figure 3.18.

If instead the right-hand side of $(**)$ is -1, then the hyperbola opens up-down. In fact take $x = c_1$; you can then solve for y and find that $y = c_2 \pm b$. So the vertices of the hyperbola are at $(c_1, c_2 - b)$ and $(c_1, c_2 + b)$. See Figure 3.19.

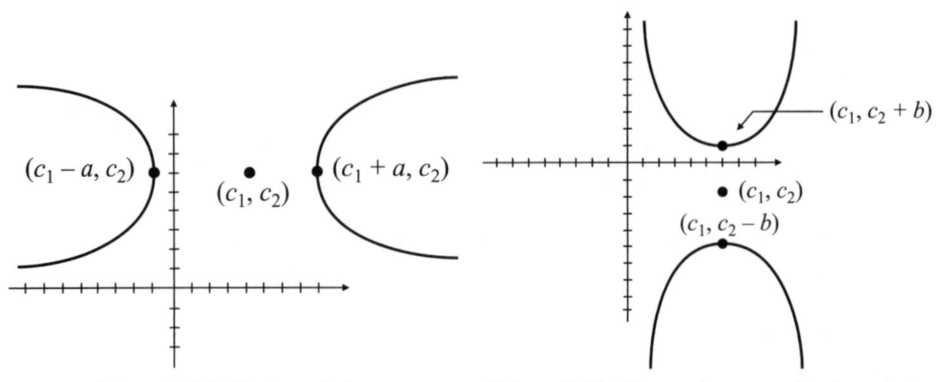

Figure 3.18. The hyperbola.

Figure 3.19. The up-down opening hyperbola.

For You to Try Discuss the hyperbola $4x^2 - 8y^2 + 8x - 16y = 12$. Does it open up-down or left-right? Can you sketch the graph?

Further Reading

Ayoub B. Ayoub, The eccentricity of a conic section, *The College Mathematics Journal* 34(2003), 116–121.

Michael Pascual, On defining the conic sections, *The American Mathematical Monthly*, 63(1956), 719–720.

Exercises

1. Let P, Q, R be three points in the plane which do not all lie on the same line. Then there is a unique circle that passes through all three of them. See Figure 3.20. There are several ways to confirm this assertion.

(a) A general circle has equation

$$x^2 + ax + y^2 + by = c .$$

Thus there are three undetermined parameters. And the three pieces of information provided by the fact that the circle must pass through $P = (p_1, p_2)$, $Q = (q_1, q_2)$, $R = (r_1, r_2)$ (and therefore these three points must satisfy the equation) will determine

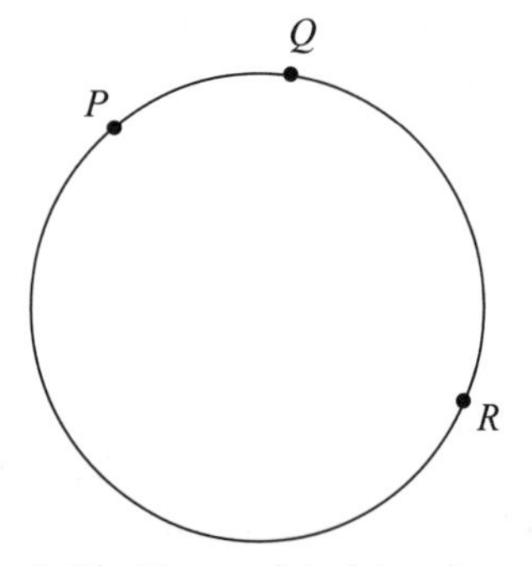

Figure 3.20. Three points determine a circle.

those parameters. Use this idea to find the unique circle that passes through $(1, 2)$, $(2, 3)$, and $(4, 9)$.

(b) There is a well-defined perpendicular bisector to the segment $\overline{PQ}$. This line represents the set of all points that are equidistant from P and Q. There is also a well-defined perpendicular bisector to the segment $\overline{QR}$. This line represents the set of all points that are equidistant from Q and R. The *intersection* of these two lines—which will be a single point C—will be the unique point that is equidistant from all three of P, Q, R. That must be the center of the circle. See Figure 3.21. The distance of C to P will be the radius. Use this idea to find the unique circle that passes through $(1, 0)$, $(0, 1)$, $(1, 1)$.

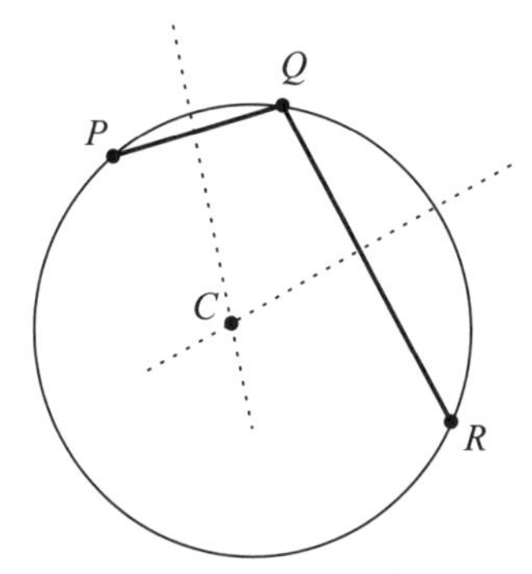

Figure 3.21. The center of the circle.

2. Consider the parabola $y = x^2$. Any ray entering the parabola from above and traveling straight down (see Figure 3.22) will bounce off the parabola and pass through the focus point $(0, 1/4)$ (the directrix is the line $y = -1/4$, as you can readily verify). Discuss this assertion in class. How would you determine the bounce of the ray? Think about the tangent line to the parabola at the point of impact. What does the tangent line have to do with the question?

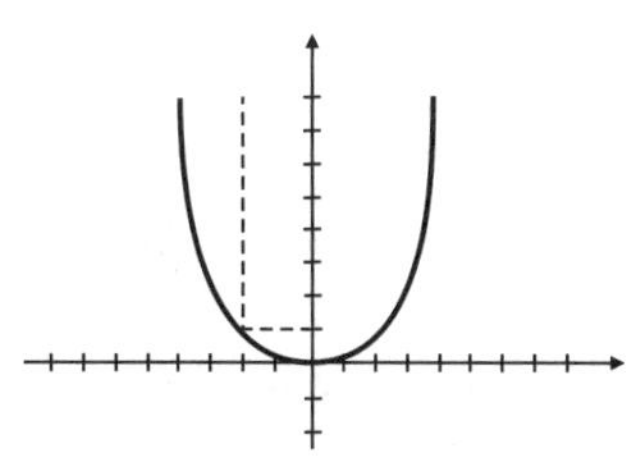

Figure 3.22. The parabola as a reflecting mirror.

3. Let $c > 0$. Fix nails at the two points $F_1 = (-c, 0)$ and $F_2 = (c, 0)$ in the plane. Imagine a string of length $2a > 2c$ that is in a closed loop. Put the loop of string over the nails. Now stretch the string taut with a pencil and move the pencil around in a closed curve. See Figure 3.23. The resulting curve will be an ellipse. You should try this yourself with two thumbtacks, a piece of string, and a real pencil.

Discuss this situation in class. Explain why the result is an ellipse. What is the length of the major axis of the ellipse? What is the length of the minor axis of the ellipse?

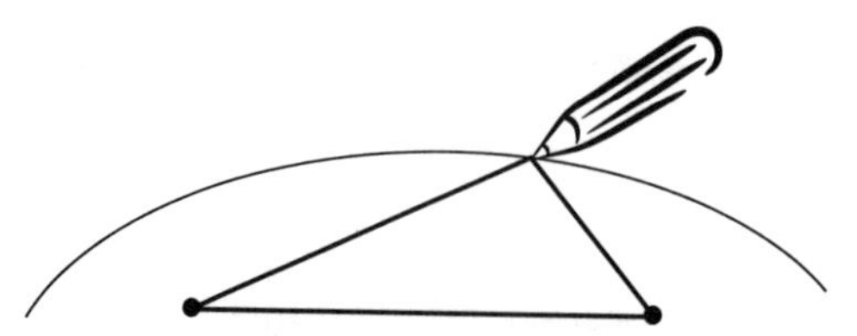

Figure 3.23. Drawing an ellipse.

4. Let $\{p_1, p_2, p_3, \ldots\}$ be an infinite collection of points in the plane. Suppose that the distance between any two of these points is an integer (different integers for different pairs of points in general). Then argue that the points must all lie on the same line. Discuss this problem in class. [*Hint:* The solution has something to do with a hyperbola!]

5. Two points in the plane do not uniquely determine a parabola. Give an example to explain why this is so. But three noncollinear points *do* uniquely determine a parabola. Explain why this is so. [*Hint:* Refer to the discussion in Exercise 1(**a**) for a clue.] What can you say for a hyperbola or an ellipse? Do you have to pre-specify the eccentricity of the ellipse?

6. Consider the line ℓ given by

$$ax + by + c = 0$$

in the plane. Let $P = (p_1, p_2)$ be a point that *does not* lie on that line. Show that the distance of P to the line ℓ is given by

$$d = \frac{|ap_1 + bp_2 + c|}{\sqrt{a^2 + b^2}} .$$

Discuss this question with your class. How does one determine the distance of a point to a line? What geometric construction is relevant?

7. Consider the parabola $y = x^2$ and the circle $x^2 + y^2 = r^2$. Is there a choice of $r > 0$ so that, at the points of intersection of the parabola and the circle, the two curves are perpendicular? [*Hint:* You can answer this question without calculating. Discuss the issue with your class.]

8. Answer Exercise 7 with the parabola $y = x^2$ replaced by the hyperbola $x^2 - y^2 = 1$.

9. Discuss the concept of tangent line to the curve $y = x^2$ at the point $(1, 1)$. What properties should it have? How could you determine this line? Discuss the issue with your class. We will consider this matter in further detail in Chapter 7.

10. Sketch the graph of the conic section

$$x^2 - 2x - 3y^2 + 6y = 10 .$$

Which type of conic section is this? How can you tell?

11. Sketch the graph of the conic section

$$x^2 + 4x - y = 15 .$$

Which type of conic section is this? How can you tell?

12. Sketch the graph of the conic section

$$4x^2 - 8x + 8y^2 + 32y = 64\,.$$

Which type of conic section is this? How can you tell?

13. Project: Imagine a billiard table in the shape of the ellipse $x^2 + 4y^2 = 4$. Place a billiard ball on the table and send it bouncing off the side. Describe the path of the ball. What does the path depend on? The book [TAB] has a delightful consideration of geometry and billiards.

14. Project: The transformation

$$x \longmapsto \frac{\sqrt{2}}{2}x - \frac{\sqrt{2}}{2}y\,,$$

$$y \longmapsto \frac{\sqrt{2}}{2}x + \frac{\sqrt{2}}{2}y$$

describes a rotation of the plane through an angle of $\pi/4$ radians (in the counterclockwise direction). Explain why this is so. Discuss the problem with your class. More generally, the transformation

$$x \longmapsto [\cos\theta]x - [\sin\theta]y\,,$$

$$y \longmapsto [\sin\theta]x + [\cos\theta]y$$

describes a rotation of the plane through an angle of θ radians (in the counterclockwise direction). Verify this assertion also.

If a quadratic equation describing a conic section—as discussed in this chapter—is subjected to one of these two changes of variable, then an equation of the form

$$Ax^2 + Bxy + Cy^2 + Dx + Ey + F = 0 \qquad (\star)$$

results. Perform the calculation and see this for yourself.

Now, if you are given an equation of the form $(\star)$, how can you tell whether it is the equation of an ellipse, a parabola, or a hyperbola? The tests that we learned in this chapter do not apply. For example,

$$x^2 + 2xy + y^2 + 1 = 0$$

describes a parabola. So how can one tell which equation corresponds to which type of curve? Try some experiments and see whether you can formulate a conjecture. Make this a project for class work. The book [BLK] has a thorough discussion of this topic.

15. Project: Refer to Exercise 14. We need a test for telling which equations of the form $(\star)$ describe which types of curves. Define the *discriminant* of the equation $(\star)$ to be

$$D = B^2 - 4AC\,.$$

It turns out that if $D = 0$ then the equation describes a parabola. If $D < 0$ then the equation describes an ellipse. And if $D > 0$ then the equation describes a hyperbola.

Test these assertions out on some familiar equations of conic sections that you know. Now rotate one of these equations, as in Exercise 14, and try the test again. You should get the same answer (because the essential nature of a conic section does not change when it is rotated). See [MOR] for more details.

16. Project: Refer to Exercises 14 and 15. Now examine the equation

$$x^2 + xy + y^2 + x + y + 1 = 0 .$$

Determine what sort of conic section it represents. Now graph the curve.

How does the curve change if $+xy$ is changed to $-xy$? Graph the new curve that has equation with this changed term. The book [MOR] is a good reference.

4

The Islamic World and the Development of Algebra

4.1 Introductory Remarks

In the early seventh century C.E., the Muslims formed a small and persecuted religious group. But by the end of that century, under the inspiration of Mohammed's leadership, they had conquered lands from India to Spain—including parts of North Africa and southern Italy. It is believed that, when Arab conquerors settled in new towns, they would contract diseases which had been unknown to them in desert life. In those days the study of medicine was confined mainly to Greeks and Jews. Encouraged by the caliphs (the local Arab leaders), these doctors settled in Baghdad, Damascus, and other cities. Thus we see that a purely social situation led to the contact between two different cultures which ultimately led to the transmission of mathematical knowledge.

Around the year 800, the caliph Haroun Al Raschid ordered many of the works of Hippocrates, Aristotle, and Galen to be translated into Arabic. Much later, in the twelfth century, these Arab translations were further translated into Latin so as to make them accessible to the Europeans. Today we credit the Arabs with preserving the grand Greek tradition in mathematics and science. Without their efforts, much of this classical work would have been lost.

4.2 The Development of Algebra

4.2.1 Al-Khwarizmi and the Basics of Algebra

There is general agreement that the rudiments of algebra found their genesis with the Hindus. Particularly Arya-Bhata in the fifth century and Brahmagupta in the sixth and seventh centuries played a major role in the development of these ideas. Notable among the developments due to these men is the summation of the first N positive integers, and also the sum of their squares and their cubes (see our discussion of these matters in Chapter 7).

But the Arab expansion two hundred years later caused the transfer of these ideas to the Arab empire, and a number of new talents exerted considerable influence on the development of these concepts. Perhaps the most illustrious and most famous of the ancient Arab mathematicians was Abu Ja'far Muhammad ibn Musa Al-Khwarizmi (780 C.E.–850 C.E.).

In 830 C.E.he wrote an algebra text that became the definitive work in the subject. Called *Kitab fi al-jabr w'al-mugabala*, it introduced the now commonly used term "algebra" (from "al-jabr"). The word "jabr" referred to the balance maintained in an equation when the same quantity is added to both sides (curiously, the phrase "al-jabr" also came to mean "bonesetter"); the word "mugabala" refers to cancelling like amounts from both sides of an equation.

Al-Khwarizmi's book *Art of Hindu Reckoning* introduced the notational system that we now call Arabic numerals: 1, 2, 3, 4, ... Al-Khwarizmi also introduced the concept, and the word, that has now come to be known as "algorithm". Further detail on the life of Al-Khwarizmi can be found in [ALD], [RAS], and [ROS].

It is worth noting, and we have made this point elsewhere in the present text, that good mathematical notation can make the difference between an idea that is clear and one that is obscure. The Arabs, like those who came before them, were hindered by lack of notation. When they performed their algebraic operations and solved their problems, they referred to everything with *words*. The modern scholars of this period are fond of saying that the Arabic notation was "rhetorical", with no symbolism of any kind (some of the examples below illustrate the Arab use of mathematical rhetoric). Moreover, the Arabs would typically exhibit their solutions to algebraic problems using geometric figures. There were particular difficulties when the solution involved a root (like $\sqrt{2}$, which can arise easily in solving a quadratic equation). They did not have an efficient method for simply writing the solution as we would today.

4.2.2 The Life of Al-Khwarizmi

Abu Ja'far Muhammad ibn Musa Al-Khwarizmi (780 C.E.–850 C.E.) was likely born in Baghdad, now part of Iraq. The little that we know about his life is based in part on surmise, and interpretation of evidence.

The "Al-Khwarizmi" in his name suggests that he came from Khwarizm, south of the Aral Sea in central Asia. But we also have this from an historian (Toomer [GIL]) of the period:

> But the historian al-Tabari gives him the additional epithet "al-Qutrubbulli", indicating that he came from Qutrubbull, a district between the Tigris and Euphrates not far from Baghdad, so perhaps his ancestors, rather than he himself, came from Khwarizm ... Another epithet given to him by al-Tabari, "al-Majusi", would seem to indicate that he was an adherent of the old Zoroastrian religion. ... the pious preface to Al-Khwarizmi's "Algebra" shows that he was an orthodox Muslim, so Al-Tabari's epithet could mean no more than that his forebears, and perhaps he in his youth, had been Zoroastrians.

We begin our tale of Al-Khwarizmi's life by describing the context in which he developed. Haroun Al Raschid became the fifth Caliph of the Abbasid dynasty on 14 September 786, at the time that Al-Khwarizmi was born. Haroun ruled in Baghdad over the Islam empire—which stretched from the Mediterranean to India. He brought culture to his court and tried to establish the intellectual disciplines which at that time were not flourishing in the Arabic world. He had two sons, al-Amin the eldest and al-Mamun the youngest. Haroun died in 809 and thus there arose (out of jealousy and ambition) a war between the two sons.

Al-Mamun won the armed struggle and al-Amin was defeated and killed in 813. Thus al-Mamun became Caliph and ruled the empire. He continued the patronage of learning started by his father and founded an academy called the House of Wisdom where Greek philosophical and scientific works were translated. He also built up a library of manuscripts, the first major library to be set up since that at Alexandria.[1] His mission was to collect important works from Byzantium. In addition to the House of Wisdom, al-Mamun set up observatories in which Muslim astronomers could build on the knowledge acquired in the past.

Al-Khwarizmi and his colleagues called the Banu Musa were scholars at the House of Wisdom in Baghdad. Their tasks involved the translation of Greek scientific manuscripts; they also studied, and wrote on, algebra, geometry, and astronomy. Certainly Al-Khwarizmi worked with the patronage of Al-Mamun; he dedicated two of his texts to the Caliph. These were his treatise on algebra and his treatise on astronomy. The algebra treatise *Kitab fi al-jabr w'al-mugabala* was the most famous and significant of all of Al-Khwarizmi's works. The title of this text is the provenance of the word "algebra". It is, in an important historical sense, the very first—and historically one of the most important—books on algebra.

Al-Khwarizmi tells us that the significance of his book is:

> ... what is easiest and most useful in arithmetic, such as men constantly require in cases of inheritance, legacies, partition, lawsuits, and trade, and in all their dealings with one another, or where the measuring of lands, the digging of canals, geometrical computations, and other objects of various sorts and kinds are concerned.

It should be remembered that it was typical of early mathematics tracts that they concentrated on, and found their motivation in, practical problems. Al-Khwarizmi's work was no exception. His motivations and his interests may have been abstract, but his presentation was very practical.

Early in the book Al-Khwarizmi describes the natural numbers in terms that are somewhat ponderous to us today. But it is easy to see that he is thereby laying the foundations of base-ten arithmetic. We must acknowledge the new abstraction and profundity of what he was doing:

> When I consider what people generally want in calculating, I found that it always is a number. I also observed that every number is composed of units, and that any number may be divided into units. Moreover, I found that every number which may be expressed from one to ten, surpasses the preceding by one unit: afterwards the ten is doubled or tripled just as before the units were: thus arise twenty, thirty, etc. until a hundred: then the hundred is doubled and tripled in the same manner as the units and the tens, up to a thousand; ... so forth to the utmost limit of numeration.

We should bear in mind that, for many centuries, the motivation for the study of algebra was the solution of equations. In Al-Khwarizmi's day these were linear and quadratic equations. His equations were composed of units, roots and squares. For example, to Al-Khwarizmi a unit was a number, a root was x, and a square was $x2$ (at least this was what

[1] The Alexandrian library was *the* great library of the ancient world. Founded around 300 B.C.E., it was the first to have books and other works from countries far and wide. It was unfortunately—at least as far as we know—destroyed by the Muslim conquest in 642 A.D. (though there are conflicting theories about when this scholarly catastrophe took place).

he seemed to be *thinking*). However, it is both astonishing and significant to bear in mind that Al-Khwarizmi did his algebra with no symbols—only words.

Al-Khwarizmi first reduces an equation (linear or quadratic) to one of six standard forms:[2]

1. Squares equal to roots.
2. Squares equal to numbers.
3. Roots equal to numbers.
4. Squares and roots equal to numbers; e.g., $x2 + 10x = 39$.
5. Squares and numbers equal to roots; e.g., $x2 + 21 = 10x$.
6. Roots and numbers equal to squares; e.g., $3x + 4 = x2$.

The reduction is carried out using the two operations of "al-jabr" and "al-mugabala". Here "al-jabr" means "completion" and is the process of removing negative terms from an equation. For example, using one of Al-Khwarizmi's own examples, "al-jabr" transforms $x2 = 40x - 4x2$ into $5x2 = 40x$. The term "al-mugabala" means "balancing" and is the process of reducing positive terms of the same power when they occur on both sides of an equation. For example, two applications of "al-mugabala" reduces $50+3x+x2 = 29+10x$ to $21 + x2 = 7x$ (one application to deal with the numbers and a second to deal with the roots).

Al-Khwarizmi then shows how to solve the six types of equations adumbrated above. He uses both algebraic methods of solution and geometric methods. We shall treat his algebraic methodology in detail below.

Al-Khwarizmi continues his study of algebra in *Kitab fi al-jabr w'al-mugabala* by considering how the laws of arithmetic extrapolate to an algebraic context. For example, he shows how to multiply out expressions such as

$$(a + bx)(c + dx)\,.$$

Again we stress that Al-Khwarizmi uses only words to describe his expressions; no symbols are used.

There seems to be little doubt, from our modern perspective, that Al-Khwarizmi was one of the greatest mathematicians of all time. His algebra was original, incisive, and profound. It truly changed the way that we think about mathematics.

The next part of Al-Khwarizmi's Algebra consists of applications and worked examples. He then goes on to look at rules for finding the area of figures such as the circle and also finding the volume of solids such as the sphere, cone, and pyramid. This section on mensuration certainly has more in common with Hindu and Hebrew texts than it does with any Greek work. The final part of the book deals with the complicated Islamic rules for inheritance but requires little from the earlier algebra beyond solving linear equations. Again, in all these aspects of the book, we see the over-arching need to justify the mathematics with practical considerations.

[2]For clarity, we continue to indulge here in the conceit of using semi-modern notation—notation that Al-Khwarizmi would never have used.

Al-Khwarizmi also wrote a treatise on Hindu-Arabic numerals. The Arabic text is lost but a Latin translation, *Algoritmi de numero Indorum* (rendered in English, the title is *Al-Khwarizmi on the Hindu Art of Reckoning*) gave rise to the word "algorithm", deriving from his name in the title. The work describes the Hindu place-value system of numerals based on 1, 2, 3, 4, 5, 6, 7, 8, 9, and 0. The first use of zero as a place holder in positional base notation was probably due to Al-Khwarizmi in this work. Methods for arithmetical calculation are given, and a method to find square roots is known to have been in the Arabic original although it is missing from the Latin version.

Another important work by Al-Khwarizmi was his work *Sindhind zij* on astronomy. The work is based in Indian astronomical works (see [SOK]):

> ...as opposed to most later Islamic astronomical handbooks, which utilized the Greek planetary models laid out in Ptolemy's *Almagest*.

The Indian text on which Al-Khwarizmi based his treatise was one which had been given to the court in Baghdad around 770 as a gift from an Indian political mission. There are two versions of Al-Khwarizmi's work which he wrote in Arabic but both are lost. In the tenth century al-Majriti made a critical revision of the shorter version and this was translated into Latin by Adelard of Bath. The main topics covered by Al-Khwarizmi in the *Sindhind zij* are calendars; calculating true positions of the sun, moon and planets, tables of sines and tangents; spherical astronomy; astrological tables; parallax and eclipse calculations; and visibility of the moon. A related manuscript, attributed to Al-Khwarizmi, concerns spherical trigonometry.

Although his astronomical work is based on that of the Indians, and most of the values from which he constructed his tables came from Hindu astronomers, Al-Khwarizmi must have been influenced by Ptolemy's work too.

Al-Khwarizmi wrote a major work on geography which gives latitudes and longitudes for 2402 localities as a basis for a world map. The book, which is based on the astronomer Ptolemy's *Geography*, lists—with latitudes and longitudes—cities, mountains, seas, islands, geographical regions, and rivers. The manuscript does include maps which on the whole are more accurate than those of Ptolemy. In particular, it is clear that where more local knowledge was available to Al-Khwarizmi, such as the regions of Islam, Africa and the Far East, then his work is considerably more accurate than that of Ptolemy; but for Europe Al-Khwarizmi seems to have used Ptolemy's data.

A number of minor works were written by Al-Khwarizmi on topics such as the astrolabe, on which he wrote two works, on the sundial, and on the Jewish calendar. He also wrote a political history containing horoscopes of prominent persons.

We have already discussed the varying views of the importance of Al-Khwarizmi's algebra which was his most important contribution to mathematics. Al-Khwarizmi is perhaps best remembered by Mohammad Kahn:

> In the foremost rank of mathematicians of all time stands Al-Khwarizmi. He composed the oldest works on arithmetic and algebra. They were the principal source of mathematical knowledge for centuries to come in the East and the West. The work on arithmetic first introduced the Hindu numbers to Europe, as the very name algorithm signifies; and the work on algebra ... gave the name to this important branch of mathematics in the European world...

4.2.3 The Ideas of Al-Khwarizmi

The ideas discussed thus far in the present chapter are perhaps best illustrated by some examples.

EXAMPLE 4.1. Solve this problem of Al-Khwarizmi:

A square and ten roots equal thirty-nine dirhems.

Solution. It requires some effort to determine what is being asked. First, a *dirhem* is a unit of money in medieval Arabic times. In modern English (we shall introduce some mathematical *notation* later), what Al-Khwarizmi is telling us is that a certain number squared plus ten times that number (by "root" he means the number that was squared—what we would call the *unknown*) equals 39. If we call this unknown number x, then what is being said is that

$$x^2 + 10x = 39 \qquad \text{or} \qquad x^2 + 10x - 39 = 0\,.$$

Of course the quadratic formula quickly tells us that

$$x = \frac{-10 \pm \sqrt{10^2 - 4 \cdot (-39) \cdot 1}}{2} = \frac{-10 \pm \sqrt{256}}{2} = \frac{-10 \pm 16}{2}\,.$$

This gives us the two roots 3 and -13.

Now the Arabs could not deal with negative numbers, and in any event Al-Khwarizmi was thinking of his unknown as the side of a square. So we take the solution

$$x = \frac{-10 + 16}{2} = 3\,.$$

Thus, from our modern perspective, this is a straightforward problem. We introduce a variable, write down the correct equation, and solve it using a standard formula.

Matters were different for the Arabs. They did not have notation, and certainly did not yet know the quadratic formula. Their method was to deal with these matters geometrically. Consider Figure 4.1. This shows the "square" mentioned in the original problem, with unknown side length that we now call x. In Figure 4.2, we attach to each side of the square a rectangle of length x and width 2.5. The reasoning here is that Al-Khwarizmi tells us to add 10 times the square's side length. We divide 10 into four pieces and thus add four times

Figure 4.1. A problem of Al-Khwarizmi.

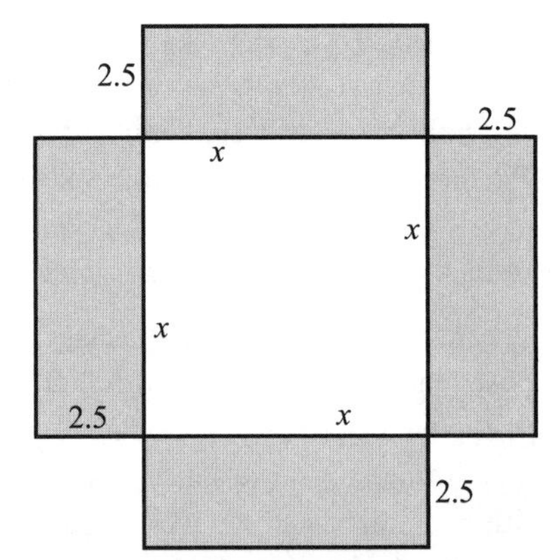

Figure 4.2. Sum of shaded areas is $10 \times x$.

"2.5 times the side length". The quantity "2.5 times the side length" is represented by an appropriate rectangle in Figure 4.2.

Now we know, according to the statement of the problem, that the sum of the areas of the square in the middle and the four rectangles around the sides is 39. We handle this situation by filling in four squares in the corners—see Figure 4.3. Now the resulting large square plainly has area equal to $39 + 2.5^2 + 2.5^2 + 2.5^2 + 2.5^2 = 64$. Since the large square has area 64, it must have side length 8. But we know that each of the squares in the four corners has side length 2.5. It must follow then that $x = 8 - 2.5 - 2.5 = 3$. And that is the correct answer.

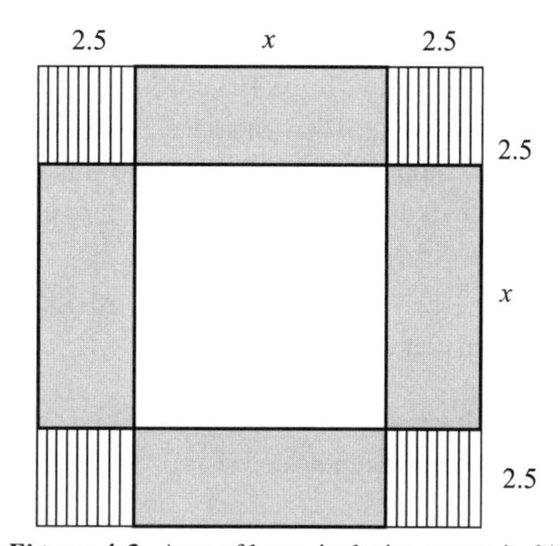

Figure 4.3. Area of large, inclusive square is 64.

For You to Try Use the method of Al-Khwarizmi to find the positive root of the quadratic equation

$$x^2 + 5x = 15.$$

In fact the method of this last example can be used to solve any quadratic equation with positive, real roots. We explore this contention in the exercises.

Now we examine another algebra problem of Al-Khwarizmi. This is in the format of a familiar sort of word problem. It has interesting social as well as mathematical content. We shall present the solution both in modern garb and in the argot of Al-Khwarizmi's time.

EXAMPLE 4.2. Solve this problem of Al-Khwarizmi:

> A man dies leaving two sons behind him, and bequeathing one-fifth of his property and one dirhem to a friend. He leaves ten dirhems in property and one of the sons owes him ten dirhems. How much does each legatee receive?

Solution. We already know that a dirhem is a unit of currency. It is curious that, in Al-Khwarizmi's time, there was no concept of "estate". A legacy could only be left to a person or people, not to an abstraction like an "estate".

However we understand what an estate is, and it helps us to solve the problem in modern language. Our solution goes as follows. The dead man's estate consists of 20 dirhems: the 10 dirhems that he has in hand and the 10 dirhems owed to him by his son. The friend receives $1/5$ of that estate plus one dirhem. Thus the friend receives $4 + 1 = 5$ dirhems. That leaves the estate with 5 dirhems in hand (the one son owing another 10 dirhems to the estate) and 10 dirhems owed to it, for a total of 15 dirhems. Thus each son is owed 7.5 dirhems. That means that the son who owes 10 dirhems should pay the estate 2.5 dirhems. Now the estate has 7.5 dirhems cash in hand. And that amount is paid to the other son.

Since Al-Khwarizmi did not have the abstraction of "estate" to aid his reasoning, he solved the problem with the following chain of logic:

> Call the amount taken out of the debt *thing*. Add this to the property. The sum is 10 dirhems plus *thing*. Subtract $1/5$ of this, since he has bequeathed $1/5$ of his property to the friend. The remainder is 8 dirhems plus $4/5$ of *thing*. Then subtract the 1 dirhem extra that is bequeathed to the friend. There remain 7 dirhems and $4/5$ of *thing*. Divide this between the two sons. The portion of each of them is three and one half dirhems plus $2/5$ of *thing*. Then you have $3/5$ of *thing* equal to three and one half dirhems. Form a complete *thing* by adding to this quantity $2/3$ of itself. Now $2/3$ of three and one half dirhems is two and one third dirhems. Conclude that *thing* is five and five sixths dirhems.

In one of the exercises we shall ask you to reconcile Al-Khwarizmi's solution of the problem with our own solution that we presented at first.

For You to Try Solve Al-Khwarizmi's preceding problem if there are three sons instead of two (and the friend still receives the indicated share).

4.2.4 Omar Khayyam and the Resolution of the Cubic

Omar Khayyam (1050–1123) is famed, and still well-remembered, for his beautiful collection of poems called *The Rubaiyat*. The words "A loaf of bread, a jug of wine, and thou beside me in the wilderness" ring down through the ages. It is perhaps less well known that Khayyam was an accomplished astronomer and mathematician. He is remembered particularly for his geometric method of solving the cubic equation (we will also discuss the cubic equation, from a somewhat more modern point of view, in Section 5.6). Here we give an example to illustrate the technique of Omar Khayyam. More on the life of Omar Khayyam may be found in [ROZ].

EXAMPLE 4.3. Consider the cubic equation

$$x^3 + Bx = C,$$

where B, C are positive constants. Find all positive, real solutions.

Solution. The first step is to choose positive numbers b, c so that $b^2 = B$ and $b^2c = C$. We know we can do this because every positive number has a square root, and every linear equation has a solution.

Thus the equation becomes

$$x^3 + b^2x = b^2c.$$

Now we construct a parabola whose *latus rectum*[3] is b. It is intuitively clear that the length

[3] The *latus rectum* of an upward-opening parabola is the horizontal line segment that begins and ends on the parabolic curve and passes through the focus—see Figure 4.4.

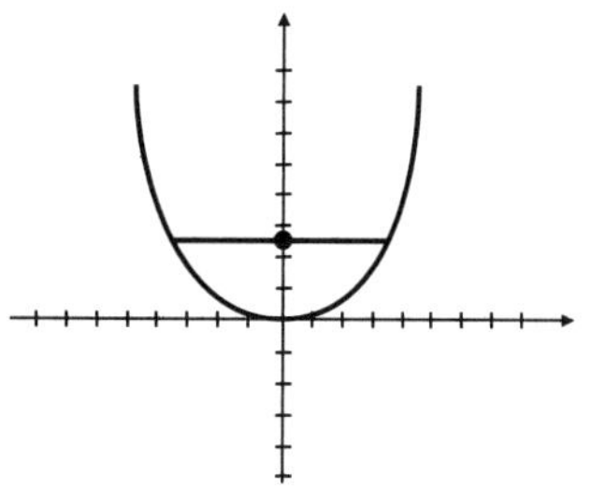

Figure 4.4. The latus rectum.

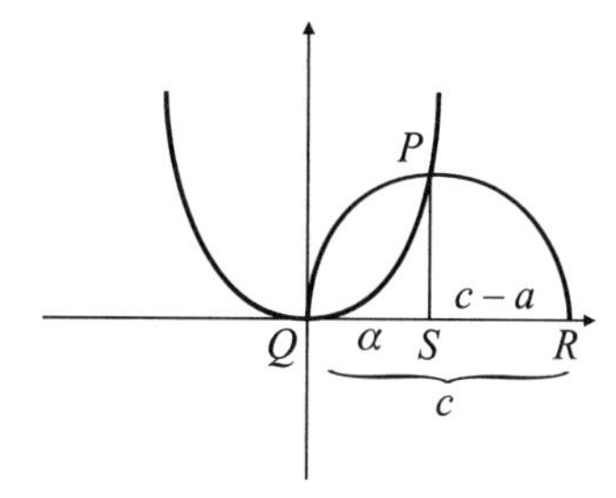

Figure 4.5. The vertex of th parabola.

of the latus rectum uniquely determines the shape of the parabola. Notice that the point Q in Figure 4.5 is the vertex of the parabola (we may take Q to be the origin if we wish). The segment QR which is shown has length c. Now consider the semicircle with diameter $\overline{QR}$. The point P is defined to be the intersection of the parabola and the semicircle. The segment PS is erected to be perpendicular to the segment QR. Then the length $\alpha = QS$ is a root of the cubic equation.

Let us explain why this last statement is true. Because the latus rectum has length b, we know that the focus of the parabola is at the point $(0, b/4)$. Moreover the directrix is the line $y = -b/4$. We can be sure (from our synthetic definition of parabola in Section 3.2) that the parabola has equation $y = x^2/b$. Thus, in Figure 4.5,

$$PS = \alpha^2/b\,. \tag{$\star$}$$

This relation may be rewritten as

$$\frac{b}{\alpha} = \frac{\alpha}{PS}\,. \tag{$*$}$$

A basic property of semicircles tells us that the triangle $\triangle QPR$ is a right triangle (with right angle at P). Since $\overline{PS}$ is an altitude of this triangle, we can be sure that

$$\frac{\alpha}{PS} = \frac{PS}{c-\alpha}\,. \tag{$**$}$$

Equations $(*)$ and $(**)$ together tell us that

$$\frac{b}{\alpha} = \frac{PS}{c-\alpha}\,. \tag{$***$}$$

But $(\star)$ tells us that

$$PS = \frac{\alpha^2}{b}\,.$$

Substituting this value for PS into $(***)$ now tells us that

$$\frac{b}{\alpha} = \frac{\alpha^2/b}{c-\alpha}\,.$$

Simplifying this last identity yields that

$$\alpha^3 + b^2\alpha = b^2 c\,.$$

Thus the positive number α solves the cubic.

We reiterate that the Arabs only understood positive, real roots of polynomial equations. Gauss's Fundamental Theorem of Algebra (Section 6.7) was centuries off. Negative numbers and certainly complex numbers were still a mystery.

Further Reading

Klaus Galda, An informal history of formal proofs: From vigor to rigor?, *The Two-Year College Mathematics Journal* 12(1981), 126–140.

Donald E. Knuth, Algorithmic thinking and mathematical thinking, *The American Mathematical Monthly* 92(1985), 170–181.

4.3 The Geometry of the Arabs

4.3.1 The Generalized Pythagorean Theorem

Arab geometry took many forms. We have already seen that they used geometry to analyze the roots of polynomial equations. The Arabs took a great interest in the parallel postulate and the existence of non-Euclidean geometries (a topic that we shall discuss later in the book), although their efforts were not very successful. We will begin our analysis of Arab geometry by considering a remarkable generalization of the Pythagorean theorem.

At this time you may wish to review our discussion of Pythagoras's theorem in Chapter 1. That result was formulated specifically for, and in fact only holds true for, right triangles. The generalization of the result that is due to Thabit ibn-Qurra in fact applies to *all* triangles.

Before we begin we must review the concept of similarity of triangles. Consider the two triangles $\triangle ABC$ and $\triangle A'B'C'$ in Figure 4.6. They appear to have the same shape. This means that the corresponding angles are equal:

- the angle at A equals the angle at A',
- the angle at B equals the angle at B',
- the angle at C equals the angle at C'.

It also means that the corresponding ratios of sides are equal. For example,

- $\dfrac{\overline{AB}}{\overline{BC}} = \dfrac{\overline{A'B'}}{\overline{B'C'}}$,
- $\dfrac{\overline{AB}}{\overline{AC}} = \dfrac{\overline{A'B'}}{\overline{A'C'}}$.

We may formulate these corresponding ratios in a slightly different fashion as follows:

- $\dfrac{\overline{AB}}{\overline{A'B'}} = \dfrac{\overline{AC}}{\overline{A'C'}}$,
- $\dfrac{\overline{BC}}{\overline{B'C'}} = \dfrac{\overline{AC}}{\overline{A'C'}}$.

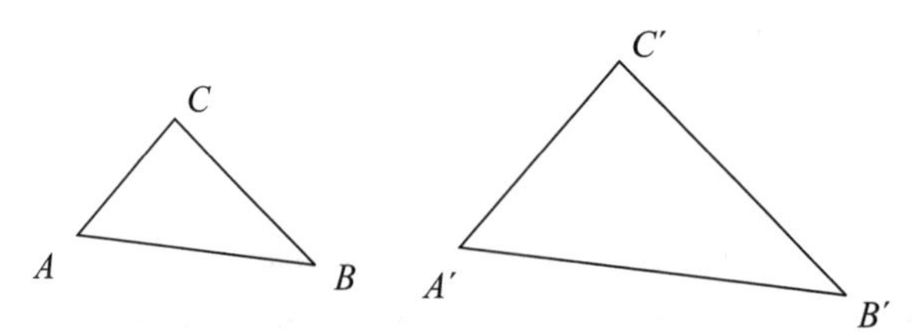

Figure 4.6. Similar triangles.

What is of particular interest is finding conditions that are *sufficient* to guarantee that two given triangles are similar. Such a condition will (unlike the concept of congruence) *not* involve equality of side lengths—after all, one triangle is *larger* than the other. In fact the most useful condition of this nature is the following:

Consider the triangles $\triangle ABC$ and $\triangle A'B'C'$ in Figure 4.7. If either

- The angle at A equals the angle at A' and the angle at B equals the angle at B'

  or

- The angle at A equals the angle at A' and the angle at C equals the angle at C'

  or

- The angle at B equals the angle at B' and the angle at C equals the angle at C'

then $\triangle ABC$ is similar to $\triangle A'B'C'$.

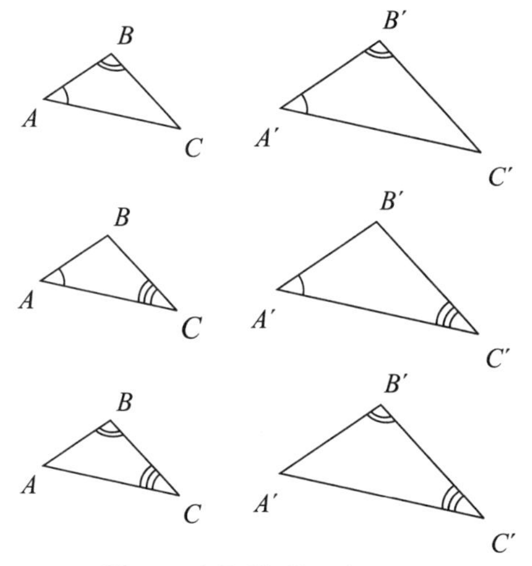

Figure 4.7. Similar triangles.

Thus, in order to test two triangles for similarity, we need only establish that two of the corresponding pairs of angles are equal. [Of course we know that the sum of the three angles in a triangle is 180°. Hence if two pairs of the angles are equal then the third pair is equal also.] So the triangles are the same shape and hence similar.

Now we may state the generalized Pythagorean theorem that was discovered by the Arabs.

Theorem. *Let $\triangle ABC$ be a planar triangle, with $\overline{BC}$ its longest side. Refer to Figure 4.8. Choose the point B' on the segment $\overline{BC}$ so that the angle $\angle B'AB$ (in dashes) is equal to angle $\angle C$ (i.e., the angle at the vertex C in the triangle). Choose the point C' on the segment $\overline{BC}$ so that the angle $\angle C'AC$ (in dots) is equal to angle $\angle B$ (the angle at vertex B in the triangle). Then*

$$\overline{AB}^2 + \overline{AC}^2 = \overline{BC} \cdot (\overline{BB'} + \overline{CC'}) .$$

For the verification of this theorem, study Figure 4.8. Choose the points B' and C' as in the statement of the theorem. We see that angle $\angle AB'B$ (marked with a single slash) equals angle $\angle CAB$ and the angle $\angle AC'C$ (marked with a double slash) equals angle $\angle BAC$.

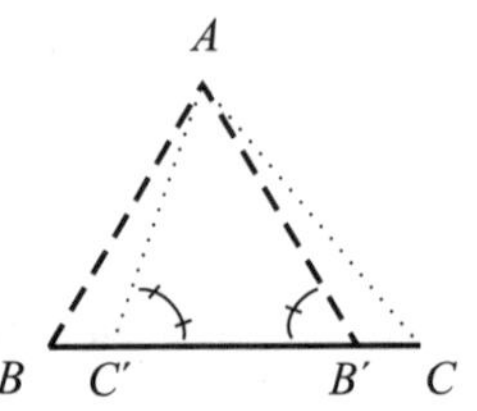

Figure 4.8. A generalized Pythagorean theorem.

It results—since $\angle AB'B = \angle CAB$ and $\angle ABB' = \angle CBA$—that triangle $\triangle B'BA$ is similar to the original triangle $\triangle ABC$. Also, by analogous reasoning, $\triangle C'AC$ is similar to the original triangle $\triangle ABC$. Thus we have the identical ratios

$$\frac{\overline{AB}}{\overline{BB'}} = \frac{\overline{BC}}{\overline{AB}}.$$

Likewise we see that

$$\frac{\overline{AC}}{\overline{CC'}} = \frac{\overline{BC}}{\overline{AC}}.$$

From the first of these equations we derive (clearing denominators) that

$$\overline{AB}^2 = \overline{BC} \cdot \overline{BB'}.$$

From the second we derive that

$$\overline{AC}^2 = \overline{BC} \cdot \overline{CC'}.$$

Adding these together yields that

$$\overline{AB}^2 + \overline{AC}^2 = \overline{BC} \cdot \overline{BB'} + \overline{BC} \cdot \overline{CC'} = \overline{BC} \cdot (\overline{BB'} + \overline{CC'}).$$

This is the desired result.

In Exercise 7 you will be asked to show that, for a right triangle, this new theorem of Thabit ibn Qurra reduces to the classical Pythagorean theorem.

4.3.2 Inscribing a Square in an Isosceles Triangle

In fact our friend Al-Khwarizmi examined a problem based on the isosceles triangle shown in Figure 4.9. We seek to inscribe a square in this triangle. Al-Khwarizmi would have used the name "thing" to refer to the side-length of the square. Now we shall emulate the analysis that he might have done more than 1000 years ago.

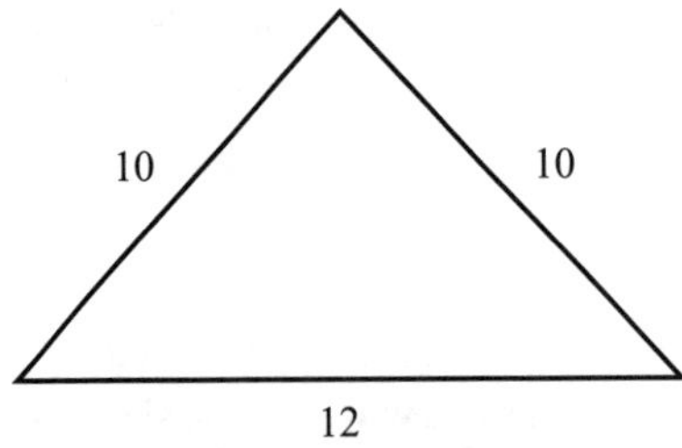

Figure 4.9. An isosceles triangle.

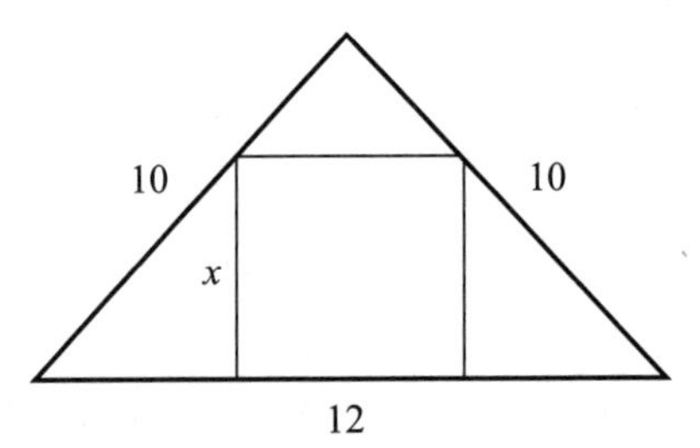

Figure 4.10. An inscribed square.

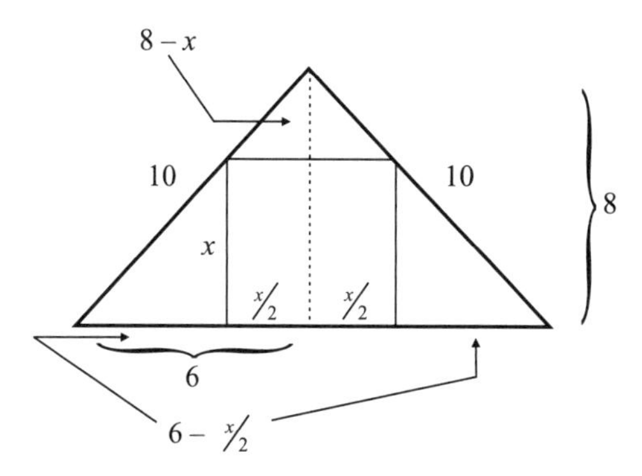

Figure 4.11. The areas of the triangles.

The area of the square is of course (thing) × (thing). Notice that, in the figures, we denote the side of the square by "x". But we call it "thing". Figure 4.11 shows how we might analyze the areas of the triangles.

The small right triangle on the left has base $6 - x/2$ and height x. Similarly for the small right triangle on the right. Thus the total area of the two triangles is $x(6 - x/2)$.

We may solve for the altitude of the large isosceles triangle using the Pythagorean theorem. It equals $\sqrt{10^2 - 6^2} = 8$. Thus the small isosceles triangle at the top of the figure has base x and height $8 - x$. We conclude that that triangle has area $[x/2] \cdot (8 - x)$.

In summary, then, the total area of the large isosceles triangle may be written in two ways. On the one hand, the triangle has base 12 and height 8. So its area is $\frac{1}{2} \cdot 12 \cdot 8 = 48$. On the other hand the area is the sum of the areas of the square and the three little triangles. So we have

$$48 = x^2 + x \cdot \left(6 - \frac{x}{2}\right) + \frac{x}{2} \cdot (8 - x) .$$

This simplifies to

$$48 = 10x$$

hence $x = 4.8$. That is the solution to Al-Khwarizmi's problem.

Further Reading

George Bruce Halsted, Non-Euclidean geometry, *The American Mathematical Monthly* 7(1900), 123–133.

Abe Shenitzer, How hyperbolic geometry became respectable, *The American Mathematical Monthly* 101(1994), 464–470.

4.4 A Little Arab Number Theory

The Arabs were fascinated by a technique that has come down through the ages as "Casting Out Nines". The basic rule for casting out nines for a positive integer N is to add its digits together. Thus

$$4873 \to 4 + 8 + 7 + 3 = 22 \to 2 + 2 = 4 .$$

We began here with the positive integer 4873. We added together its digits: $4+8+7+3 = 22$. Then we again added together the digits of 22—$2 + 2 = 4$—to obtain 4. Part of the rule of casting out nines is that if we ever encounter a 9 then we set it equal to 0. Thus if we cast out nines on the number 621 we obtain $6 + 2 + 1 \to 9 \to 0$.

The remarkable thing about casting out nines is that the process respects addition and multiplication. If we let "c.o.n." stand for casting out nines, then we have

$$\text{c.o.n.}[k + m] = \text{c.o.n.}(k) + \text{c.o.n.}(m)$$

and

$$\text{c.o.n.}[k \cdot m] = \text{c.o.n.}(k) \cdot \text{c.o.n.}(m)\,.$$

Thus casting out nines can be used to check arithmetic problems. We illustrate the idea with some examples.

EXAMPLE 4.4. Using casting out nines, check whether

$$693 \times 42 = 29206\,.$$

Solution. Casting out nines on the left gives

$$6 + 9 + 3 = 18 \to 9 \to 0$$

and

$$4 + 2 = 6\,.$$

Therefore

$$693 \times 42 \to 0 \times 6 = 0\,.$$

Casting out nines on the right gives $2 + 9 + 2 + 0 + 6 = 19 \to 10 \to 1$.

Thus the result of casting out nines gives $0 \times 6 = 0$ on the left and 1 on the right. These do not match. Thus the multiplication is incorrect. In fact checking with a calculator gives that $693 \times 42 = 29106$.

Casting out nines does not provide a failsafe method for checking arithmetic problems. For example, casting out nines on 6×8 gives 14 and then 5. Casting out nines on 23 also gives 5. Yet it certainly is not the case that $6 \times 8 = 23$. What is true is this: If casting out nines *does not work* then the original arithmetic problem is incorrect. If casting out nines *does work* then it is likely that the original arithmetic problem was correct. But it is not guaranteed.

EXAMPLE 4.5. Check whether the addition

$$385 + 2971 + 1146 = 4502 \qquad (\star)$$

is correct.

Solution. Casting out nines on the left gives

$$3 + 8 + 5 = 16 \to 7$$

and

$$2 + 9 + 7 + 1 = 19 \to 10 \to 1$$

and

$$1 + 1 + 4 + 6 = 12 \to 3\,.$$

Casting out nines on the right yields

$$4 + 5 + 0 + 2 = 11 \to 2\,.$$

Altogether then, applying casting out nines to the equation ($\star$) gives the result

$$7 + 1 + 3 \cong 2\,,$$

where we use the notation $\cong$ to indicate equivalence under casting out nines. Casting out nines on the left yields $11 \cong 2$ or $2 = 2$.

Thus the casting out nines checks out. This does *not* guarantee that our original addition was correct. But it provides strong evidence that it is.

What is interesting for us is why the method of casting out nines works. And the answer, in our modern language, is simplicity itself:

> Casting out nines is nothing other than arithmetic modulo 9. And arithmetic modulo 9 respects addition and multiplication.

Modular arithmetic will be discussed in greater detail later in the text (Chapters 10 and 20). Suffice it to say for now that we do arithmetic modulo 9 by subtracting from any number all the multiples of 9 that we possibly can. Thus

$$17 \bmod 9 = 8\,,$$

$$94 \bmod 9 = 4\,,$$

$$87 \bmod 9 = 6\,,$$

and

$$-5 \bmod 9 = 4\,.$$

We can perform addition and multiplication modulo 9. For instance,

$$[23 + 35] \bmod 9 = 58 \bmod 9 = 4 \bmod 9\,.$$

This may also be performed by first reducing the summands modulo 9:

$$23 \bmod 9 + 35 \bmod 9 = 5 \bmod 9 + 8 \bmod 9 = 13 \bmod 9 = 4 \bmod 9\,.$$

Matters are similar with multiplication:

$$[12 \bmod 9] \cdot [15 \bmod 9] = [3 \bmod 9] \cdot [6 \bmod 9] = 18 \bmod 9 = 0 \bmod 9\,.$$

To understand now why casting out nines works, first note that

$$1 \bmod 9 = 1\,,$$

$$10 \bmod 9 = 1\,,$$

$$100 \bmod 9 = 1\,,$$

$$1000 \bmod 9 = 1\,,$$

and so forth. Now let us look at a specific example.

Consider the number 5784. Then

$$\begin{aligned}
5784 \bmod 9 &= [5000 + 700 + 80 + 4] \bmod 9 \\
&= 5 \cdot [1000 \bmod 9] + 7 \cdot [100 \bmod 9] + 8 \cdot [10 \bmod 9] + 4 \cdot [1 \bmod 9] \\
&= (5 \cdot 1 + 7 \cdot 1 + 8 \cdot 1 + 4 \cdot 1) \bmod 9 \\
&= (5 + 7 + 8 + 4) \bmod 9 .
\end{aligned}$$

In other words, we see rather directly that casting out nines on the number 5794 consists of just adding the digits together. If the result is greater than 9, we just add the digits together again. If the digit 9 occurs then we replace it by 0 (which is consistent with arithmetic modulo 9).

Of course the Arabs did not have modular arithmetic at their disposal. Their reasoning was more indirect. But they nonetheless gave us a useful and fascinating arithmetical tool.

Further Reading

David Eugene Smith and Clara C. Eaton, Rithmomachia, the great medieval number game, *The American Mathematical Monthly* 18(1911), 73–80.

David Eugene Smith and Jekuthial Ginsburg, Rabbi Ben Ezra and the Hindu-Arabic Problem, *The American Mathematical Monthly* 25(1918), 99–108.

Exercises

1. Use the method of Example 4.1 to solve the quadratic equation

$$x^2 + x - 2 = 0 .$$

Can you find both roots, or just the positive root?

2. What goes wrong if we attempt to use the method of Example 4.1 to solve the quadratic equation

$$x^2 + x + 4 = 0 ?$$

(Note that this quadratic equation has complex roots.) Discuss in class.

3. What goes wrong if we try to use the method of Example 4.1 to solve the quadratic equation

$$x^2 + 3x + 2 = 0 ?$$

Discuss in class.

4. Explain why the modern solution and Al-Khwarizmi's solution of Example 4.2 are consistent.

5. Solve the following algebra problem of Al-Khwarizmi:

> A man marries while in his final illness and pays a marriage settlement of his entire property in the amount of 100 dirhems, 10 dirhems of which was his

wife's dowry. His plans are upset, however, as his wife dies first, leaving one-third of her property to a third party, after which the husband dies. There are then three sets of claimants to the 100 dirhems: **(1)** the third party, **(2)** the wife's direct heirs (her family), and **(3)** the husband's direct heirs (his children or parents). How is the estate to be divided up?

Certainly discuss this question in class.

6. The Persian mathematician Nasir-Eddin proved that if you roll a circle of radius r around the inside edge of a circle of radius $2r$, and if the smaller circle has a dot on the edge, then the dot will trace out a diameter of the larger circle. Draw a picture to illustrate this result. Now try to verify it. Discuss this problem in class.

7. Suppose that $\triangle ABC$ is a right triangle with right angle at A. Apply Thabit ibn-Qurra's generalization of the Pythagorean theorem and show that, in this case, his result reduces to the standard Pythagorean theorem.

8. This problem comes from Al-Khwarizmi's *Algebra*. You should solve it. A woman dies and leaves her daughter, her mother, and her husband. She bequeathes to some person as much as the share of her mother and to another person as much as one-ninth of her entire capital. Find the share of each person. [*Note:* It is known, from Arab legal principles of the time, that the mother's share would be $2/13$ and the husband's share $3/13$.]

9. Abu Kamil (850 C.E.–930 C.E.) wrote a commentary on Al-Khwarizmi's *Algebra*. In it, he contributed a number of ingenious algebra problems. Solve the following one: The number 50 is divided by a certain other number. If the divisor is increased by 3, then the quotient decreases by $3\,3/4$. What is the divisor?

10. The method of "Casting out Elevens" is mathematically equivalent to doing arithmetic modulo 11 (just as we learned in the text for casting out nines). Casting out elevens is performed on a positive integer by **(i)** adding up the digits in the odd positions, **(ii)** adding up the digits in the even positions, and **(iii)** subtracting the second sum from the first. Explain why this method works. Use the method of casting out elevens to check the examples in the text.

11. A pair of positive integers is *amicable* if each is equal to the sum of the proper divisors of the other. For example, the numbers 220 and 284 are amicable. For notice that the proper divisors (sometimes called the *aliquot divisors*) of 220 are $1, 2, 4, 5, 10, 11, 20, 22, 44, 55, 110$ and these sum to 284; also the proper divisors of 284 are $1, 2, 4, 71, 142$ and these sum to 220. Thabit ibn Qurra found the following formula for generating pairs of amicable numbers. If n is a positive integer then set

$$h = 3 \cdot 2^n - 1 \quad t = 3 \cdot 2^{n-1} - 1 \quad s = 9 \cdot 2^{2n-1} - 1\,.$$

If h, t, s are all prime numbers then $2^n \cdot h \cdot t$ and $2^n \cdot s$ are amicable. Verify that, for $n = 2$, Thabit ibn Qurra's formula gives the pair of amicable numbers that we just discussed. Also check that, for $n = 4$, this formula gives a new pair of amicable numbers. Discuss your results in class. Today about $6,262,871$ pairs of amicable numbers have been identified. Nobody knows whether this formula will generate infinitely many pairs of amicable numbers. Explain what this last statement means in the context of what went before.

12. Heron's formula for finding the area of a triangle was known both to the Hindus and the Arabs. It says this. Let a, b, c be the side lengths of a given triangle. Let $s = (a + b + c)/2$ be the semi-perimeter. Then the area A of the triangle is given by

$$A = \sqrt{s(s-a)(s-b)(s-c)}\,.$$

Verify Heron's formula for some triangles that you know. Discuss in class why Heron's formula might be true. [*Hint:* Think about the symmetric roles of a, b, c.]

13. Refer to Exercise 12. Why does Heron's formula imply the Pythagorean theorem?

14. Project: There are many different "generalized Pythagorean theorems". We discuss one such here.

The generalized Pythagorean theorem is valid not only for right triangles but also for any kind of triangle. Refer to Figure 4.12.

$$\begin{aligned} a^2 &= b^2 + c^2 \pm 2bp\,, \\ b^2 &= a^2 + c^2 \pm 2cq\,, \\ c^2 &= a^2 + b^2 \pm 2ar\,, \end{aligned}$$

with

$$\begin{aligned} p &= \text{projection of } c \text{ onto } b\,, \\ q &= \text{projection of } a \text{ onto } c\,, \\ r &= \text{projection of } b \text{ onto } a\,. \end{aligned}$$

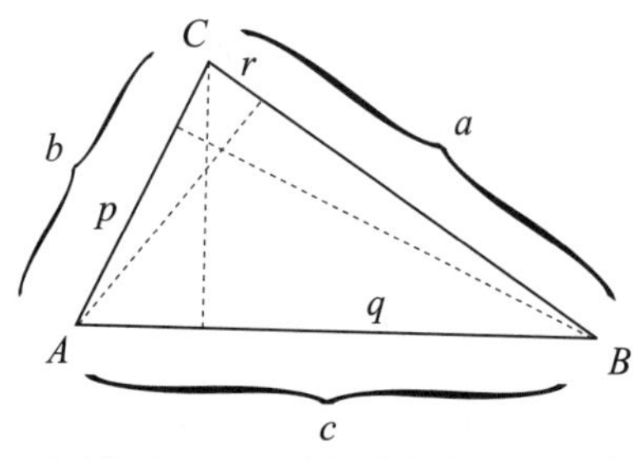

Figure 4.12. The generalized Pythagorean theorem.

Find a synthetic proof of this version of the Pythagorean theorem. The reference [MAO] gives considerable background and context for Pythagoras's theorem.

15. Project: We understand negative numbers better than the Arabs did. Find a way to modify the Arab method of solving quadratic equations that will take into account negative roots. See [KAT] for more on this matter.

16. Project: The Arabs were interested in the question of finding the center of a sphere if you are given finitely many points on that sphere. Recall that, for a circle in the plane, three distinct points on the circle uniquely determine its center. How many points are needed to determine the center of a sphere? The article [SMA] discusses this and related problems.

5

Cardano, Abel, Galois, and the Solving of Equations

5.1 Introduction

Ever since the eighth century among Arab scholars, algebra has exerted a profound influence on modern mathematics. One of the prevailing themes has been the solving of equations—especially polynomial equations. Early on, mathematicians realized that some equations, such as

$$2x + 3 = 9\,,$$

can be solved by elementary manipulation. One writes

$$[2x + 3] - 3 = 9 - 3\,,$$

then

$$2x = 6$$

hence

$$x = 3\,.$$

More interesting are the higher-order equations. An equation like

$$x^2 - 5x + 6 = 0$$

can be factored as

$$(x - 2)(x - 3) = 0$$

and a complete solution (namely $x = 2, x = 3$) obtained. Other equations, such as

$$x^2 + 1 = 0\,,$$

do not admit real solutions. This was certainly one of the motivations for the invention of the complex number system. In the present chapter we shall concentrate our attention on *real* solutions of polynomials. Later on (Chapter 9) we shall consider the complex number system.

A mathematical program of long standing was to determine *which* polynomial equations are solvable. Particularly, which ones are solvable by a procedure of finitely many operations of arithmetic and taking roots? And which are not? For those which are solvable, what is an algorithm or methodology for finding the solution(s)? Can one write an explicit formula for the solution(s)? Work on these problems absorbed many centuries.

Further Reading

Jerry L. Kazdan, Solving equations, an elegant legacy, *The American Mathematical Monthly* 105(1998), 1–21.

Bill Bompart, An alternate method for solving radical equations, *The Two-Year College Mathematics Journal* 13(1982), 198–199.

5.2 The Story of Cardano

We begin our saga with an account of the life of Girolamo Cardano (1501 C.E.–1576 C.E.). Hieronymus Cardano is the name that he used in his writings. But he is sometimes known by the English version of his name: Jerome Cardan. Further details of the life of Girolamo Cardano may be found in [FIE].

He was the illegitimate child of Fazio Cardano and Chiara Micheria. His father was a lawyer in the Italian city of Milan. But his father knew quite a lot of mathematics; he was actually consulted by Leonardo da Vinci on questions of geometry. In addition to practicing law, Fazio Cardano lectured on geometry at the University of Pavia and the Piatti foundation in Milan.

Girolamo Cardano's mother was struck by the plague when she was pregnant with him. She went to Pavia for safety, and stayed with wealthy friends of Fazio. Her other children died of the plague, but Girolamo survived.

After Girolamo Cardano grew up, he became his father's assistant. But his health was very poor, and he required assistance from two nephews in order to perform some of the more arduous tasks.

Over his father's objections, Girolamo Cardano ended up entering the University of Pavia and studying medicine (his father wanted him to study law, of course). War broke out and interrupted Cardano's studies. He was forced to transfer to the University of Padua. Cardano was always outspoken and politically oriented. He campaigned to become rector of the university, and he succeeded.

Being a fiery and irrepressible personality, Cardano squandered the small bequest that his father left when he died. He ended up supporting himself by gambling—at cards, dice, and chess. Being a mathematician by nature, he understood probability theory rather well. So he was better equipped to gamble than most, and he won more than he lost. He managed to support himself with gambling, and his addiction to the pastime persisted for years.

Cardano did succeed in earning his medical degree in 1525 C.E.. He applied to join the College of Physicians in Milan. But his difficult personality turned out to be a problem for the admissions committee. When they learned that he was a bastard child, they found grounds to decline his application.

Cardano then went to the small village of Sacco near Padua. There he was able to set up a medical practice. Cardano subsequently married, but his modest practice did not give him the resources to support a family. He moved to Gallarate, near Milan. He was again turned down by the College of Physicians, and he found himself unable to practice medicine. He reverted again to gambling, and he also hocked many family valuables. Things went from bad to worse, and the Cardanos ended up in the poorhouse.

Cardano was finally able to assume his father's position as lecturer at the Piatti Foundation in Milan. This allowed him some free time, and he was able to treat some patients. He had such success as a practicing physician that he was able to build a coterie of backers. Cardano continued to be resentful that he could not gain admission to the College of Physicians. As a result, in 1536 he then published a book attacking the College's medical abilities and also it's character. A passage from the book gives a sense of its quality:

> The things which give most reputation to a physician nowadays are his manners, servants, carriage, clothes, smartness and caginess, all displayed in a sort of artificial and insipid way ...

This broadside aggravated the College of Physicians even further, and they continued to rebuff Cardano's applications. However, in 1539, Cardano's admirers convinced the college to modify the clause excluding illegitimate children. Cardano was finally admitted. In that same year, Cardano's first two mathematical books were published. He subsequently published numerous other books on mathematics; he wrote on topics as diverse as medicine, philosophy, astronomy, and theology (as well as mathematics).

In 1539 Cardano approached Tartaglia (1500 C.E.–1557 C.E.), who had achieved fame in winning a contest on solving cubics; he tried to convince Targaglia to divulge his methods. It should be understood that it was not common for scientists in those days to publish their results or their methods. Much was kept secret. He finally convinced Tartaglia to share his ideas, on the condition that Cardano would not publish the ideas until he (Tartaglia) himself had published them. In fact Cardano's oath was

> I swear to you, by God's holy Gospels, and as a true man of honour, not only never to publish your discoveries, if you teach me them, but I also promise you, and I pledge my faith as a true Christian, to note them down in code, so that after my death no one will be able to understand them.

Cardano spent the next six years in intense study of the solution of cubic and quartic equations.

One of Cardano's difficulties with this study was that he often was forced to confront roots of negative numbers. Complex numbers were not an established tool for mathematicians of the age. Even though the ultimate solution of the problem at hand was usually a genuine real number, the complex numbers came up as tools along the way. Cardano wrote to Tartaglia on August 4, 1539:

> I have sent to enquire after the solution to various problems for which you have given me no answer, one of which concerns the cube equal to an unknown plus a number. I have certainly grasped this rule, but when the cube of one-third of the coefficient of the unknown is greater in value than the square of one-half of the number, then, it appears, I cannot make it fit into the equation.

Ultimately Cardano exhibited the means for dealing with this difficulty, and as a result Tartaglia was jealous. He regretted revealing his methods to Cardano. He endeavored to confuse Cardano with his reply to the letter. Thus arose a raging feud between the two mathematicians of lifelong duration.

In 1540 Cardano resigned his post at the Piatti Foundation. The vacancy was filled by Cardano's assistant Ferrari, who had brilliantly solved quartic equations by radicals. From

1540 to 1542 Cardano spent his time gambling and playing chess all day. From 1543 to 1552 Cardano lectured on medicine at the Universities of Milan and Pavia.

In 1543 Cardano realized that Tartaglia had *not* been the first to solve cubics by radicals. He therefore felt justified in publishing what he knew on the subject. Thus in 1545 he published his masterpiece *Ars Magna*. This book contains, among other important facts, the very first calculations with complex numbers.

Although Cardano's wife died in 1546, he was not much taken aback by this loss. He had achieved considerable fame with his writings, and had finally been elected rector of the College of Physicians. He was, by some measures, the most famous physician in the world. He received offers from heads of state all over Europe to tend to their medical needs.

In 1552 Cardano was asked by the Archbishop of St. Andrews to treat his asthma. Although Cardano had routinely declined invitations of this sort, he found time to accept this one. He undertook the considerable journey, and was able to treat and to cure the Archbishop's illness. He was paid over 2000 gold crowns as an honorarium, and his considerable reputation was even more enhanced. On his return to Italy, Cardano was appointed Professor of Medicine at the University of Pavia.

Unfortunately, it was at this point in time that Cardano's life was struck by his profoundest tragedy. It affected him deeply, and led to his decline and to his death.

Cardano's eldest son Giambatista had studied medicine and qualified as a physician in 1557. But, meanwhile, he had married a woman of whom his father disapproved. In fact he characterized her as "a worthless, shameless woman." The elder Cardano supported his son financially and the young couple kept house with her parents. But the young woman's parents seemed to be scheming to extort money from Giambatista and his wealthy father. And she mocked her new husband for not being the father of her three children.

Giambatista took the situation badly, and ended up poisoning his wife. The young man confessed the crime and was ultimately brought to trial. The judge demanded, as part of the settlement, that Girolamo Cardano make peace with his son's wife's parents. They demanded a payment which was far beyond Cardano's means. So Giambatista was kept in prison and tortured. His left hand was cut off. On April 13, 1560, Giambatista Cardano was executed.

The elder Cardano never recovered from these circumstances. He tormented himself for failing to rescue his son. Since he was now the father of a convicted murderer, he became a hated man. He had to move, and obtained a Professorship of Medicine at the University of Bologna (the oldest university in the Western world). But his time in Bologna was plagued by controversy. His arrogant manner and questionable reputation combined to alienate him from his colleagues. At one point they conspired to have him dismissed from his post.

Cardano had additional problems with his children. His remaining son Aldo was a compulsive gambler who spent his time with low life. Cardano wrote in his autobiography of the four greatest disappointments in his life:

> The first was my marriage; the second, the bitter death of my son; the third, imprisonment ; the fourth, the base character of my youngest son.

In fact, in 1569, young Aldo gambled away all of his clothes and possessions as well as a notable portion of his father's assets. He even broke into his father's house and stole jewelry, cash, and valuables. Cardano was forced to report Aldo to the authorities, and the miscreant was banished from Bologna.

In 1570 Girolamo Cardano himself was jailed for heresy. He had cast a horoscope of Jesus Christ and written a book in praise of Nero (tormenter of the Christian martyrs). Evidently this was an attempt to pump up his notoriety and perpetuate his name. But this made him obvious fodder for the Spanish Inquisition, and he suffered accordingly.

Fortunately for Cardano, he was given lenient treatment (in part because public opinion had come full circle and there was actually considerable sympathy for Cardano in those days). He only served a short time in prison. But he was banned from the university and forbidden from publishing his work.

At this point in his life Cardano moved to Rome. There he received a surprisingly warm reception. He was granted membership in the College of Physicians. The Pope gave Cardano a pardon, and granted him a pension. It was at this time that Cardano wrote his autobiography and published it in Paris and Amsterdam.

One of the legends of Girolamo Cardano is that he predicted the exact date of his own death. But he fulfilled this prophecy by committing suicide.

In addition to Cardano's significant contributions to algebra he also made important contributions to probability, hydrodynamics, mechanics, and geology. He wrote a number of important and influential books, and he was the first ever to write on the subject of probability and its applications to gambling. He even wrote two encyclopedias of natural science, which were comprehensive compendia of all the scientific knowledge of the day.

Girolamo Cardano was a multi-talented individual who made profound contributions to the development of mathematics. His chaotic personal life certainly cut into, and in the end cut short, his scientific activities and contributions. But he will long be remembered for his significant ideas.

Further Reading

Janet Bellcourt Pomeranz, The dice problem—then and now, *The College Mathematics Journal* 15(1984), 229–237.

Blair K. Spearman, Kenneth S. Williams, Characterization of solvable quintics x^5+ax+b, *The American Mathematical Monthly* 101(1994), 986–992.

5.3 First-Order Equations

Girolamo Cardano is best remembered for the solution of algebraic (especially polynomial) equations. Thus we will concentrate here on topics of that nature.

Of course solving a linear equation, one of the form

$$ax + b = c\,,$$

is trivial. One engages in simple manipulations, such as

$$\begin{aligned} ax &= c - b \\ \frac{1}{a}\cdot ax &= \frac{1}{a}\cdot(c-b) \\ x &= \frac{c-b}{a}\,. \end{aligned}$$

to find the complete solution. And this method works on *all* linear (or first-order) polynomial equations.

For You to Try Solve the equation

$$3x - 9 = 15\,.$$

Further Reading

Sylvan Burgstahler, An algorithm for solving polynomial equations, *The American Mathematical Monthly* 93(1986), 421–430.

Dan Kalman, A generalized logarithm for exponential-linear equations, *The College Mathematics Journal* 32(2001), 2–14.

5.4 Rudiments of Second-Order Equations

Second-order, or quadratic, equations are slightly more subtle than first-order equations. A quadratic equation has the form

$$ax^2 + bx + c = 0\,.$$

Here a, b, c are real constants which can be positive, negative, or zero.

In the special case that $b = 0$, we have

$$ax^2 + c = 0$$

or

$$ax^2 = -c\,.$$

Division by a yields

$$x^2 = -\frac{c}{a}$$

hence

$$x = \pm\sqrt{-\frac{c}{a}}\,.$$

In summary, the quadratic equation

$$ax^2 + c = 0$$

has the two solutions $x = \sqrt{-c/a}$ and $x = +\sqrt{-c/a}$. This is all correct provided that $-c/a \geq 0$, so that the square root operation is valid.

For You to Try Solve the quadratic equation

$$8x^2 - 4 = 0\,.$$

You should find two solutions.

The philosophy for solving a *general* quadratic equation is a time-honored one in mathematics: to reduce the general case to the special case that we have already understood. We do this by the method of completing the square. Let us first acquaint ourselves with that technique before we proceed.

Further Reading

Chris Christensen, Newton's method for resolving affected equations, *The College Mathematics Journal* 27(1996), 330–340.

R. S. Varga, Iterative methods for solving matrix equations, *The American Mathematical Monthly* 72(1965), 67–74.

5.5 Completing the Square

Consider the square expression

$$A = (x + \alpha)^2 .$$

Formulas like this are common in elementary algebra. We frequently want to multiply it out so that we can manipulate it more effectively. In fact we may write

$$A = (x + \alpha)(x + \alpha) = x \cdot (x + \alpha) + \alpha \cdot (x + \alpha) = x \cdot x + x \cdot \alpha + \alpha \cdot x + \alpha \cdot \alpha .$$

Combining terms finally gives

$$A = (x + \alpha)(x + \alpha) = x^2 + 2\alpha x + \alpha^2 . \qquad (*)$$

Now it is also worthwhile to be able to look at a quadratic expression and recognize when it is a square. In examining $(*)$, we observe that it has a special feature:

$$A = x^2 + 2\underline{\alpha}x + \underline{\alpha}^2 . \qquad (\star)$$

The number whose square gives the constant term (namely, α) is just half of the coefficient of the x-term. Let us examine this feature in the context of a specific example.

EXAMPLE 5.1. Consider the expression

$$A = x^2 + 8x + 16 .$$

Observe that half the coefficient of x is 4; and $4^2 = 16$—which is the constant term. This matches equation $(\star)$ exactly. So A must be a perfect square. Indeed it must be that $\alpha = 4$, hence

$$A = x^2 + 8x + 16 = (x + 4)^2 .$$

EXAMPLE 5.2. Examine the polynomial

$$C = x^2 - 6x + 9 .$$

Notice that half the coefficient of x is -3; and $(-3)^2 = 9$—which is the constant term. According to our analysis of equation $(\star)$, we know that C is a perfect square. Note that $\alpha = -3$, hence

$$C = x^2 - 6x + 9 = (x - 3)^2 .$$

EXAMPLE 5.3. Let us determine whether

$$D = x^2 - 20x + 140$$

is a perfect square. We see that half the coefficient of x is -10 and $(-10)^2 = 100 \neq 140$. So the square of half the coefficient of x is *not* the constant term. Thus D is *not* a perfect square. Put in other words: the coefficient of the x-term forces $\alpha = -10$; but the square of this α does not match the constant term.

Even in this circumstance, we may rewrite D in terms of a square. Using the fact that $(-10)^2 = 100$, we rewrite D as

$$D = [x^2 - 20x + 100] + 40 .$$

According to our calculations, the expression in brackets is actually a perfect square. So we finally may write

$$D = (x - 10)^2 + 40 .$$

Further Reading

O. Fraser and B. Gordon, On representing a square as the sum of three squares, *The American Mathematical Monthly* 76(1969), 922–923.

R. Rozen and A. Sofo, Area of a parabolic region, *The College Mathematics Journal* 16, No. 5(1985), 400–402.

5.6 The Solution of a Quadratic Equation

Let us now use the philosophy of completing the square, combined with the methodology for solving $ax^2 + c = 0$, to solve an arbitrary quadratic equation. It is natural to build on the ideas in Sections 5.4 and 5.5 in order to treat the general case.

Now our general quadratic equation is

$$ax^2 + bx + c = 0 .$$

We may assume that $a \neq 0$, otherwise the equation has no quadratic term and it is in fact linear. Let us then divide out by a:

$$\frac{1}{a}\left[ax^2 + bx + c\right] = 0 .$$

This reduces to

$$x^2 + \frac{b}{a}x + \frac{c}{a} = 0 .$$

Write this as

$$\left[x^2 + \frac{b}{a}x\right] + \frac{c}{a} = 0 .$$

We know from the last example that the expression in brackets may be turned into a perfect square by the following device: We divide the coefficient of x by two and square it, then add the result on as our constant term. Thus we need

$$\left(\frac{1}{2} \cdot \frac{b}{a}\right)^2 = \frac{b^2}{4a^2} .$$

This is what we must add to the expression in brackets. But if we add a number to one side of the equation then of course we must add it to the other side (this is the Arab philosophy of keeping the equation balanced, or *al-jabr*, that we encountered in Chapter 4). The result is

$$\left[x^2 + \frac{b}{a}x + \frac{b^2}{4a^2}\right] + \frac{c}{a} = \frac{b^2}{4a^2}$$

or

$$\left[x + \frac{b}{2a}\right]^2 = \frac{b^2}{4a^2} - \frac{c}{a}.$$

It is convenient to put the right-hand side over a common denominator so that we have

$$\left[x + \frac{b}{2a}\right]^2 = \frac{b^2 - 4ac}{4a^2}.$$

The equation we have now is quite analogous to the sort of quadratic equation that we solved in Section 5.4: something squared equals a constant. The natural thing to do now is take the square root of both sides. We must remember, of course, that a positive real number has both a positive square root and a negative square root. The result is

$$x + \frac{b}{2a} = \pm\sqrt{\frac{b^2 - 4ac}{4a^2}}$$

or

$$x + \frac{b}{2a} = \pm\frac{\sqrt{b^2 - 4ac}}{2a}.$$

Now a little algebraic manipulation allows us to rewrite our result as

$$x = -\frac{b}{2a} \pm \frac{\sqrt{b^2 - 4ac}}{2a}$$

or

$$x = \frac{-b \pm \sqrt{b^2 - 4ac}}{2a}. \qquad (\dagger)$$

Of course this is the familiar quadratic formula that we all learn in high school algebra. There is evidence that the Egyptians dealt with quadratic equations (in the so-called Berlin Papyrus). The Babylonian tablet *Plimpton 322* contains many fascinating sets of data about such problems. There are even more definite indications that the ancient Greeks and Hindus knew the quadratic formula (around 500 B.C.E.). It was almost certainly then passed on to the Arabs.

EXAMPLE 5.4. Find all the roots of the quadratic equation

$$x^2 + 3x - 10 = 0.$$

Solution. This equation is in the standard form of a quadratic equation, with $a = 1, b = 3,$

and $c = -10$. According to the quadratic formula (†),

$$\begin{aligned} x &= \frac{-3 \pm \sqrt{3^2 - 4 \cdot 1 \cdot (-10)}}{2 \cdot 1} \\ &= \frac{-3 \pm \sqrt{49}}{2} \\ &= \frac{-3 \pm 7}{2} \\ &= \begin{cases} 2 \\ -5\,. \end{cases} \end{aligned}$$

EXAMPLE 5.5. Find all the roots of the quadratic equation

$$x^2 + 3x - 7 = 0\,.$$

Solution. Of course this equation fits our paradigm for a quadratic equation with $a = 1$, $b = 3$, $c = -7$. According to the quadratic formula (†),

$$\begin{aligned} x &= \frac{-3 \pm \sqrt{3^2 - 4 \cdot 1 \cdot (-7)}}{2 \cdot 1} \\ &= \frac{-3 \pm \sqrt{9 + 28}}{2} \\ &= \frac{-3 \pm \sqrt{37}}{2} \\ &= \begin{cases} \dfrac{-3 + \sqrt{37}}{2} \\ \dfrac{-3 - \sqrt{37}}{2}\,. \end{cases} \end{aligned}$$

EXAMPLE 5.6. Find all the roots of the quadratic equation

$$5x^2 - 8x + 2 = 0\,.$$

Solution. This equation certainly fits our paradigm for a quadratic equation with $a = 5$, $b = -8$, $c = 2$. The solution is therefore

$$\begin{aligned} x &= \frac{-(-8) \pm \sqrt{(-8)^2 - 4 \cdot 5 \cdot 2}}{2 \cdot 5} \\ &= \frac{8 \pm \sqrt{64 - 40}}{10} \\ &= \frac{8 \pm \sqrt{24}}{10} \\ &= \begin{cases} \dfrac{8 + \sqrt{24}}{10} \\ \dfrac{8 - \sqrt{24}}{10}\,. \end{cases} \end{aligned}$$

For You to Try Find all roots of the quadratic equation

$$x^2 + x + 1 = 0 .$$

Knowing the roots, factor the equation.

For You to Try Find all roots of the quadratic equation

$$x^2 - x + 10 = 0 .$$

Now factor the equation.

Further Reading

Henry Heaton, A method of solving quadratic equations, *The American Mathematical Monthly* 3(1896), 236–237.

E. John Hornsby, Jr., Geometrical and graphical solutions of quadratic equations, *The College Mathematics Journal* 21(1990), 362–369.

5.7 The Cubic Equation

It is Girolamo Cardano (1501 C.E.–1576 C.E.) who deserves the credit for finally taming the cubic equation. Cardano also solved the quartic, or fourth-degree; equation. Both solutions appeared in Cardano's important treatise *Ars magna*. It should be noted that Cardano's work was in some ways anticipated by work of Scipione del Ferro (1465–1526), Niccolo Tartaglia (1500–1557), and Lodovico Ferrari (1522–1565). We shall only treat the cubic equation in this text. The analysis of the quartic equation is similar, but much more complicated. In fact one solves the quartic by reducing it to a cubic, and then solves the cubic by reducing it to a quadratic. This is how mathematics works.

5.7.1 A Particular Equation

Let us begin by examining a particular cubic equation—one that Cardano wrote about in *Ars magna* in 1545. Of course Cardano did not have our modern notation, and we are rendering his ideas in contemporary language. That is the equation

$$x^3 + 6x = 20 . \qquad (*)$$

This may seem rather special, but in fact it is quite typical. And the general case may be reduced to it.

Cardano's idea is to introduce two new variables u and v. In fact we let $u^3 - v^3 = 20$ and $uv = 2$.[1] As a result, we may rewrite the equation $(*)$ as

$$x^3 + (3uv)x = u^3 - v^3 . \qquad (**)$$

[1] Remember that we saw a trick like this when we studied Pythagorean triples in Chapter 1.

Now be forewarned that Cardano's solution method is a bag of tricks. His idea now is to observe—just by educated guessing—that $x = u - v$ solves this new equation $(**)$. Let us verify this claim:

$$(u - v)^3 + (3uv)(u - v) \stackrel{(?)}{=} u^3 - v^3 ,$$

$$[u^3 - 3u^2v + 3uv^2 - v^3] + (3u^2v - 3uv^2) \stackrel{(?)}{=} u^3 - v^3 ,$$

$$u^3 - v^3 = u^3 - v^3 .$$

Thus we may write $x = u - v$. It is our job, then, to determine u and v.

But we know that

$$u^3 = 20 + v^3 = 20 + \frac{8}{u^3} \tag{$\star$}$$

because $uv = 2$ so $(uv)^3 = 8$ hence $v^3 = 8/u^3$. Now it is convenient to let $\alpha = u^3$. Then equation $(\star)$ becomes

$$\alpha = 20 + \frac{8}{\alpha}$$

or (multiplying through by α)

$$\alpha^2 = 20\alpha + 8 ;$$

we may rearrange this to read

$$\alpha^2 - 20\alpha - 8 = 0 . \tag{$\ddagger$}$$

Of course equation $(\ddagger)$ is a quadratic equation, and we may solve it. We find that

$$\begin{aligned} \alpha &= \frac{20 \pm \sqrt{(-20)^2 - 4 \cdot 1 \cdot (-8)}}{2 \cdot 1} \\ &= \frac{20 \pm \sqrt{400 + 32}}{2} \\ &= 10 \pm \sqrt{108} \\ &= \begin{cases} 10 + \sqrt{108} \\ 10 - \sqrt{108} . \end{cases} \end{aligned}$$

So $u^3 = \alpha = 10 \pm \sqrt{108}$.

Now we must unwind our construction. Since $u^3 - v^3 = 20$, we know that $v^3 = u^3 - 20 = -10 \pm \sqrt{108}$. Now we have two cases:

The Case $u^3 = 10 + \sqrt{108}$. In this situation, $v^3 = -10 + \sqrt{108}$. Taking roots, we find that $u = \sqrt[3]{10 + \sqrt{108}}$ and $v = \sqrt[3]{-10 + \sqrt{108}}$. In conclusion, $x = u - v = \sqrt[3]{10 + \sqrt{108}} - \sqrt[3]{-10 + \sqrt{108}}$.

The Case $u^3 = 10 - \sqrt{108}$. In this situation, $v^3 = -10 - \sqrt{108}$. Taking roots, we find that $u = \sqrt[3]{10 - \sqrt{108}}$ and $v = \sqrt[3]{-10 - \sqrt{108}}$. In conclusion, $x = u - v = \sqrt[3]{10 - \sqrt{108}} - \sqrt[3]{-10 - \sqrt{108}} = \sqrt[3]{10 + \sqrt{108}} - \sqrt[3]{-10 + \sqrt{108}}$.

For the last equality, distribute the minus signs.

We see that we have discovered the same root of our polynomial equation twice. We invite the reader to actually plug this value for x into the expression $x^3 + 6x$ and confirm that the result is 20.

There is one remaining thing to consider. We expect that a cubic equation will factor as three linear factors. So we expect there to be three roots. But we have only found one root. Where are the other two hiding?

It turns out that the other two roots of $x^3 + 6x = 20$ are complex. We are not going to get into the complex numbers at this time (but see Section 9.1), so we shall content ourselves (just as Cardano did) with just one root for the polynomial equation.

5.7.2 The General Case

At the beginning of this section, we made the bold assertion that *any* cubic equation can be reduced to the one that we have just studied. Let us now see why that is so. Consider a cubic equation

$$x^3 + ax^2 + bx + c = 0 .$$

We make the change of variable $x = t - a/3$. The result is

$$\left(t - \frac{a}{3}\right)^3 + a\left(t - \frac{a}{3}\right)^2 + b\left(t - \frac{a}{3}\right) + c = 0$$

or

$$\left(t^3 - 3 \cdot t^2 \cdot \frac{a}{3} + 3 \cdot t \cdot \frac{a^2}{3^2} - \frac{a^3}{3^3}\right) + a\left(t^2 - 2 \cdot \frac{a}{3} \cdot t + \frac{a^2}{3^2}\right) + b\left(t - \frac{a}{3}\right) + c = 0 .$$

Now we regroup the left-hand side in powers of t. The result is

$$\begin{aligned} &t^3 + \left(-3 \cdot \frac{a}{3} + a\right)t^2 + \left(3 \cdot \frac{a^2}{9} - \frac{2a^2}{3} + b\right)t + \left(-\frac{a^3}{27} + \frac{a^3}{9} - \frac{ab}{3}\right) \\ &\quad = t^3 + \left(-\frac{a^2}{3} + b\right)t + \left(\frac{2a^3}{27} - \frac{ab}{3}\right) = 0 . \end{aligned}$$

Observe that what we have accomplished is that our polynomial now has no square term.

If we assume for the moment that $b - a^2/3 > 0$, then[2] we may make the change of variable $t = \sqrt{(b - a^2/3)/6}\,u$. The result is

$$\left[\sqrt{(b - a^2/3)/6}\,u\right]^3 + \left(-\frac{a^2}{3} + b\right)\left[\sqrt{(b - a^2/3)/6}\,u\right] + \left(\frac{2a^3}{27} - \frac{ab}{3}\right) = 0 .$$

Doing the algebra, and simplifying, we find that our polynomial has become

$$u^3 + 6u + \frac{2a^3/27 - ab/3}{[(b - a^2/3)/6]^{3/2}} = 0 .$$

This is Cardano's polynomial equation, with -20 replaced by a somewhat different constant. But Cardano's technique still applies, and the solution may be found. Then one can resubstitute t for u, and then resubstitute x for t, and find the root of the original polynomial equation.

---

[2]Things are a bit more complicated if $b - a^2/3 < 0$ and we shall not discuss that situation here. The situation $b - a^2/3 = 0$ is trivial since then the equation reduces to $t^3 + ([2a^3/27] - [ab/3]) = 0$ or $t = ([ab/3 - [2a^3/27])^{1/3}$.

For You to Try Use Cardano's method to find a root of the polynomial equation

$$x^3 + 9x^2 + 33x + 35 = 0 .$$

For You to Try Use Cardano's method to find a root of the polynomial equation

$$x^3 - 6x^2 + 18x - 24 = 0 .$$

Further Reading

R. S. Luthar, Luddhar's method of solving a cubic equation with a rational root, *The Two-Year College Mathematics Journal* 11(1980), 107–110.

H. A. Nogrady, A New Method for the Solution of Cubic Equations, *The American Mathematical Monthly* 44(1937), 36–38.

5.8 Fourth-Degree Equations and Beyond

Cardano's method can be extended to fourth-degree equations. That situation is fairly complicated, and we shall not discuss it here. It was an open problem for a long time—nearly 240 years—to determine whether there was a formula, or a technique, for solving fifth-degree (or higher) equations. Both Niels Henrik Abel (1802–1829) and Évariste Galois (1812–1832) thought that they found a means, but then discovered their own error. It was Abel himself, at the age of twenty-two, who finally proved the *impossibility of solving a quintic equation with a formula involving only arithmetic operations and roots.* There is an advanced theorem, called the *Implicit Function Theorem* (see the reference [KRP]), that allows one to solve a quintic equation using transcendental functions (like sine and cosine and logarithm, for example). But Abel showed that there was no elementary formula.

Further Reading

Abraham Arcavi and Maxim Bruckheimer, Reading Bombelli's x-purgated algebra, *The College Mathematics Journal* 22(1991), 212–219.

H. T. R. Aude, Notes on quartic curves, *The American Mathematical Monthly* 56(1949), 165–170.

5.8.1 The Brief and Tragic Lives of Abel and Galois

Niels Henrik Abel (1802–1829) lived an all-too-brief life that was dominated by poverty and deprivation. At the time, the Norwegian economy was suffering a blockade by the British, and there were also political difficulties with Denmark and Sweden. The entire country of Norway was in a bad way, and poverty was widespread.

Both Abel's father and grandfather were ministers. His father was also involved in politics and in fact held office in the national legislative body, the *Storting*. Niels Henrik was the second of seven children. In those hard times, the young man's parents had difficulty putting food on the table. In addition, it is suspected that Niels Henrik's father was a drunk and his mother a woman of lax morals. More can be learned about Abel's life in [ORE1].

In 1815 the young genius was sent to the Cathedral School in Christiana. Once a distinguished academy, this institution had lost all its good teachers to the staffing of the university. So education was in a bad state when the young man arrived. He was uninspired by the instruction, but exhibited some talent for mathematics and physics. It was Niels Henrik Abel's good fortune that a new instructor, Bernt Holmboe, arrived at the Cathedral School in 1817. He immediately recognized Abel's talent and encouraged him to study university-level mathematics. The young student bloomed under this attention, and he advanced rapidly. Tragedy struck, however, when Abel's father died in 1820.

Abel's father had ended his political career in disgrace because he had made false charges against his fellow members of the Storting. His excessive drinking led to his dismissal from the Storting, and it followed that the family was in dire straits when the old man died. Niels Henrik feared that he would have to quit school in order to support his family.

But his teacher Holmboe arranged a scholarship so that Abel could attend the University of Christiania. He also raised money from his colleagues to help support the young scholar. Abel did manage to graduate from the University in 1822.

During his final year in school, Abel worked on the solution of the quintic equation. In 1821 he believed that he had found the solution, and he submitted a paper to the Danish mathematician Ferdinand Degen for publication by the Royal Society of Copenhagen. Degen questioned Abel closely about his work, and led him to find an error. Degen also encouraged Abel to develop an interest in elliptic integrals.

Abel was also fortunate at this time to have found a new mentor, Christopher Hansteen, at the University of Christiania. In fact Hansteen's wife took Abel under her wing, and treated the young fellow as her own son. Abel was able to publish papers in a new scientific journal (*Magazin for Naturvidenskaberne*) that had been started by Hansteen. In particular, he produced the first known solution of an integral equation.

At this time Abel won a small grant that enabled him to visit Degen in Copenhagen. In Copenhagen he met Christine Kemp, who became his fiancée. Abel had ambitions to visit the leading mathematical scholars in France and Germany in order to be able to discuss and develop his work. But he did not have the funds and did not speak the languages, so he instead obtained more modest funds to stay in Christiania and study. In 1824 he succeeded in proving the impossibility of solving the quintic equation by radicals. He published the work in French, as a pamphlet, at his own expense. This decision was motivated by a desire to get into print quickly so that he would have an impressive piece of work to bring with him when he engaged in his planned travels. In order to save printing costs, he reduced his proof to fit on half a folio sheet (six pages).

Abel sent his pamphlet to a number of distinguished mathematicians of the day, including Carl Friedrich Gauss. He intended to visit Göttingen when he engaged in his travels. In 1825 he obtained a scholarship from the Norwegian government that finally made his planned European sojourn possible. Reaching Copenhagen, Abel was disappointed to learn that Degen had died. He decided not to go to Paris but instead to stay with his traveling companions and proceed to Berlin.

Abel had obtained a letter of introduction to Crelle. He then met Crelle in Berlin, and the two men became fast friends. At the time, Crelle was developing a new journal (*Die Journal für die Reine und Angewandte Mathematik*—The Journal of Pure and Applied Mathematics) which was to become a very distinguished showcase for mathematical research. Today it is the oldest extant mathematics journal. Crelle encouraged Abel to develop a more detailed version of his ideas about the unsolvability of quintic equations, and to publish it in his new journal. That Abel did, and his paper appears in the very first volume of the journal. In fact a total of seven of his papers appear in that volume.

Abel began to dedicate himself to the development of the rigorous foundations of mathematical analysis and to publish more papers in Crelle's journal. He was disappointed to learn that the only open professorship at the only university in Norway had been given to Holmboe. Abel had had plans to go with Crelle to Paris and to visit Gauss in Göttingen along the way. But Gauss, who was notorious for being unsupportive of bright, young mathematicians, had evinced displeasure with Abel's pamphlet on the nonsolvability of the quintic. This may at first seem rather odd, as Abel's pamphlet was later found still in the envelope and unopened among Gauss's papers. But it is believed that Gauss attached no significance to the explicit solution of particular equations. Recall that Gauss was the one who proved the Fundamental Theorem of Algebra, which says that *any* polynomial has a complex root. That is an abstract, nonconstructive result—the sort of theorem that Gauss favored. In any event, Gauss's lack of support deeply affected Abel.

When Abel finally got to Paris he was upset to find that the leading French mathematicians had little interest in his work. Cauchy, in particular, had no time for him. He wrote to Holmboe that

> The French are much more reserved with strangers than the Germans. It is extremely difficult to gain their intimacy, and I do not dare to urge my pretensions as far as that; finally every beginner had a great deal of difficulty getting noticed here. I have just finished an extensive treatise on a certain class of transcendental functions to present it to the Institute which will be done next Monday. I showed it to Mr. Cauchy but he scarcely deigned to glance at it.

Abel had important new results on elliptic integrals—some of which far surpassed earlier work of Euler—but he could find no interest for them. He was running out of money, could only afford one meager meal per day, and was becoming emaciated, despondent, and tired.

But Abel doggedly continued his work on elliptic integrals. He ultimately left Paris and returned to Berlin. There he borrowed some money so that he could continue his work on elliptic functions. But his health was in a poor state. Crelle continued to be Abel's staunch supporter. He attempted to land a professorship for the young scholar, and also offered him the editorship of his journal. But Abel determined to return to his homeland. Abel finally reached Christiania in 1827 and obtained a very small grant from the university. He tutored school children to make ends meet, and his fiancée was employed as a governess.

At this time Hansteen received a major grant to investigate the Earth's magnetic field in Siberia. Thus Abel was hired to replace him as a Professor at the University. This improved Abel's circumstances slightly.

In 1828 Abel became aware of work of Jacobi on transformations of elliptic integrals. These ideas were a revelation to Abel, and he realized that they fit into the context of what he had been studying. He quickly wrote several papers which transformed the subject, and which finally gained the attention of Adrien-Marie Legendre (1752 C.E.–1833 C.E.) among others.

While his health deteriorated, Abel continued to produce first-class work on elliptic functions. He spent the summer of 1828 with his fiancée in Froland. He had submitted his masterpiece on elliptic function theory to the Paris Academy, but they had somehow lost the manuscript. This was long before the days of photocopying, so Abel had to produce the manuscript from scratch again.

Abel traveled by sled to visit his fiancée in Froland for Christmas of 1828. On that trip he became seriously ill. Crelle, ever his friend and mentor, redoubled his efforts to obtain better circumstances for Abel. He finally succeeded in obtaining a professorial appointment for Abel in Berlin. Crelle wrote to him on April 8, 1829 to tell of the great news. But it was too late; Abel had already died.

After Abel's untimely death, Cauchy (after much searching) found his Paris manuscript. It was printed in 1841 but again somehow vanished. It did not surface again until 1952!—some 123 years after Abel's death. It was in fact found in Florence, Italy. Another manuscript found after Abel's death—he continued working to the very end, even on his deathbed—gave important results about the solution of polynomial equations. These anticipated seminal results that would later be proved by Galois.

Evariste Galois (1811–1832) also lived a painfully brief life. His demise was not brought on by abject poverty, but rather by personal chaos and, in the end, death by a gunshot wound. Details of the sadly brief life of Galois may be found in [RIG].

Galois's family consisted of intelligent and well-educated people. His mother was his only teacher until the age of 12, and she taught him classical languages and religion. There is no evidence of mathematical talent in the family before Evariste Galois himself.

Galois lived in times of great political turmoil in France. The storming of the Bastille took place in 1789, and set the tone of unrest and foment in which the young Galois grew up. His school itself—the Lycée of Louis-le-Grand—was marked by rebellion among the students.

The year 1827 was a turning point for Galois, because he had his first mathematics class from M. Vernier. He quickly became absorbed by the subject and excelled dramatically. His director of studies wrote of him

> It is the passion for mathematics which dominates him, I think it would be best for him if his parents would allow him to study nothing but this, he is wasting his time here and does nothing but torment his teachers and overwhelm himself with punishments.

Young Galois's school reports described him repeatedly as "singular, bizarre, original, and closed." Since Galois is today remembered as one of the most *original* mathematicians who ever lived, it is remarkable that his originality was at first taken to be a liability.

In 1828 Galois applied to the École Polytechnique, the most distinguished technical school in France (analogous to M.I.T. in the United States today). His interests in the school were of course academic, but he was also interested in the powerful political movements that existed among the students. He failed the entrance exam, and was not admitted.

Galois disappointedly returned to Louis-le-Grand, where he took mathematics from Louis Richard. But the young man concentrated more on his own interests (the work of Adrien-Marie Legendre and J. L. Lagrange) and less on his classwork. In 1829 he published his first research paper. Two more papers quickly followed. Unfortunately, Galois's father committed suicide later that year, and of course young Galois took this event very badly.

His second application to the École Polytechnique, as a result, failed. Galois instead entered the École Normale, which was an annex to the Louis-le-Grand school.

Galois always had trouble formulating and expressing his mathematical ideas, and this may have contributed to his failure to pass the entrance exam to the École Polytechnique. In order to enter the École Normale, he had to pass Baccalaureate examinations. His examiner in mathematics reported

> This pupil is sometimes obscure in expressing his ideas, but he is intelligent and shows a remarkable spirit of research.

As a counterpoint, his examiner in literature said

> This is the only student who has answered me poorly, he knows absolutely nothing. I was told that this student has an extraordinary capacity for mathematics. This astonishes me greatly, for, after his examination, I believed him to have but little intelligence.

Galois sent some of his work to Cauchy at this time, and was informed that it overlapped with work of Abel. He subsequently read Abel's papers, and this changed the course of his research. He began to study elliptic functions and abelian integrals. Galois had submitted his work to Fourier, the secretary of the French Academy, for consideration for their Grand Prize in Mathematics. The prize was subsequently awarded to Abel and Jacobi, certainly disappointing Galois. They seem to have lost his submission.

France experienced considerable political unrest in 1830, and Galois became involved. This certainly distracted from his mathematics. He only published two more papers in 1831, and these were to be his last. Sophie Germain (discussed elsewhere in this book) noted in a letter that Galois was suffering because his mentor Fourier had died. Galois was without money, dispirited, and distracted by radical politics. He was expelled from the École Normale.

In 1831 Galois was arrested for making public threats against the King, Louis-Phillipe. Testimony revealed that there was confusion about what Galois had actually said, and no reliable witness could be brought against him. He was acquitted.

Not long after, Galois was found carrying loaded weapons on Bastille day, and he was arrested again. While in prison, he got word of the rejection of his latest mathematical memoir. He attempted suicide while incarcerated, but the other prisoners wrested the dagger from him.

During a cholera epidemic in March, 1832 the prisoners, including Galois,were transferred to the pension Sieur Faultrier. There he seems to have fallen in love with Stephanie-Felice du Motel, daughter of the resident physician. After his release in April, he pursued a correspondence with Stephanie, but she distanced herself from the relationship.

Galois fought a duel with Perscheux d'Herbinville on May 30, 1832. Although the specific reasons for the duel have been lost to history, it seems clear that the issue was related to Stephanie. According to legend, Galois knew that he had no skills related to dueling, and was convinced that he would die in this confrontation. So he spent the night before writing out all that he knew about group theory. In any event, he was wounded in the duel and was abandoned by d'Herbinville and his own second. Later a peasant found him and arranged for him to be taken to Cochin hospital. When Galois was taken to the hospital with his fatal wounds, his brother waited there weeping at his bedside. Galois said, "Don't cry. I need all my courage to die at twenty." Galois died on May 31, 1832.

Galois's brother and his friend Chevalier copied out Galois's mathematical papers and sent them to Gauss. There is no record that Gauss ever studied them, but the papers found their way to Liouville. Liouville did study them, and subsequently announced to the French Academy that Galois had found a complete solution of the problem of when a polynomial may be solved by radicals. These papers contain the foundations of what is now known as *Galois theory*—one of the central cornerstones of modern number theory.

Further Reading

Philip J. Davis, What do I know? A study of mathematical self-awareness, *The College Mathematics Journal* 16(1985), 22–41.

Michael Rosen, Abel's theorem on the lemniscate, *The American Mathematical Monthly* 88(1981), 387–395.

5.9 The Work of Abel and Galois in Context

As you can see from their dates, both Galois and Abel led tragically short lives. Abel was a relatively happy person, but was burdened with the support of his six-member family and ultimately was defeated by his poverty. He died of consumption at the age of 26. Galois turned out to be his own worst enemy. He was tormented by his ill fortune and the lack of recognition that his work had received. He turned to radical politics amid social upheaval in order to expiate his frustrations. He ended up involved in a self-destructive duel that he knew he would lose. He spent the night before the duel recording, as best he could, his many brilliant ideas. Then he went out the next morning and died from a bullet shot. He was only twenty.

One of the astonishing theorems of mathematics, that was proved by Carl Friedrich Gauss (1777-1855) in his thesis, is the fundamental theorem of algebra. This theorem asserts that every nonconstant polynomial has a (complex) root. It does *not* give a formula or a method for finding that root. But it does assert that one exists. However, there is a catch. Consider the polynomial

$$p(x) = x^2 + 1 .$$

There is no *real* value for x which makes this polynomial equal to 0. Why not? Well, for any real x, $x^2 \geq 0$ so $x^2 + 1 \geq 1$. Thus the polynomial cannot take the value 0. And that is all there is to it.

So how can Gauss's fundamental theorem be true? The answer is that it is true in a larger number system—the complex numbers. We shall consider the complex numbers in Chapter 9.

Further Reading

Sharon Barrs, James Braselton, and Lorraine Braselton, A rational root theorem for imaginary roots, *The College Mathematics Journal* 34 (2003), 380–382.

George Mackiw, Computing in abstract algebra , *The College Mathematics Journal* 27 (1996), 136–142.

Michael I. Rosen, Niels Hendrik Abel and equations of the fifth degree, *The American Mathematical Monthly* 102 (1995), 495–505.

Exercises

1. Find all solutions of the equation $3x - 7 = 4$. Explain why you have found *all* the roots. Discuss this problem in class.

2. Find all solutions of the quadratic equations $x^2 + x - 4 = 0$. Explain why you have found *all* the roots. Discuss this problem in class.

3. Apply the quadratic formula to the equation $x^2 + x + 4$. What difficulty do you encounter? Does this equation have any real roots? Draw the graph of $p(x) = x^2 + x + 4$ and discuss in class why there are no real roots.

4. If a polynomial with real coefficients of odd degree is to have real roots then its graph must cross the x-axis. Discuss in class why this is true. Then discuss why a cubic equation will *always* have at least one real root while a quadratic equation may not. In fact complex roots always occur in conjugate pairs. Discuss this result, defining terms as needed, and prove it.

5. Discuss in class whether a quartic (i.e., a fourth-degree) equation will always have a real root. Look at some examples. What about $p(x) = x^4 + 1$? What about $x^4 - 2x^2 + 1$?

6. Use Cardano's method to find a root of the polynomial $x^3 - x - 6$.

7. Use Cardano's method to find a root of the polynomial $3x^3 - 10x^2 + 9$.

8. Can you write a polynomial whose roots are $-1, 3, 5$?

9. Can you give an example of a polynomial of degree 2 that has no real roots? How about degree 4?

10. Explain why a polynomial with real coefficients of odd degree at least 1 will always have at least one real root.

11. **Project:** The only real root of the quintic equation

$$x^5 - 5x^4 - 10x^3 - 10x^2 - 5x - 1 = 0 \qquad (*)$$

is $\alpha = 1 + \sqrt[5]{2} + \sqrt[5]{4} + \sqrt[5]{8} + \sqrt[5]{16}$. Verify with a calculation that this is true.

Further, show that this is the only real root. You can divide the equation $(*)$ by $x - \alpha$ and obtain a quartic polynomial. It will have only complex roots. What can you say about those complex roots? See [VAN] and [PRA] for more on solving algebraic equations.

12. **Project:** The quintic equation

$$x^5 - x^4 - x + 1$$

can be completely solved by factorization. Perform this procedure and find all the roots. See [PRA].

13. Project: Find all roots x and y of the equation

$$(x - y)^3 - (x - y) + 1 = 0 .$$

The solution set will be a collection of ordered pairs (x, y) in $\mathbb{C}^2 = \mathbb{C} \times \mathbb{C}$. What can you say about this set? What will be its dimension? What are its geometric characteristics? See [FUL] for further details on this topic.

6

René Descartes and the Idea of Coordinates

6.0 Introductory Remarks

The idea of coordinates is an old one. Apollo'nius of Per'ga (about 200 B.C.E.) set up a special coordinate system on the cone in order to study conic sections (Figure 6.1). Hipparchus (about 150 B.C.E.) and Marinus of Tyre (about 150 C.E.) used a version of latitude and longitude for purposes of navigation and astronomy. The idea of locating the real numbers on a number line is also an old one.

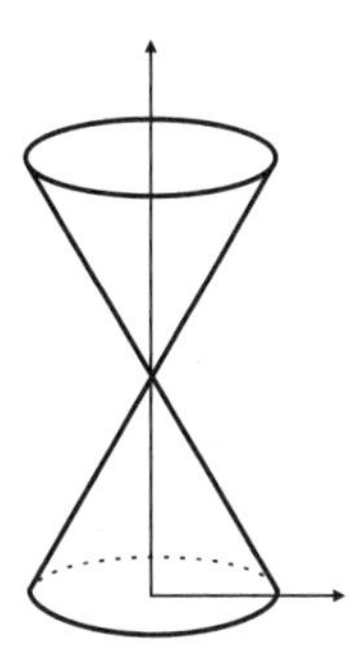

Figure 6.1. Coordinates of Apollo'nius.

But it was René Descartes who conceived the idea of unifying algebra and geometry with a rectangular coordinate system on the plane.[1] In particular, it was Descartes who created the idea of *graphing a function*. John Stuart Mill (1806–1873) said that this was "the greatest single step ever made in the exact sciences." Certainly the idea of rectangular coordinates has had a profound influence on all of modern science, engineering, and mathematics.

Today, in analytical thinking, we use many types of coordinate systems. For some types of problems, the traditional (rectangular) cartesian coordinates are well-suited. For others, a coordinate system with some circular symmetry (such as polar coordinates or cylindrical coordinates or spherical coordinates[2]) are more appropriate. For certain problems in cos-

[1] In fact legend has it that Descartes was lying on his back in bed, staring at the shadow that a window screen cast on the ceiling, when the idea for his coordinates struck him.

[2] The Schwarzchild model for general relativity is calculated in spherical coordinates.

mology and higher-dimensional geometry, more abstract curvilinear coordinate systems are what best suits the task at hand.

In the present chapter we shall learn about coordinate systems, and about the synthesis between geometry and algebra that Descartes created with his profound idea.

6.1 The Life of René Descartes

René Descartes was born on March 31, 1596 in La Haye, Touraine, France. In fact the town is now named "Descartes" in his honor. He died on February 11, 1650. Descartes was educated at the Jesuit College of La Flèche in Anjou. In fact he was only eight years of age when he entered the college, just a few months after it opened its doors. Young René studied there from 1604 until 1612. He concentrated on classics, logic, and traditional Artistotelian philosophy. His health was poor in those days, and he obtained special permission to remain abed each day until 11:00 AM. He maintained that habit for his entire adult life, and usually spent the morning hours in bed thinking. The remarkable life of René Descartes is recounted in loving detail in [HAL] and [VRO].

School impressed on Descartes how little he knew. Of all his areas of study, he found mathematics to be the most satisfying, as it gave him some sense of closure. His mathematical studies became the basis for all his future investigations—in mathematics, in philosophy, and in the natural sciences.

After the Jesuit College, Descartes spent some time in Paris—primarily keeping his own counsel. Then he studied at the University of Poitiers, where he received a law degree in 1616. After that he enlisted in the military school at Breda. After two years he began studying mathematics in earnest under the direction of the Dutch scientist Isaac Beeckman. Descartes's goal was to find a unified science of nature.

In 1619 Descartes joined the Bavarian Army. From 1620 to 1629 Descartes traveled throughout Europe. He spent time in Bohemia, Hungary, Germany, Holland, and France. In 1623 Descartes found himself in Paris, where he was able to spend time with Mersenne. The latter proved to be an important liaison who kept Descartes abreast of scientific developments for many years.

By 1628 Descartes was tired of traveling and determined to settle down. He chose Holland for his residence. This turned out to be a good choice, and he immediately began work on his physics treatise entitled *Le Monde, ou Traité de la Lumière*. This ambitious work was near completion when Descartes received word of Galileo's house arrest (for his scientific ideas about the planets). Descartes decided on the basis of this news not to risk publication, and in fact his book on physics was published, and only in part, after his death. Descartes decided then to concentrate his efforts on more abstract issues (which were less likely to upset the powers that be). He used these words to express his thoughts:

> ...in order to express my judgment more freely, without being called upon to assent to, or to refute the opinions of the learned, I resolved to leave all this world to them and to speak solely of what would happen in a new world, if God were now to create ...and allow her to act in accordance with the laws He had established.

Descartes maintained a number of scientific contacts while in Holland, and was actually quite cordial with many of them. Among these were Mersenne (in Paris), Mydorge, Hortensius, Huygens, and Frans van Schooten (the elder). These allies encouraged Descartes to publish his ideas. He was firm in not wishing to publish *Le Monde*, but he instead published a tract on science with the title *Discourse de la méthode pour bien conduire sa raison et chercher la vérité dans les sciences*. This book had three important Appendices entitled La Dioptrique, Les Météores, and La Géométrie. Descartes's book was published in Leiden in 1637.

The first of the Appendices is a work on optics, and the second a work on meteorology. Although Descartes's scientific method was flawed, and he made a number of incorrect assertions, he nonetheless laid the foundations for future work in these fields.

Certainly the third Appendix, on geometry, is the most important. In this tract Descartes lays the foundations for the theory of geometric invariants, and particularly for the connections between algebra and planar geometry. Although Descartes's thoughts are inspired by Oresme, there is much here that is original.

Descartes's first major philosophical work, entitled *Meditations on First Philosophy*, was published in 1641. The book consisted of six "meditations":

- Of the Things that we may doubt;
- Of the Nature of the Human Mind;
- Of God: that He exists;
- Of Truth and Error;
- Of the Essence of Material Things;
- Of the Existence of Material Things and of the Real Distinction between the Mind and the Body of Man.

Unfortunately many prominent scientists, including Arnauld, Hobbes, and Gassendi, were opposed to the ideas expressed in this book.

René Descartes most comprehensive work was *Principia Philosophiae*, published in Amsterdam in 1644. The book has four parts:

- The Principles of Human Knowledge;
- The Principles of Material Things;
- Of the Visible World;
- The Earth.

Following the philosophical principles that Descartes put in place when he was a student, he attempted in this work to put the entire universe on a mathematical foundation, reducing the study to one of mechanics. Descartes's study had some strange features. He did not believe in action at a distance. Therefore he could not account for gravity. Descartes believed instead that the universe is filled with matter which, due to some initial motion, has settled down into a system of vortices which carry the sun, the stars, the planets, and the comets in their paths. Descartes's theories held sway for more than one hundred years, even after Isaac Newton showed that the theory was impossible and replaced it with his universal law of gravitation.

In the year 1644, the date of publication of his *Meditations*, René Descartes visited France. He returned again to France in 1647, when he established contact with Blaise Pascal. In fact he tried to convince Pascal that a vacuum could not exist (again bearing out his idea that no force can act at a distance). He returned once more to France in 1648.

Descartes was a solitary figure with many eccentricities. He was a short (5′0″ dripping wet), irascible Frenchman who was also one of our greatest philosophers and mathematicians. He thought very highly of himself and his abilities, and he had little patience along with a blazing temper. He enjoyed staying in bed naked each day until late in the morning. He would think about philosophical and mathematical issues during his sojourns abed. In fact he conceived his ideas about coordinates in the plane during one of his bed sessions.

Descartes gave up mathematics when he was still a young man because, he said, he'd gone as far in mathematics as a human being could go. He read many romance novels and sagas of chivalry; Descartes had an active fantasy life. He had a particular fetish for cross-eyed women.

Among his more unusual beliefs was the contention that animals were "senseless machines". Even so, Descartes had a pet dog named "Monsieur Grat" or "Mr. Scratch" of which he was very fond.

Descartes used to play cards and gamble with his friend Blaise Pascal (1623 C.E.–1662 C.E.). It is said that Descartes made a lot of money thereby.

Descartes's ideas were widely read and highly influential. In 1649, Queen Christina of Sweden persuaded him to travel to Stockholm to tutor her. One of the Queen's eccentricities was that she wanted to draw tangents at 5:00 AM. Descartes reluctantly broke his lifetime habit of sleeping late. Unfortunately the new routine of walking to the palace so early every morning, in the dark and cold, led to Descartes contracting pneumonia. He died as a result.

Further Reading

B. F. Finkel, Biography: René Descartes, *The American Mathematical Monthly* 5(1898), 191–195.

Dwight D. Freund, A genuine application of synthetic division, Descartes' rule of signs, and all that stuff, *The College Mathematics Journal* 26(1995), 106–110.

6.2 The Real Number Line

Now we shall study some of the mathematical ideas of René Descartes. We begin by laying out the integers in a linear pattern on a fixed straight line (Figure 6.2). Notice that numbers to the left of 0 are negative, and the further left we go the more negative the numbers become. Likewise, numbers to the right of 0 are positive, and the further right we go the more positive the numbers become.

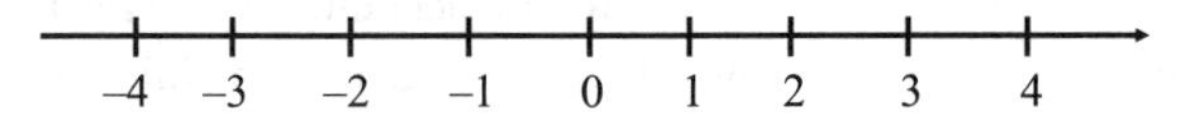

Figure 6.2. Beginnings of the real number line.

Now it makes sense to interpolate rational numbers in between the integers. Of course a fraction, such as $2/3$, is easy to locate because it is just two thirds of the way from 0 to 1. We exhibit a couple of rational numbers in Figure 6.3.

For practical purposes, in everyday life, the rational numbers will suffice. When we speak of quantities to the butcher or the baker or the doctor, we use rational numbers—sometimes as decimals and sometimes as fractions. We ask for 2.5 pounds of beef, or

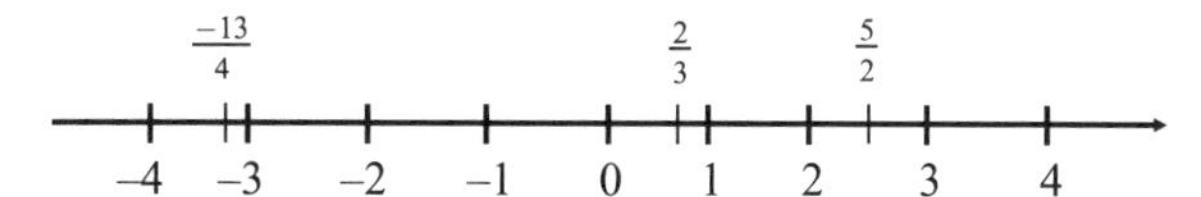

Figure 6.3. Location of some rational numbers.

10 gallons of gas, or a pint of blood. Numbers like $\sqrt{2}$ or π never come up in ordinary conversation. Why are they needed at all?

We learned in Chapter 2 that there are very concrete numbers, that come up in ordinary measurement, that are not rational. For example, the diagonal of a square of side 1 has length $\sqrt{2}$. The circumference of a circle of diameter 1 is π. These strange, irrational numbers, really exist and they apply to quite tangible, tactile, quantities.

We can picture irrational numbers on the number line by using their decimal approximations. Your pocket calculator will tell you that $\sqrt{2} \approx 1.414$ and $\pi \approx 3.14159$. A few irrational numbers are depicted on the number line in Figure 6.4.

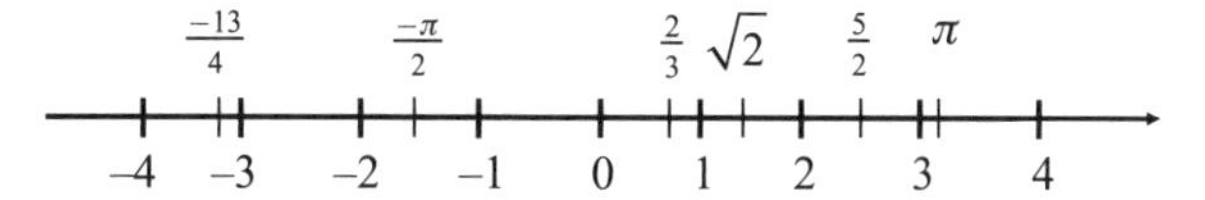

Figure 6.4. Location of some irrational numbers.

The real number line is a useful mnemonic for picturing the relative locations of real numbers. It is particularly helpful when we try to understand *sets* of real numbers. For example, the set

$$S = \{x \in \mathbb{R} : -3 \leq x < 2\}$$

is shown in Figure 6.5. The slightly more subtle set

$$T = \{x \in \mathbb{R} : |x - 1| < 2\}$$

is shown in Figure 6.6. Observe that we use a solid dot to indicate that an endpoint is included in the set; a hollow dot denotes that the endpoint is excluded.

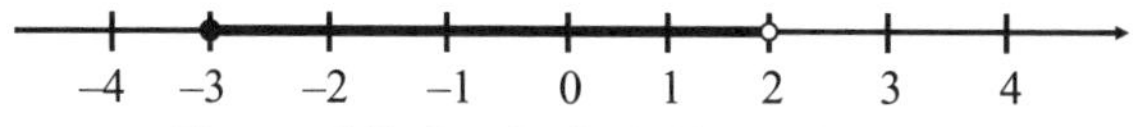

Figure 6.5. Graph of a half-open interval.

Figure 6.6. Graph of an open interval.

It is important now to understand Descartes's contribution in the context of the history of mathematics. The great driving force in the subject, since the time of the ancient Greeks, had been geometry in the plane. That is, people had been studying the geometry of triangles and circles and other planar figures for a long, long time. Euclid's axiomatization of geometry, which is the blueprint for the way that we do mathematics today, was an effort to put this geometry on a rigorous footing. Cartesian coordinates injected an entirely new set of tools into this great tradition. It provided a unification of algebra (the other great theoretical flow in mathematical development) and geometry. And it provided a technique for graphing and picturing functions. In the next section we begin to explore this new circle of ideas.

Further Reading

R H Bing, Properties of the real numbers, *The American Mathematical Monthly* 67(1960), 35–37.

Paul Fjelstad and Peter Hammer, A picture for real arithmetic, *The College Mathematics Journal* 31(2000), 56–60.

6.3 The Cartesian Plane

Consider the layout of two perpendicular coordinate lines as shown in Figure 6.7.

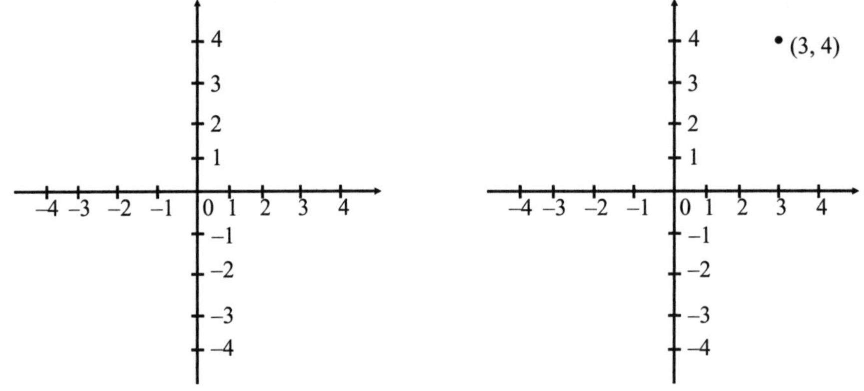

Figure 6.7. The basis for cartesian coordinates.

Figure 6.8. Cartesian coordinates of a point.

We locate a point in the plane by specifying its position in the left-to-right direction and then its position in the up-and-down direction. Put in other words, we write down an ordered pair of numbers consisting of the displacement from the vertical axis followed by the displacement from the horizontal axis. See Figure 6.8. The exhibited point has coordinates $(3, 4)$.

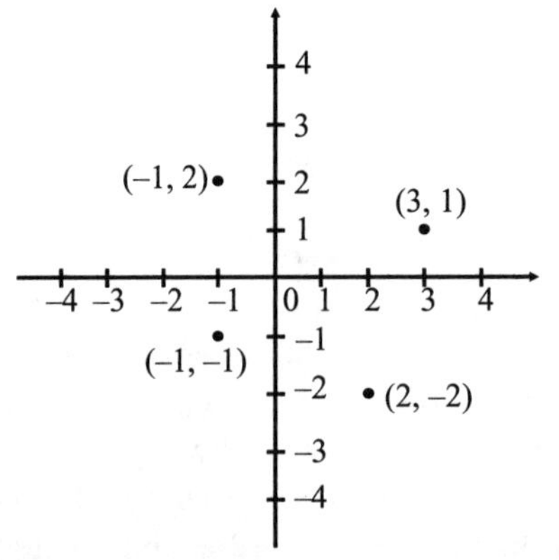

Figure 6.9. Several points plotted in the cartesian plane.

The first coordinate of a point is called the x-*coordinate* (or *abscissa*). The second coordinate is called the y-*coordinate* (or *ordinate*). The point in Figure 6.8 has x-coordinate 3 and y-coordinate 4. Figure 6.9 exhibits several points on a cartesian coordinate plane. Notice that the points with negative x-coordinate lie to the *left* of the y-axis. The points with negative y-coordinate lie *below* the x-axis.

We conclude this section by plotting a simple locus. Consider the equation

$$y = 3x + 1\,.$$

It is useful to form a chart of values (Table 6.10). We plot each of these points on the same set of axes and connect them in a plausible manner. The result is the line exhibited in Figure 6.11.

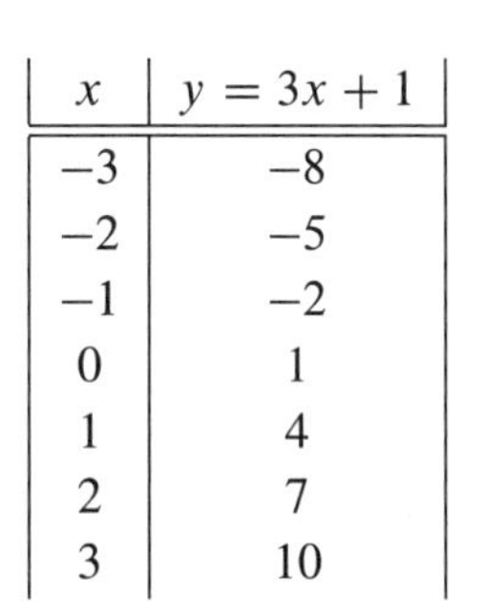

x	$y = 3x + 1$
−3	−8
−2	−5
−1	−2
0	1
1	4
2	7
3	10

Table 6.10. Chart of Values for $y = 3x + 1$.

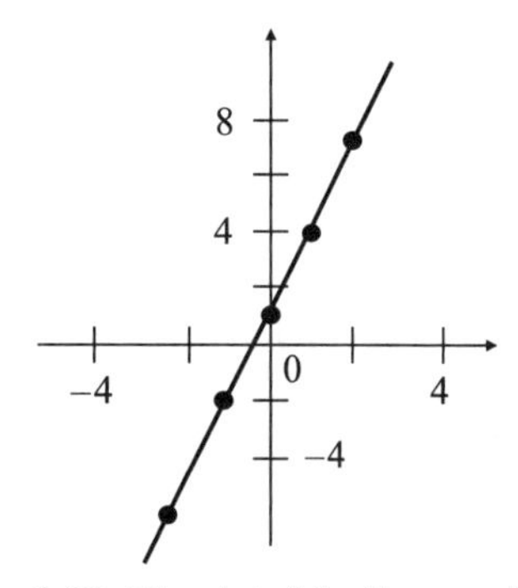

Figure 6.11. The plot of the line $y = 3x + 1$.

EXAMPLE 6.1. One of Descartes's great insights was that his new coordinate system could be used to envision the *graph* of a function. As a simple example, plot the graph of $f(x) = x^2$.

Solution. We begin with a table of values (Table 6.12) for the function. Plotting these values on a set of axes, we obtain Figure 6.13.

x	$f(x) = x^2$
−3	9
−2	4
−1	1
0	0
1	1
2	4
3	9

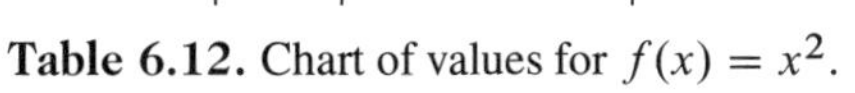

Table 6.12. Chart of values for $f(x) = x^2$.

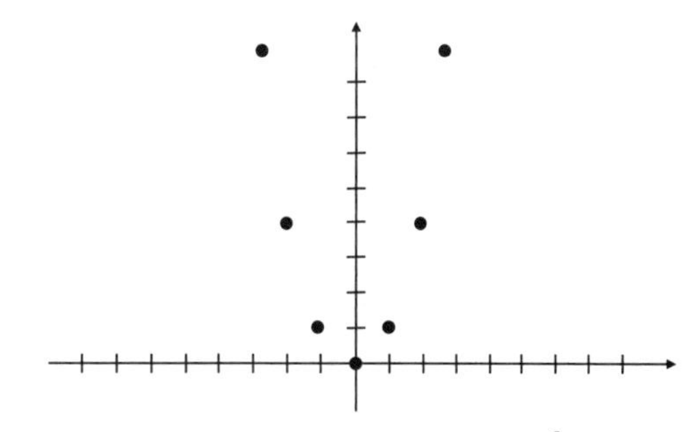

Figure 6.13. Plot for $f(x) = x^2$.

Connecting these points in a plausible manner gives the familiar graph of a parabola—Figure 6.14.

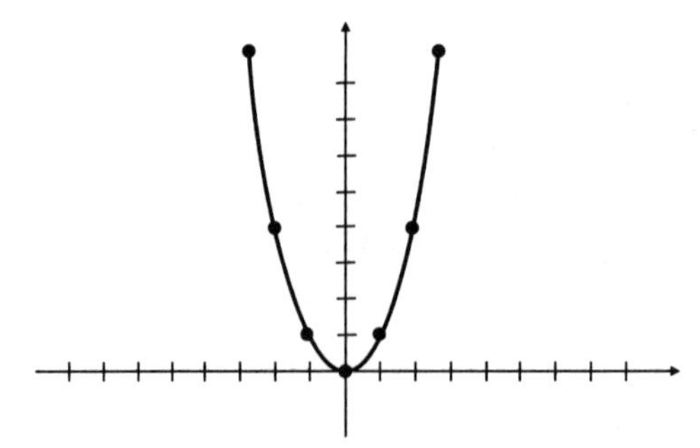

Figure 6.14. Graph of $f(x) = x^2$.

Further Reading

James R. Boone, The Moise plane, *The College Mathematics Journal* 27(1996), 182–185.
E. B. Leach, Finding the Cartesian equation of a locus, *The American Mathematical Monthly* 63(1956), 661–663.

6.4 The Use of Cartesian Coordinates to Study Euclidean Geometry

One of the big innovations that René Descartes introduced was the idea of studying questions of classical Euclidean geometry using his new coordinate system. We illustrate this idea with a few examples.

EXAMPLE 6.2. Consider the triangle shown in Figure 6.15. The horizontal segment I halfway up the figure connects the midpoints of the two sides $\overline{AB}$ and $\overline{AC}$. We claim that the length of I is half the length of $\overline{BC}$.

In fact it is not difficult to prove this assertion synthetically, by classical methods of Euclidean geometry. But our purpose here is to see how to use cartesian coordinates to achieve the result. Glance at Figure 6.16, where we have placed the triangle on a set of coordinate axes and labeled the coordinates of A, B, and C respectively. Notice that $A = (0, a)$, $B = (b, 0)$, and $C = (c, 0)$. Then the endpoints of I are $(b/2, a/2)$ and $(c/2, a/2)$. It follows that the length of I is $c/2 - b/2 = [c - b]/2$. But this is just half the length of BC, which is $c - b$. That is the result that we wished to establish.

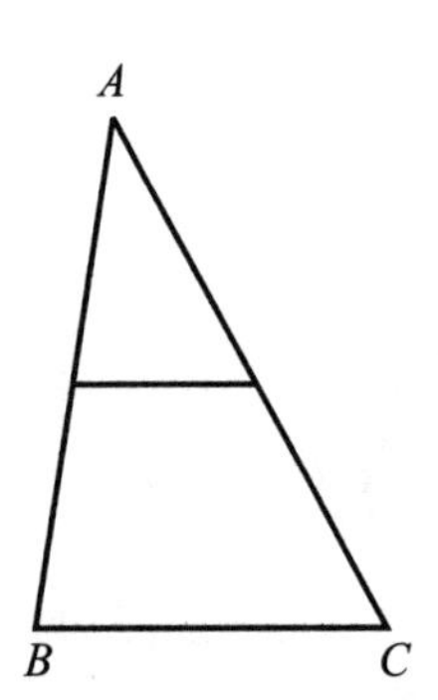

Figure 6.15. The median segment of a triangle.

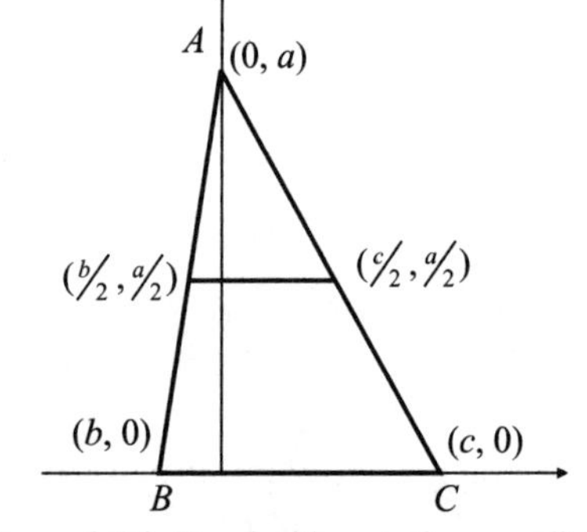

Figure 6.16. Proof with cartesian coordinates.

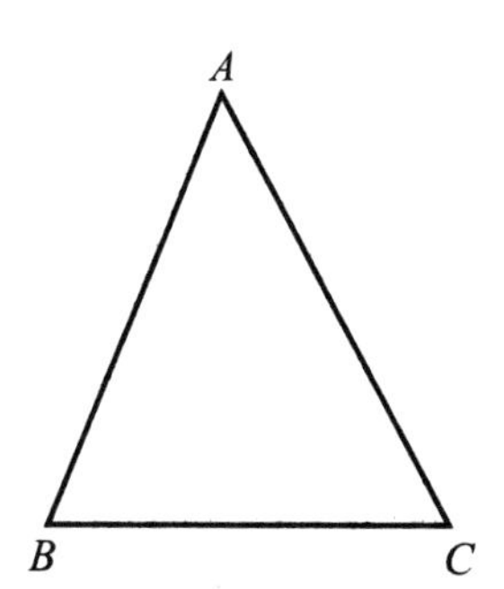

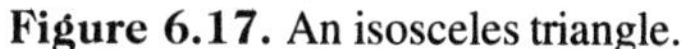
Figure 6.17. An isosceles triangle.

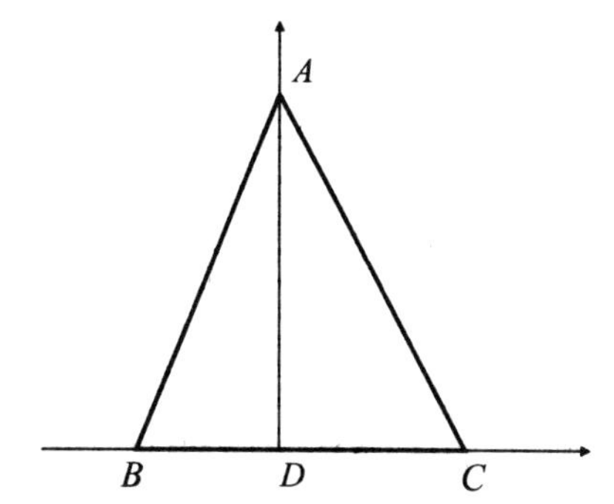

Figure 6.18. Triangle configured on a pair of axes.

EXAMPLE 6.3. Let $\triangle ABC$ be an isosceles triangle as in Figure 6.17. So $AB = AC$. Let $\overline{AD}$ be the median passing between the two equal sides. Then $\overline{AD}$ is perpendicular to the base $\overline{BC}$. Prove this result using cartesian coordinates.

Solution. We configure the triangle on a pair of axes as in Figure 6.18. Coordinates are assigned to each of the relevant points. Notice that the point D is at the origin.

Now the line determined by points B and C has slope 0. And the line determined by A and D has slope ∞. Thus the two lines are perpendicular, as was to be proved.

EXAMPLE 6.4. It is a classical result of Euclidean geometry that the three medians of a triangle (here a median is the segment connecting a vertex to the midpoint of the opposite side) intersect at a single point. Use cartesian coordinates to give a proof of this fact.

Solution. Examine the triangle in Figure 6.19 that is exhibited on a set of coordinate axes. The vertices are $A = (0, a)$, $B = (b, 0)$, and $C = (c, 0)$. Now the endpoints of the medians are $\alpha = (b/2 + c/2, 0)$, $\beta = (c/2, a/2)$, and $\gamma = (b/2, a/2)$.

The line through α and A has slope

$$m_1 = \frac{a - 0}{0 - (b/2 + c/2)} = \frac{-2a}{b + c}.$$

The corresponding median then has equation

$$y - a = \frac{-2a}{b + c} \cdot (x - 0).$$

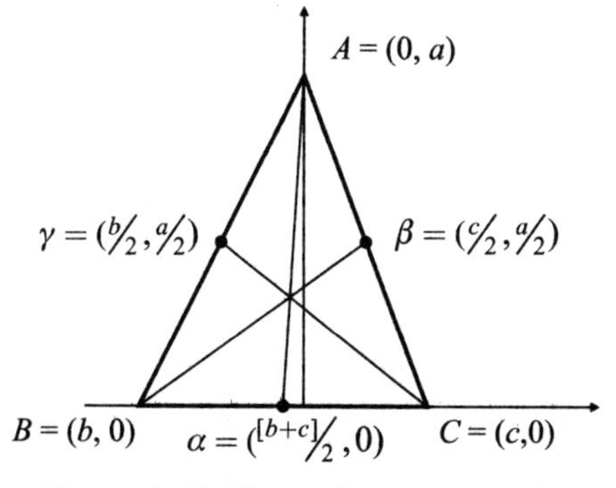

Figure 6.19. The medians of a triangle.

The line through β and B has slope

$$m_2 = \frac{a/2 - 0}{c/2 - b} = \frac{a}{c - 2b}.$$

The corresponding median then has equation

$$y - 0 = \frac{a}{c - 2b} \cdot (x - b).$$

The line through γ and C has slope

$$m_3 = \frac{0 - a/2}{c - b/2} = \frac{a}{b - 2c}.$$

The corresponding median then has equation

$$y - 0 = \frac{a}{b - 2c} \cdot (x - c).$$

Now, by elementary algebra, the first two lines intersect at the point

$$x = \frac{b + c}{3}, \quad y = \frac{a}{3}.$$

The second two lines also intersect at this point.

We conclude that the unique point of intersection of the three medians is

$$P = \left(\frac{b + c}{3}, \frac{a}{3}\right).$$

For You to Try Use cartesian coordinates to demonstrate that a rhombus (a quadrilateral with sides of equal length) has diagonals that are perpendicular.

For You to Try Use cartesian coordinates to demonstrate that an isosceles triangle has two medians of equal length.

Further Reading

Larry Cannon, Geometry of the rational plane, *The College Mathematics Journal* 17(1986), 392-402.

Marvin J. Greenberg, Euclidean and non-Euclidean geometries without continuity, *The American Mathematical Monthly* 86(1979), 757–764.

Adolf Mader, A Euclidean model for Euclidean geometry, *The American Mathematical Monthly* 96(1989), 43–49.

6.5 Coordinates in Three-Dimensional Space

To locate a point in the 2-dimensional plane requires two coordinates. By analogy, to locate a point in 3-dimensional space requires three coordinates. Figure 6.20 indicates how this is done. There are three axes: the x-axis, the y-axis, and the z-axis. The indicated point has coordinates (x, y, z). The first coordinate indicates displacement along the direction of the x-axis. The second coordinate indicates displacement along the direction of the y-axis. And the third coordinate indicates displacement along the direction of the z-axis.

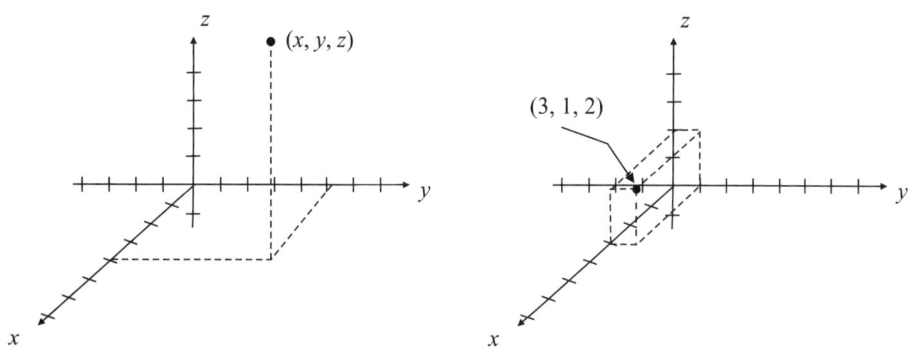

Figure 6.20. Three coordinates to locate a point in space.

Figure 6.21. Use of a box to situate a point in space.

EXAMPLE 6.5. Sketch the point $(3, 1, 2)$ on a 3-dimensional set of axes.

Solution. Examine Figure 6.21. You will see that we have drawn a box to show how the point is situated in space. The box aids in our perspective of the geometry. The side lengths of the box indicate the magnitude of each coordinate, and the orientation of the box shows the sign of each coordinate.

EXAMPLE 6.6. Sketch the point $(-2, -4, 3)$ on a 3-dimensional set of axes.

Solution. Examine Figure 6.22. You will see that we have drawn a box to show how the point is situated in space. The side lengths of the box indicate the magnitude of each coordinate, and the orientation of the box shows the sign of each coordinate.

You know that, when we graph the locus of points in the 2-dimensional plane that satisfies a given equation, then the result is usually a curve in the plane. This makes good intuitive sense, since the imposition of a condition given by one equation removes 1 degree of freedom, hence removes 1 dimension. Since we begin with 2 dimensions, the result is a 1-dimensional object—or a curve.

Likewise, if we impose an equation on 3-dimensional space, then we remove 1 degree of freedom. Hence there should be a loss of 1 dimension, and the result should be a 2-dimensional surface.

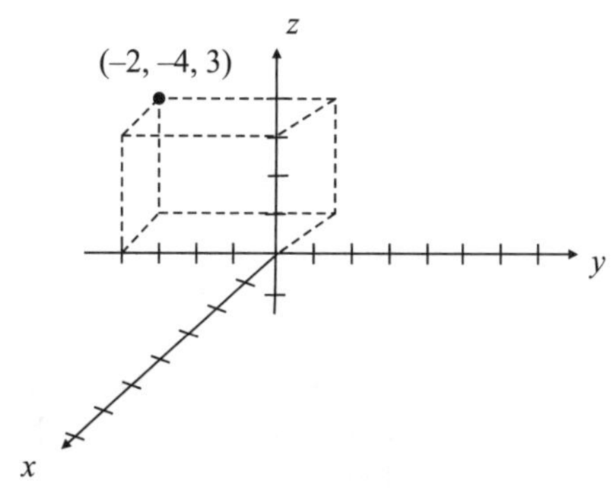

Figure 6.22. Use of a box to picture a point in space.

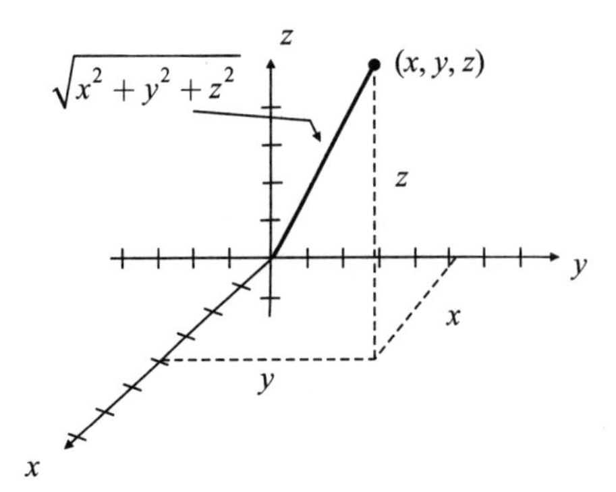

Figure 6.23. Distance of a point to the origin.

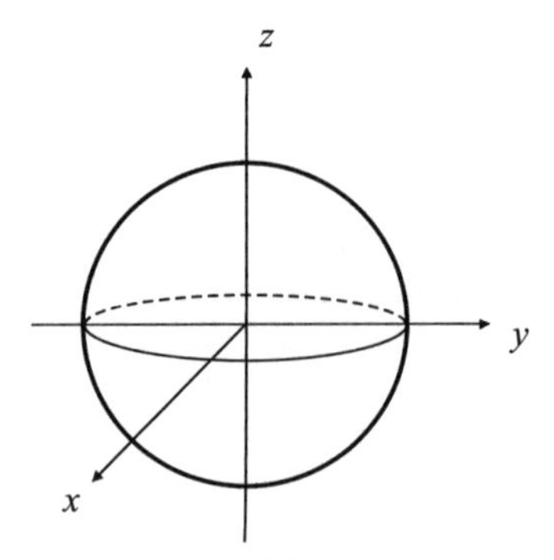

Figure 6.24. A sphere.

EXAMPLE 6.7. Sketch the surface in 3-dimensional space that is defined by the equation

$$x^2 + y^2 + z^2 = 1 . \qquad (*)$$

Solution. Examine Figure 6.23. It shows that the distance of the point (x, y, z) to the origin is $\sqrt{x^2 + y^2 + z^2}$. Write $X = (x, y, z)$ and $0 = (0, 0, 0)$. Then we set

$$d\big(X, 0\big) = \sqrt{x^2 + y^2 + z^2} .$$

The equation $(*)$ may then be written schematically as

$$[d(X, 0)]^2 = 1 ,$$

or

$$d(X, 0) = 1 .$$

Thus we see that our equation describes the set of all points X in space that have distance 1 from the origin. This is a sphere. The surface is shown in Figure 6.24.

EXAMPLE 6.8. Sketch the surface in 3-dimensional space that is defined by the equation

$$x + y + z = 4 . \qquad (\star)$$

Solution. Recall that, in the plane, an equation of the form

$$ax + by = c$$

gives rise to a line. It is plausible, therefore, that the equation $(\star)$ will describe a "linear" object. In fact observe that if (x, y, z) is a point that satisfies $(\star)$, then also the point $(x - 2t, y + t, z + t)$ for any t will satisfy $(\star)$. Likewise, $(x + t, y - 2t, z + t)$ will satisfy $(\star)$ for any t. And $(x + t, y + t, z - 2t)$ will satisfy $(\star)$ for any t.

We see, therefore, that three lines pointing in three different directions all lie in the surface defined by $(\star)$. We conclude that $(\star)$ describes a plane. Notice that the points $(4, 0, 0)$ and $(0, 4, 0)$ and $(0, 0, 4)$ all satisfy the equation and hence all must lie on the plane. The resulting picture is shown in Figure 6.25.

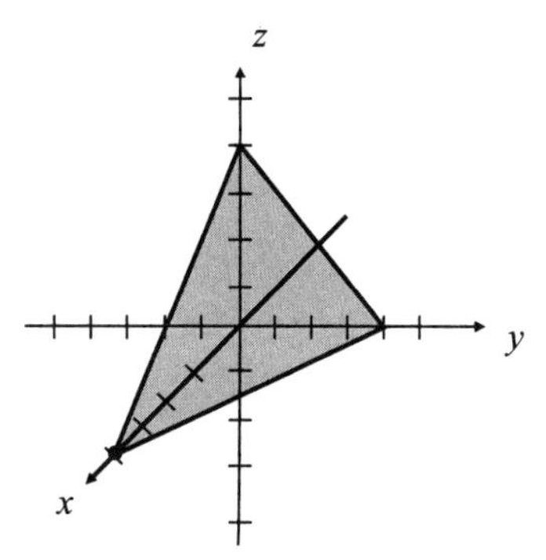

Figure 6.25. A plane in space.

Further Reading

Roland Brossard, Metric postulates for space geometry, *The American Mathematical Monthly* 74(1967), 777–788.

Daniel Pedoe, In love with geometry, *The College Mathematics Journal*, 29(1998), 170–188.

Exercises

1. On a single set of axes, sketch the points in the plane that satisfy

$$|x| + |y| = 1\,.$$

2. On a single set of axes, sketch the points in the plane that satisfy

$$|x + y| = 1\,.$$

3. On a single set of axes, sketch the points in the plane that satisfy

$$|x| - |y| = 1\,.$$

4. Sketch the points in 3-dimensional space that satisfy

$$(x-1)^2 + (y-2)^2 + (z-3)^2 = 4\,.$$

5. Sketch the points in 3-dimensional space that satisfy

$$x - 2y + 5z = 10\,.$$

6. Use cartesian coordinates to verify that if $\triangle ABC$ is an isosceles triangle with equal sides AB and AC, then the median from the vertex A to the side BC will be perpendicular to the segment $\overline{BC}$.

7. Use cartesian coordinates to verify that three noncollinear points will uniquely determine a circle.

8. On a single set of axes, sketch those points in the plane that satisfy

$$x^2 = y^3\,.$$

9. Sketch the points in 3-dimensional space that satisfy

$$z^2 = x^2 + y^2 .$$

10. Find the equation of the line in the plane that passes through the points $(1, 2)$ and $(-3, 1)$.

11. Find the equation of the plane in 3-dimensional space that passes through the points $(1, 0, 0)$, $(0, 1, 0)$ and $(0, 0, 2)$.

12. Describe in words the set of points in 3-dimensional space given by

$$\{(x, y, z) : |x| < 1, |y| < 1, |z| < 1\} .$$

13. Project: In the study of relativity we make use of 4-dimensional space-time. The coordinates are (x, y, z, t), where x, y, z are the usual Cartesian coordinates of 3-dimensional space and t is time. The inner product, or dot product, on space-time is

$$\langle\langle (x, y, z), (x', y', z') \rangle\rangle \equiv xx' + yy' + zz' - tt' .$$

Explain why it makes sense to have a negative signature in front of the time component in this dot product.

The set

$$\{(x, y, z, t) : x^2 + y^2 + z^2 - t^2 < 0, t > 0\}$$

is called the *forward light cone*. It is of particular interest in relativity theory. Explain why this should be so. The book [MTW] contains chapter and verse on the forward light cone.

14. Project: A famous formula of Euler says that, for α a real number,

$$e^{i\alpha} = \cos\alpha + i \sin\alpha .$$

One may verify this formula by writing out the power series expansion of the exponential, of sine, and of cosine, gathering terms, and comparing. You should do so.

Use this formula to derive the standard trigonometric formulas for $\sin(\alpha + \beta)$ and $\cos(\alpha + \beta)$. Use the formula to derive a new formula for rotating Cartesian coordinates through an angle θ.

The formula is also valid when α is complex. Use this fact to give a definition of $\cos\alpha$ and $\sin\alpha$ when α is complex. See [KRA1] and [GRK] for the full story of Euler's formula.

7

Pierre de Fermat and the Invention of Differential Calculus

7.1 The Life of Fermat

Pierre de Fermat (1601 C.E.–1665 C.E.) was one of the most remarkable mathematicians who ever lived. He spent his entire adult life as a magistrate or judge in the city of Toulouse, France. His career was marked by prudence, honesty, and scrupulous fairness. He led a quiet and productive life. His special passion was for mathematics. Fermat was perhaps the most talented amateur mathematician in history. He led a full and productive life which is recounted in [MAH].

Fermat is commemorated today by a large statue that is in the basement of the Hôtel de Ville in Toulouse. The statue depicts Fermat, dressed in formal attire, and seated. There is a sign, etched in stone and part of the statue, that says, "Pierre de Fermat, the father of differential calculus." Seated in Fermat's lap is a scantily clad muse showing her ample appreciation for Fermat's powers.

Pierre Fermat had a brother and two sisters and was almost certainly brought up in the town (Beaumont-de-Lomagne) of his birth. Although there is little evidence concerning his school education, it must have been at the local Franciscan monastery.

He attended the University of Toulouse before moving to Bordeaux in the second half of the 1620s. In Bordeaux he began his first serious mathematical research and in 1629 he gave a copy of his restoration of Apollonius's *Plane loci* to one of the mathematicians there. Certainly in Bordeaux he was in contact with Jean Beaugrand (1590–1640), an influential French mathematician of the time. During this period he produced important work on maxima and minima which he gave to Étienne d'Espagnet who clearly shared mathematical interests with Fermat.

From Bordeaux, Fermat went to Orléans where he studied law at the University. He received a degree in civil law and he purchased the offices of councillor at the parliament in Toulouse. So by 1631 Fermat was a lawyer and government official in Toulouse and because of the office he now held he became entitled to change his name from Pierre Fermat to Pierre de Fermat.

For the remainder of his life he lived in Toulouse but, as well as working there, he also worked in his home town of Beaumont-de-Lomagne and in the nearby town of Castres. The plague struck the region in the early 1650s, meaning that many of the older men died.

Fermat himself was struck down by the plague and in 1653 his death was wrongly reported, then corrected:

> I informed you earlier of the death of Fermat. He is alive, and we no longer fear for his health, even though we had counted him among the dead a short time ago.

The period from 1643 to 1654 was one when Fermat was out of touch with his scientific colleagues in Paris. There are a number of reasons for this. First, pressure of work kept him from devoting so much time to mathematics. Secondly the Fronde, a civil war in France, took place and from 1648 Toulouse was greatly affected. Finally there was the plague of 1651 which must have had great consequences both on life in Toulouse and of course its near fatal consequences on Fermat himself. However it was during this time that Fermat worked on the theory of numbers.

Fermat is best remembered for this work in number theory, in particular for Fermat's Last Theorem. This theorem states that the equation

$$x^n + y^n = z^n$$

has no nonzero integer solutions x, y and z when the integer exponent $n > 2$. Fermat wrote, in the margin of Bachet's translation of Diophantus's *Arithmetica*, a note that included the claim

> I have discovered a truly remarkable proof which this margin is too small to contain.

These marginal notes only became known after Fermat's death, when his son Samuel published in 1670 an edition of Bachet's translation of Diophantus's *Arithmetica* with his father's notes.

It is now believed that Fermat's "proof" was wrong although it is impossible to be completely certain. The truth of Fermat's assertion was claimed to be proved in June, 1993 by the British mathematician Andrew Wiles, but Wiles withdrew the claim when problems emerged later in 1993. In November, 1994 Wiles (in collaboration with his student Richard Taylor) again claimed to have a correct proof which has now been accepted. Unsuccessful attempts to prove the theorem over a 300 year period led to the discovery of commutative ring theory and a wealth of other mathematical developments. Wiles himself has said in a public lecture that he thinks that Fermat probably made a mistake in claiming that he could prove the "last theorem". He allows, however, that Fermat made few mistakes.

Fermat's correspondence with the Paris mathematicians restarted in 1654 when Blaise Pascal wrote to him to ask for confirmation about his ideas on probability. Blaise Pascal knew of Fermat through his father, who had died three years before, and was well aware of Fermat's outstanding mathematical abilities. Their short correspondence set up the theory of probability and from this they are now regarded as joint founders of the subject.

It was Fermat's habit to solve problems and then pose them to the community of mathematicians. Some of these were quite deep and difficult, and people found them aggravating. One problem that he posed was to prove that the sum of two cubes cannot be a cube (a special case of Fermat's Last Theorem which may indicate that by this time Fermat realized that his proof of the general result was incorrect), that there are exactly two integer solutions of $x^2 + 4 = y^3$, and that the equation $x^2 + 2 = y^3$ has only one integer solution. He posed problems directly in correspondence with the English. Everyone failed to see that

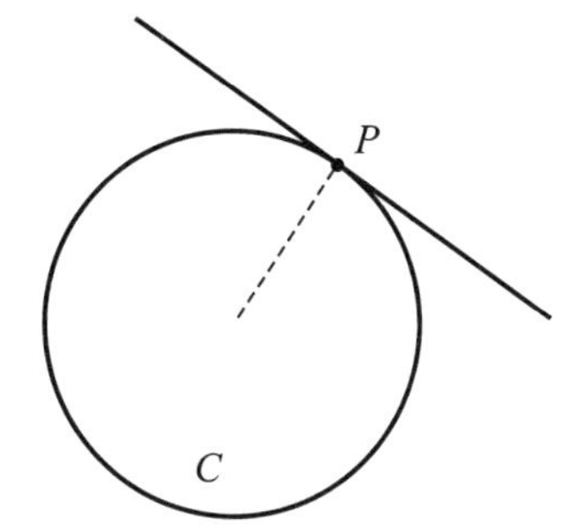

Figure 7.1. The tangent line to a curve.

Fermat had been hoping his specific problems would lead them to discover, as he had done, deeper theoretical results.

Fermat has been described by some historical scholars as

> Secretive and taciturn, he did not like to talk about himself and was loath to reveal too much about his thinking. ... His thought, however original or novel, operated within a range of possibilities limited by that time [1600 -1650] and that place [France].

Further Reading

Hidefumi Katsuura, From Euler to Fermat, *The College Mathematics Journal* 30(1999), 118–119.

David A. Cox, Introduction to Fermat's last theorem, *The American Mathematical Monthly* 101(1994), 3–14.

7.2 Fermat's Method

One of the fundamental ideas of calculus is to find the tangent line to a given curve. Figure 7.1 exhibits the familiar idea of the tangent line to a *circle*. This is a particularly simple situation. In classical geometry texts, we are told that the tangent line to a circle C at a point P of the circle is that line which passes through P and is perpendicular to the radius at P. The figure amply illustrates this idea.

For a more general curve—say the graph of a function—we have an intuitive idea of what the tangent line to a point on the curve might be (Figure 7.2), but it is hard to define the idea precisely. *How can we say analytically what the tangent line is supposed to be?* For the curve in the figure, there is no notion of radius. The only thing that we know about the tangent line is that it passes through P and "touches" the curve at P. How can we come up with a precise formulation of "touches"

Fermat's idea, a precursor of the full-bore version of calculus that Newton and Leibniz developed some years later, was this: the tangent line has the special feature that it only intersects the curve at one point. This idea is not foolproof—for example, the tangent line to the curve in Figure 7.2 actually intersects the curve in two points. Nonetheless, in many examples Fermat's idea gives us just what we are looking for.

In order to actually implement Fermat's idea, we shall need the concept of *slope*. Recall that if we are given a line ℓ in the plane and two points (p_1, q_1) and (p_2, q_2) on that line,

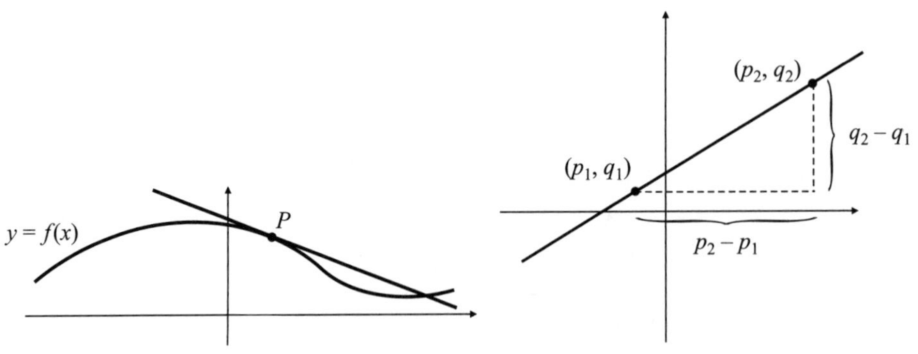

Figure 7.2. Intuitive idea of the tangent.

Figure 7.3. The idea of slope.

then the *slope* of the line is

$$m = \frac{q_2 - q_1}{p_2 - p_1}.$$

Figure 7.3 illustrates the idea of slope.

EXAMPLE 7.1. Use Fermat's idea to find the tangent line to the curve $y = x^2$ at the point $(2, 4)$.

Solution. Refer to Figure 7.4 as you read along. Let us consider the equation of a line passing through $(2, 4)$. Say that it has slope m. Then the line is

$$y - 4 = m(x - 2). \tag{†}$$

We calculate the intersection of the line with the curve $y = x^2$. Our equations are then

$$y = 4 + m(x - 2),$$
$$y = x^2.$$

Equating the two expressions for y, we find that

$$x^2 = 4 + m(x - 2).$$

In the more familiar format of a quadratic equation, this is

$$x^2 - mx + (2m - 4) = 0.$$

Using the quadratic formula, we find that

$$x = \frac{m \pm \sqrt{m^2 - 4 \cdot 1 \cdot (2m - 4)}}{2}.$$

We may rewrite this as

$$x = \frac{m \pm \sqrt{m^2 - 8m + 16}}{2}.$$

We are in luck. The expression under the square root sign is a perfect square: the square of $(m - 4)$. Thus our solution becomes

$$x = \frac{m \pm (m - 4)}{2}. \tag{*}$$

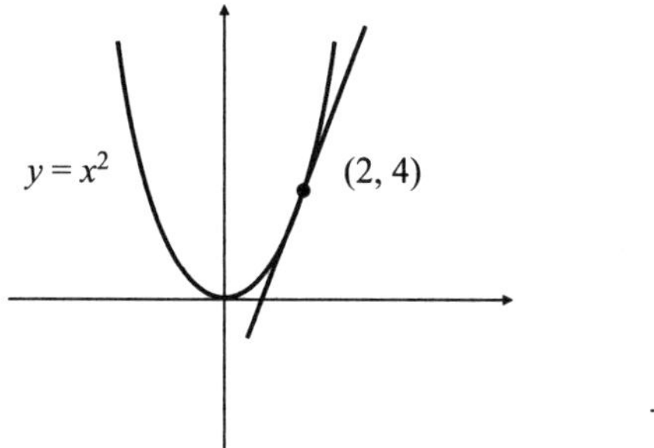

Figure 7.4. The line and the curve.

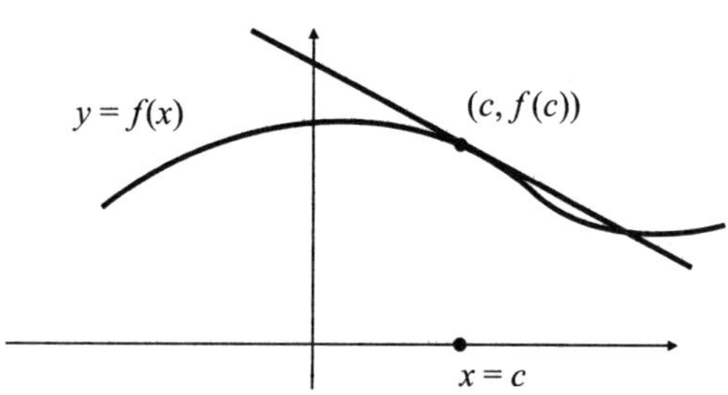

Figure 7.5. The slope of the tangent line.

Now we are looking for a choice of m so that the line only intersects the curve in one point. So we want the system of equations to have only one solution. But $(*)$ certainly looks like *two* solutions. The only way to make this reduce to just one is to have the expression coming from the square root go away. In other words, we want $(m - 4) = 0$. In conclusion, we want to choose $m = 4$.

What we have learned is that the only line that passes through the point $(2, 4)$ and intersects the curve $y = x^2$ just once is the line with slope $m = 4$ (recall our discussion in connection with equation $(\dagger)$). It has equation

$$y - 4 = 4(x - 2).$$

The line and the curve are exhibited in Figure 7.4.

Further Reading

Carl B. Boyer, The history of the calculus, *The Two-Year College Mathematics Journal* 1(1970), 60–86.

J. L. Coolidge, The story of tangents, *The American Mathematical Monthly*, 58(1951), 449–462.

7.3 More Advanced Ideas of Calculus: The Derivative and the Tangent Line

There is little doubt that Fermat's work was one of the seminal inspirations for the huge subject that today is known as differential calculus. Thanks to his efforts, and to the work of Descartes, Newton, Leibniz, and many others, we now have a body of mathematical machinery for calculating tangents, finding maxima and minima of functions, and performing many other important operations in analysis and mechanics.

We now give an idea of the general approach provided by calculus for calculating the tangent line to the graph of a function f at a point $P = (p, f(p))$ on its graph. As we saw in the example of the last section, what this comes down to is finding the *slope* of the tangent line. Examine Figure 7.5.

Now let us consider slope. Look at the graph of the function $y = f(x)$ in Figure 7.5. We wish to determine the "slope" of the graph at the point $x = c$. This is the same as

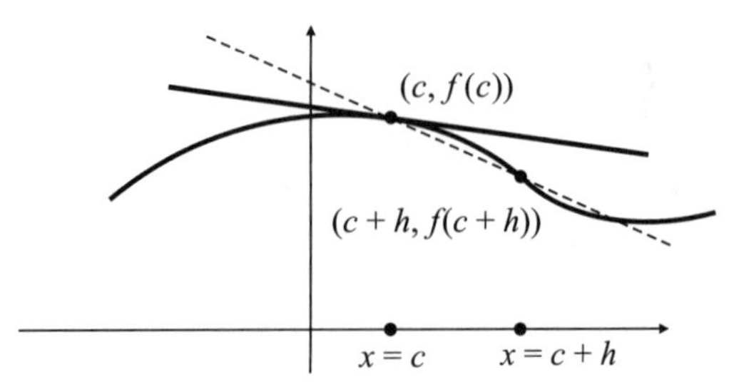

Figure 7.6. The secant (which is a dotted line).

determining the slope of the *tangent line* to the graph of f at $x = c$, where the tangent line is the line that best approximates the graph at that point. See Figure 7.6. What could this mean? After all, it takes two points to determine the slope of a line, yet we are only given the point $(c, f(c))$ on the graph. One reasonable interpretation of the slope at $(c, f(c))$ is that it is the limit of the slopes of secant lines[1] determined by $(c, f(c))$ and nearby points $(c + h, f(c + h))$. See the dotted line in Figure 7.6. When we say "limit", we mean to consider the behavior of the expression as h tends to 0. Let us calculate this limit:

$$\lim_{h\to 0} \frac{f(c+h) - f(c)}{(c+h) - c} = \lim_{h\to 0} \frac{f(c+h) - f(c)}{h}.$$

Now this last limit is what we shall call the *derivative* of f at c. We denote the derivative by $f'(c)$. When the limit exists, we say that the function f is *differentiable* at c.

Notice that the definition of "derivative" involves the important limiting process. We calculate the limit of the so-called *Newton quotient*

$$\frac{f(c+h) - f(c)}{h}.$$

This means that we consider the behavior of the quotient as h tends to zero. The theory of the limit is deep and subtle. It was considered by the ancients, more than two thousand years ago, and they never got it right. Isaac Newton himself used limits (with trepidation), but he never really understood them. It is only in the past 150 years that we have developed an accurate and rigorous way to think about limits. In the present book we treat limits intuitively. As you will see in the first example, we write out every step of our calculation so that the procedure of taking the limit becomes transparent.

Now let us return to the concept of derivative. We have learned the following:

> Let f be a differentiable function on an interval (a, b). Let $c \in (a, b)$. Then the slope of the tangent line to the graph of f at c is $f'(c)$.

EXAMPLE 7.2. Calculate the slope of the tangent line to the graph of $y = f(x) = x^3 - 3x$ at $x = -2$. Write the equation of the tangent line. Draw a figure illustrating these ideas.

[1] In simple terms, a "secant line" is a line connecting two different points on the curve.

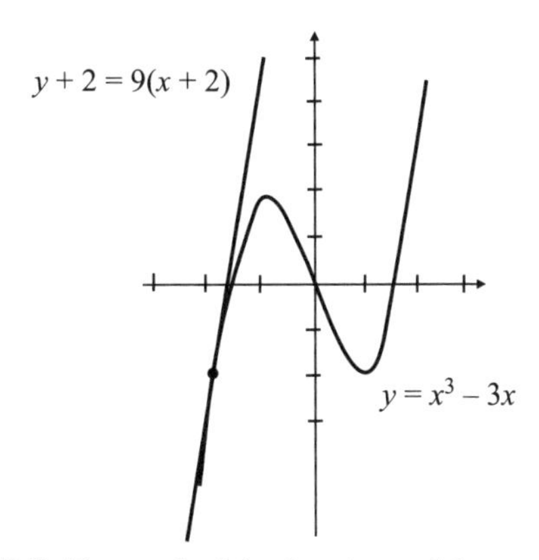

Figure 7.7. The graph of the function and the tangent line.

Solution. We know that the desired slope is equal to $f'(-2)$. We calculate

$$\begin{aligned}
f'(-2) &= \lim_{h\to 0} \frac{f(-2+h) - f(-2)}{h} \\
&= \lim_{h\to 0} \frac{[(-2+h)^3 - 3(-2+h)] - [(-2)^3 - 3(-2)]}{h} \\
&= \lim_{h\to 0} \frac{[(-8+12h-6h^2+h^3) + (6-3h)] + [2]}{h} \\
&= \lim_{h\to 0} \frac{h^3 - 6h^2 + 9h}{h} \\
&= \lim_{h\to 0} [h^2 - 6h + 9] \\
&= 9\,.
\end{aligned}$$

Notice that, in the last equality, we have observed that h^2 tends to 0 and $6h$ tends to 0 as $h \to 0$.

We conclude that the slope of the tangent line to the graph of $y = x^3 - 3x$ at $x = -2$ is 9. The tangent line passes through $(-2, f(-2)) = (-2, -2)$ and has slope 9. Thus it has equation

$$y - (-2) = 9(x - (-2))\,.$$

The graph of the function and the tangent line are exhibited in Figure 7.7.

For You to Try Calculate the tangent line to the graph of $f(x) = 4x^2 - 5x + 2$ at the point where $x = 2$.

The process that we have been describing has a number of important interpretations. We have already considered slope. Another one of these is in terms of velocity. Suppose that the position of a moving body at time t is given by $p(t)$. This position could be measured, for example, in feet. And time t could be measured in seconds (of course other choices are possible). Now the *average velocity* over a time interval $[t, t+h]$ is given by "change in position" divided by "change in time". This quantity is

$$(\text{average velocity}) = \frac{p(t+h) - p(t)}{h}.$$

The limit as $h \to 0$ of this quantity is declared to be the *instantaneous velocity* at time t. Of course this limit is just the derivative $p'(t)$. We conclude the following:

> If the position of a moving body is represented by the differentiable function $p(t)$, then the instantaneous velocity of the motion at time t is $p'(t)$.

EXAMPLE 7.3. Calculate the instantaneous velocity at time $t = 5$ of the moving body whose position at time t seconds is given by $g(t) = t^3 + 4t^2 + 10$ feet.

Solution. We know that the required instantaneous velocity is $g'(5)$. We calculate

$$\begin{aligned}
g'(5) &= \lim_{h\to 0} \frac{g(5+h) - g(5)}{h} \\
&= \lim_{h\to 0} \frac{[(5+h)^3 + 4(5+h)^2 + 10] - [5^3 + 4\cdot 5^2 + 10]}{h} \\
&= \lim_{h\to 0} \frac{[(125 + 75h + 15h^2 + h^3) + 4\cdot(25 + 10h + h^2) + 10)}{h} \\
&\qquad - \frac{(125 + 100 + 10)}{h} \\
&= \lim_{h\to 0} \frac{115h + 19h^2 + h^3}{h} \\
&= \lim_{h\to 0} 115 + 19h + h^2 \\
&= 115.
\end{aligned}$$

We conclude that the instantaneous velocity of the moving body at time $t = 5$ is $g'(5) = 115$ ft/sec.

Remark Since position (or distance) is measured in feet, and time in seconds, then we measure velocity in "feet per second". □

EXAMPLE 7.4. A rubber balloon is losing air steadily. At time t seconds the balloon contains $75 - 10t^2 + t$ cubic inches of air. What is the rate of loss of air in the balloon at time $t = 1$?

Solution. Let $\psi(t) = 75 - 10t^2 + t$. Of course the rate of loss of air is given by $\psi'(1)$. We therefore calculate

$$\begin{aligned}
\psi'(1) &= \lim_{h\to 0} \frac{\psi(1+h) - \psi(1)}{h} \\
&= \lim_{h\to 0} \frac{[75 - 10(1+h)^2 + (1+h)] - [75 - 10\cdot 1^2 + 1]}{h} \\
&= \lim_{h\to 0} \frac{[75 - (10 + 20h + 10h^2) + (1+h)] - [66]}{h} \\
&= \lim_{h\to 0} \frac{-19h - 10h^2}{h} \\
&= \lim_{h\to 0} -19 - 10h \\
&= -19 .
\end{aligned}$$

In conclusion, the rate of air loss in the balloon at time $t = 1$ is $\psi'(1) = -19$ cu. in./sec. Observe that the negative sign in this answer indicates that the change is *negative*, i.e., that the quantity is decreasing.

For You to Try The amount of water in a leaky tank is given by $W(t) = 50 - 5t^2 + t$ gallons. Here time t is measured in minutes. What is the rate of leakage of the water at time $t = 2$?

Remark. We have noted that the derivative may be used to describe a rate of change and also to denote the slope of the tangent line to a graph. These are really two different manifestations of the same thing, for a slope is the rate of change of rise with respect to run (see the discussion of Figure 7.3).

□

Further Reading

Irl C. Bivens, What a tangent line is when it isn't a limit, *The College Mathematics Journal* 17(1986), 133–143.

H. A. Thurston, On the definition of a tangent-line, *The American Mathematical Monthly* 71(1964), 1099–1103.

7.4 Fermat's Lemma and Maximum/Minimum Problems

Fermat's lemma is based on a simple geometric observation about differentiable functions. Examine the graph exhibited in Figure 7.8. The points P and Q on the graph are special. Notice that, if we compare P to points nearby on the graph, then we see that the point P is vertically *higher* than its neighbors (see the blowup in Figure 7.9). We say that P is a *local maximum*. Likewise, if we compare Q to points nearby on the graph, then we see that the point Q is vertically *lower* than its neighbors (see the blowup in Figure 7.10). We say that Q is a *local minimum*.

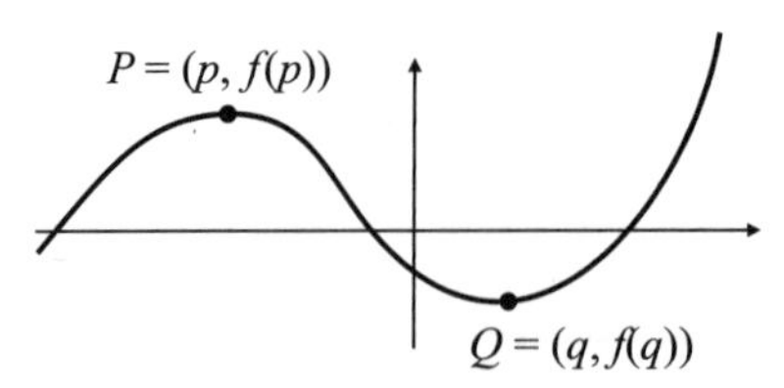

Figure 7.8. A differentiable function.

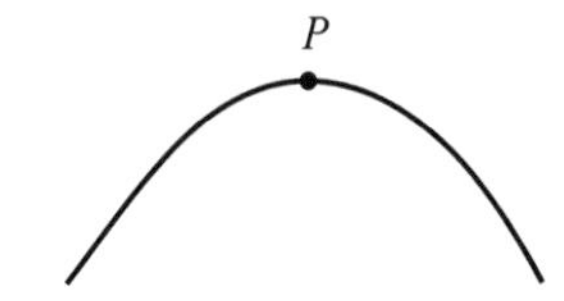

Figure 7.9. Blowup of the figure.

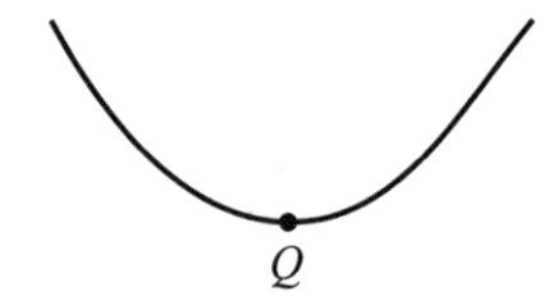

Figure 7.10. Another blowup.

From the point of view of calculus, what is special about the point P is that the graph goes neither uphill nor downhill there. In other words, *the tangent line is horizontal*. That means that its slope is zero. Thus

> The derivative of a function at a point of differentiability where the function assumes a local maximum is 0.

From the point of view of calculus, what is special about the point Q is that the graph goes neither uphill nor downhill there. In other words, *the tangent line is horizontal*. That means that its slope is zero. Thus

> The derivative of a function at a point of differentiability where the function assumes a local minimum is 0.

These two displayed rules are the content of Fermat's lemma. For the sake of the present discussion, let us (inspired by Fermat) call a point x a *critical point* for the function f if $f'(x) = 0$. We illustrate, with a simple example, how Fermat's lemma can be used to gain important information about a function.

In fact it is worth considering this matter in a bit more detail. Let f be a function and x a point of its domain. Calculate the derivative $f'(x)$. If $f'(x) > 0$, then this says that the approximating quotients (or *Newton quotients*)

$$\frac{f(x+h) - f(x)}{h}$$

are *positive*. As Figure 7.11 shows, the graph is going *uphill* at x. If instead $f'(x) < 0$ then we see that the approximating quotients (or *Newton quotients*)

$$\frac{f(x+h) - f(x)}{h}$$

are *negative*. As Figure 7.12 shows, the graph is going *downhill* at x. Finally, if $f'(x) = 0$, then we have seen that the graph is (instantaneously) horizontal—*neither* uphill nor downhill—at x. These simple observations will be useful in our discussions below.

EXAMPLE 7.5. Sketch the graph of the function

$$f(x) = 2x^3 - 3x^2 - 12x + 4\,.$$

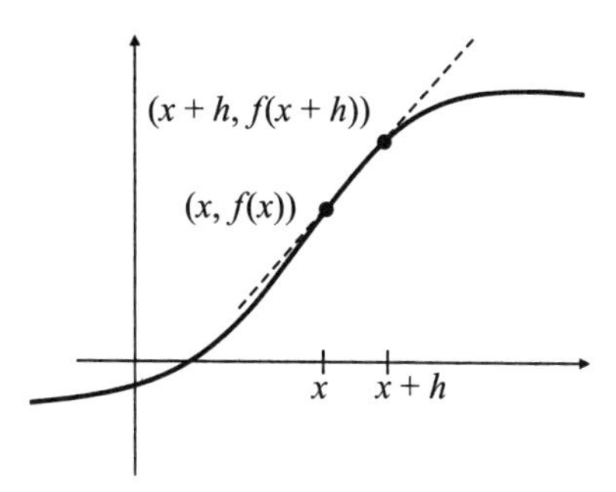

Figure 7.11. The graph goes uphill.

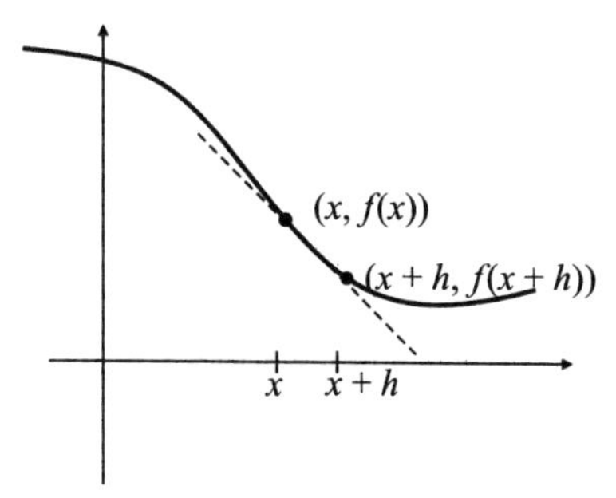

Figure 7.12. The graph goes downhill.

Solution. We have seen cubic curves like this one in the past. When the leading coefficient is positive, the graph will go up, then down, then up. So it will have a local maximum and a local minimum. If we can find those two special points, then we can draw a useful and compelling graph of f. We use the derivative, and Fermat's lemma, to do so.

We calculate, for any point $(x, f(x))$ on the curve, that

$$
\begin{aligned}
f'(x) &= \lim_{h\to 0} \frac{f(x+h) - f(x)}{h} \\
&= \lim_{h\to 0} \frac{[2(x+h)^3 - 3(x+h)^2 - 12(x+h) + 4] - [2x^3 - 3x^2 - 12x + 4]}{h} \\
&= \lim_{h\to 0} \frac{[(2x^3 + 6x^2h + 6xh^2 + 2h^3) - (3x^2 + 6xh + 3h^2) - (12x + 12h) + 4]}{h} \\
&\qquad - \frac{[2x^3 - 3x^2 - 12x + 4]}{h} \\
&= \lim_{h\to 0} \frac{(6x^2h + 6xh^2 + 2h^3) - (6xh + 3h^2) - (12h)}{h} \\
&= \lim_{h\to 0} \left(6x^2 + 6xh + 2h^2 - 6x - 3h - 12\right) \\
&= 6x^2 - 6x - 12 .
\end{aligned}
$$

We are interested in points where $f'(x) = 0$. So we must solve the equation

$$0 = f'(x) = 6x^2 - 6x - 12 .$$

In fact the quadratic equation factors:

$$0 = 6(x+1)(x-2) .$$

So we find that $x = -1$ or $x = 2$.

Let us examine points on either side of $x = -1$. Since $f'(-3/2) = 21/2$ and $f'(-1/2) = -15/2$, we see that the graph is going uphill to the *left* of $x = -1$ and downhill to the *right* of $x = -1$. Thus $x = -1$ is the location of a local maximum.[2] See Figure 7.13.

[2]Some might say that the maximum is at $(-1, f(-1))$. But most calculus books simply say that the maximum is at $x = -1$.

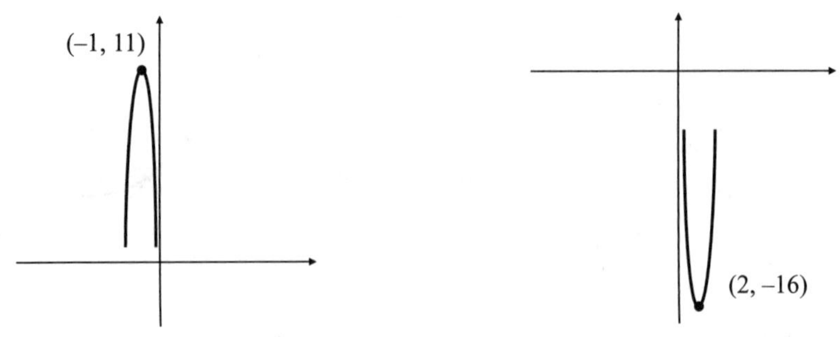

Figure 7.13. A local maximum. **Figure 7.14.** A local minimum.

Now let us examine points on either side of $x = 2$. Likewise, $f'(3/2) = -15/2$ and $f'(5/2) = 21/2$. So we see that the graph is going downhill to the *left* of $x = 2$ and uphill to the *right* of $x = 2$. Thus $x = 2$ is the location of a local minimum. See Figure 7.14.

Noting that $f(-1) = 11$ and $f(2) = -16$, we can assemble all our information and produce the graph shown in Figure 7.15.

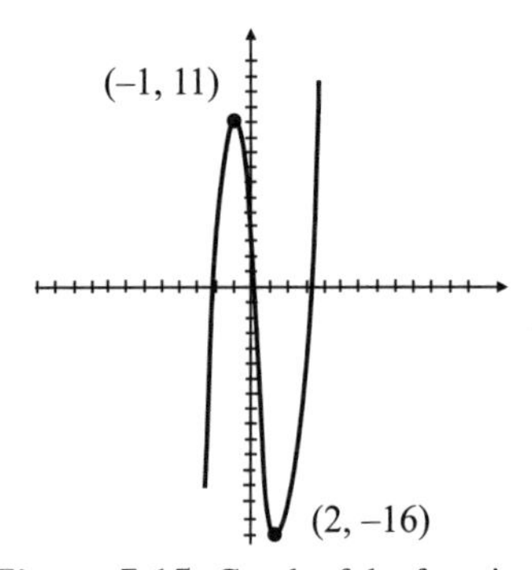

Figure 7.15. Graph of the function.

EXAMPLE 7.6. A box is to be made from a sheet of cardboard that measures $12'' \times 12''$. The construction will be achieved by cutting a square from each corner of the sheet and then folding up the sides (see Figure 7.16). What is the box of greatest volume that can be constructed in this fashion?

Solution. It is important to introduce a variable. Let x be the side length of each of the squares that are to be cut from the sheet of cardboard. Then the side length of the resulting box will be $12 - 2x$ (see Figure 7.17). Also the *height* of the box will be x. As a result, the volume of the box will be

$$V(x) = x \cdot (12 - 2x) \cdot (12 - 2x) = 144x - 48x^2 + 4x^3 .$$

Our job is to maximize this function V.

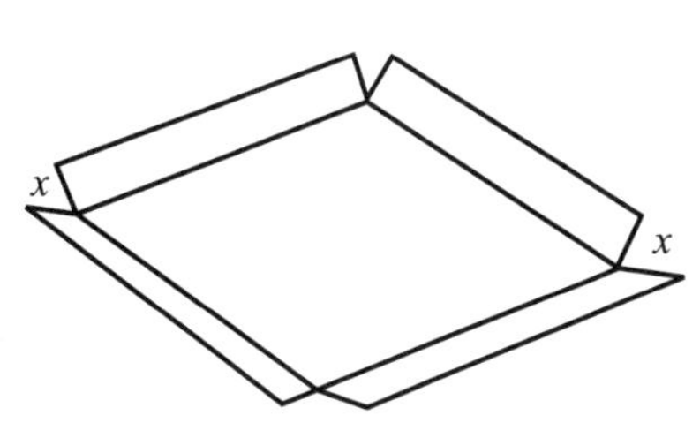

Figure 7.16. Construction of a box.

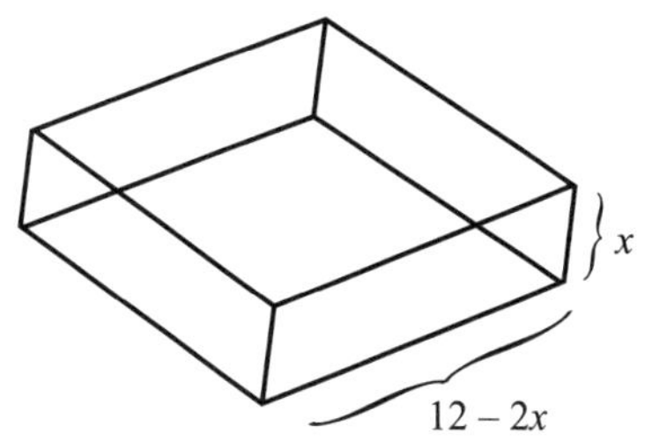

Figure 7.17. Side length of the box.

Now

$$\begin{aligned}
V'(x) &= \lim_{h\to 0} \frac{[144(x+h) - 48(x+h)^2 + 4(x+h)^3] - [144x - 48x^2 + 4x^3]}{h} \\
&= \lim_{h\to 0} \frac{[(144x + 144h) - (48x^2 + 96xh + 48h^2)}{h} \\
&\quad + \frac{(4x^3 + 12x^2h + 12xh^2 + 4h^3)] - [144x - 48x^2 + 4x^3]}{h} \\
&= \lim_{h\to 0} \frac{144h - 96xh - 48h^2 + 12x^2h + 12xh^2 + 4h^3}{h} \\
&= \lim_{h\to 0} 144 - 96x - 48h + 12x^2 + 12xh + 4h^2 \\
&= 144 - 96x + 12x^2 .
\end{aligned}$$

In summary, the derivative of the volume function is $V'(x) = 144 - 96x + 12x^2$. We may solve the quadratic equation

$$144 - 96x + 12x^2 = 0$$

to find the critical points for this problem. Using the quadratic formula, we find that $x = 2$ and $x = 6$ are the points that we seek (i.e., the potential maximum or minimum).

Of course $V'(2) = 0$. A little to the left of 2, we see that $V'(1.5) = 27$. A little to the right of 2, we see that $V'(2.5) = -21$. We conclude that $x = 2$ is a maximum.

We conclude that if squares of side 2″ are cut from the sheet of cardboard, then a box of maximum volume will result.

Observe in passing that if squares of side 6″ are cut from the sheet, then (there will be no cardboard left!) the resulting box will have zero volume. This value for x gives a minimum for the problem.

EXAMPLE 7.7. A rectangular garden is to be constructed against the side of a garage. The gardener has 100 feet of fencing, and will construct a three-sided fence; the side of the garage will form the fourth side. What dimensions will give the garden of greatest area?

Solution. Look at Figure 7.18. Let x denote the side of the garden that is perpendicular to the side of the garage. Then the resulting garden has width x feet and length $100 - 2x$ feet.

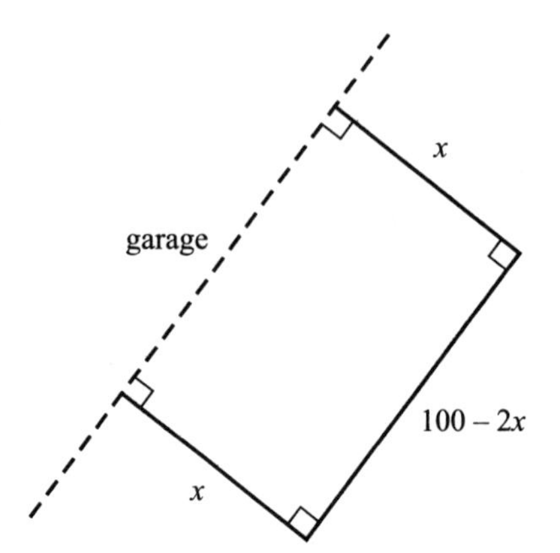

Figure 7.18. Dimensions of the garden.

The area of the garden is

$$A(x) = x \cdot (100 - 2x) = 100x - 2x^2 .$$

We calculate

$$\begin{aligned} A'(x) &= \lim_{h \to 0} \frac{[100(x+h) - 2(x+h)^2] - [100x - 2x^2]}{h} \\ &= \lim_{h \to 0} \frac{[100x + 100h - 2x^2 - 4xh - 2h^2] - [100x - 2x^2]}{h} \\ &= \lim_{h \to 0} \frac{100h - 4xh - 2h^2}{h} \\ &= \lim_{h \to 0} 100 - 4x - 2h \\ &= 100 - 4x \end{aligned}$$

and solve the equation $0 = A'(x) = 100 - 4x$. We find that the only critical point for the problem is $x = 25$. By inspection, we see that the graph of A is a downward-opening parabola. So $x = 25$ must be the global maximum that we seek. The optimal dimensions for the garden are

$$\text{width} = 25\,\text{ft.} \qquad \text{length} = 50\,\text{ft.}$$

For You to Try A cylindrical tin can is to be designed to hold 96 cubic inches of stewed tomatoes. What dimensions will minimize the amount of material used to construct the can?

For You to Try The sum of two numbers is 100. How can we choose them so as to maximize their product?

Further Reading

L. H. Lange, Maximize $x(a - x)$, *The Two-Year College Mathematics Journal* 5(1974), 22–24.

Arthur Rosenthal, The history of calculus, *The American Mathematical Monthly* 58(1951), 75–86.

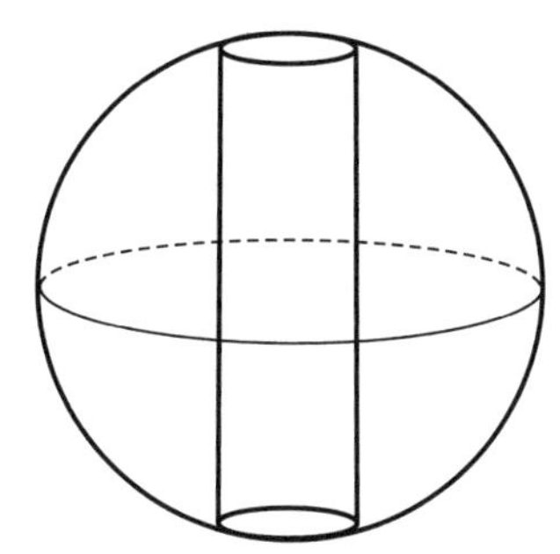

Figure 7.19. Cylinder inscribed in the sphere.

Exercises

1. What is the slope of the curve $f(x) = 3x^2 - 4x + 7$ at the point where $x = 2$?

2. What is the slope of the curve $g(x) = 4x^3 - x$ at the point $(-2, -30)$?

3. The height in feet of a falling body at time t seconds is given by $p(t) = -16t^2 + 20t + 34$. At what rate is the body falling when $t = 1$? At what time t does the body hit the ground? What is its velocity at that time?

4. Write the equation of the tangent line to the curve in Exercise 2 at the given point.

5. Write the equation of the line in Exercise 2 that is *perpendicular* to the given curve at the given point. Recall that two lines are perpendicular if the product of their slopes is -1.

6. Find all local maxima and minima of the curve $h(x) = -3x^3 + 6x^2 - 4x + 6$.

7. Of all the rectangles with perimeter 20, find the one of greatest area.

8. A cylinder is to be inscribed inside a sphere of radius 5, as shown in Figure 7.19. What dimensions of the cylinder will result in the cylinder of greatest volume?

9. If you try to calculate the slope of the tangent line to the curve $y = \sqrt{4 - x^2}$ at $x = 2$, you get an unpleasant answer. What does it mean? What is the geometric significance of your answer? You should be able to answer this question without doing any calculations.

10. An arrow is shot into the air, and its path describes a parabolic arc. The equation for the height in feet of the arrow at time t in seconds is $h(t) = -16t^2 + 42t + 100$. What is the greatest height that the arrow reaches? At what time t does the arrow hit the ground? What is the terminal velocity of the arrow?

11. The volume of a ball of radius r inches is $V = [4/3]\pi r^3$ cubic inches. What is the rate of change of volume with respect to the radius when $r = 4$?

12. The position of a moving body at time t seconds is given by $p(t) = 4t^3 - 7t^2 + 18t - 5$ feet . Focus on the time range $0 \leq t \leq 4$. At what time t is the velocity greatest?

13. **Project:** Let $y = f(x)$ be a differentiable curve. If $x = c$ is in the domain of f, then the *tangent line* to the graph of f at c is the line with slope $f'(c)$ and passing through the point $(c, f(c))$.

Explain why the tangent line at $(c, f(c))$ is the line of best approximation to the curve at the point $(c, f(c))$. This will entail two things: **(i)** finding a rigorous definition of "line of best approximation" and **(ii)** proving that the tangent line satisfies the rigorous definition in **(i)**. This approach to calculus is explored particularly cogently in [FAD].

14. **Project:** One of the mistakes that classical mathematicians of the nineteenth century made was that they assumed that if $f_1, f_2, \ldots$ are differentiable functions on an interval (a, b), then the sum

$$f(x) = \sum_{j=1}^{\infty} f_j(x)$$

will also be differentiable on (a, b). Provide an example to show that this statement is incorrect. What additional hypothesis is required on the f_j to make the conclusion correct? The book [GRA] gives a thorough grounding in the history of these ideas.

15. **Project:** Let f be a differentiable function on the interval (a, b) and f' its derivative. Will f' be continuous at each point of (a, b)? Why or why not? What sort of discontinuities can f' have? These ideas are considered in [KRA1].

8

The Great Isaac Newton

8.1 Introduction to Newton

Isaac Newton (1642–1727) is widely recognized to have been the greatest scientist of all time and certainly one of the three greatest mathematicians (along with Archimedes and Carl Friedrich Gauss). He is one of the inventors of calculus, he virtually created the subject of mechanics, he laid the groundwork for significant parts of modern physics, he developed the theory of infinite series, and made many other contributions as well. Today the Isaac Newton Institute for Mathematical Sciences in Cambridge, England memorializes his stature as a scholar and scientist. Newton's remarkable career is recounted in [BRE] and [WES]. The more recent [GLE] has new information about Newton's youth, and the time up to his invention of the calculus.

Isaac Newton's father was a man of some wealth and property, but (even though he fathered one of the greatest thinkers of all time) he could not sign his own name. He died three months before Isaac was born in the manor house of Woolsthorpe, near Grantham in Lincolnshire, England. His mother remarried—her second husband was Barnabas Smith, the minister of the church at North Witham—but part of the marital agreement was that little Isaac would *not* be part of the household. As a result, Isaac Newton was raised by his grandmother Margery Ayscough. Young Newton grew up feeling like an orphan, and his childhood was not a happy one.

When Isaac Newton was just ten years old his stepfather Smith died and bequeathed to young Isaac a rather large notebook. This was Isaac Newton's first access to paper, and he was thus able to begin calculating and working out his original ideas.

Newton's record in school was undistinguished. But Master Stokes of Newton's high school thought he was university material. He provided extra tutoring for Newton, and helped him gain admission to Trinity College at Cambridge University in 1661. Although his mother had considerable financial resources from her two marriages, Newton entered Cambridge as a "sizar". This was a special status for those in financial need. Newton acted as a manservant to the other students in exchange for funds to subvene his food and lodging and tuition.

Newton's original goal at the university was to study law. But he was somehow exposed to geometry, particularly the ideas of Descartes, and ultimately became fascinated with Euclidean and analytic geometry. Isaac Barrow, later to be Newton's teacher and mentor, arrived at Cambridge in 1663.

A seminal event in Newton's life is that the university was closed in Summer of 1665 on account of the plague. Medicine was a very primitive art in those days, but there was some understanding of contagion. So it was thought best for the students to disburse. Newton used his two years away from the university to develop his ideas about calculus, mechanics, optics, and astronomy. This was one of the most revolutionary periods in the history of science. Newton recorded some of his profoundly original ideas in *De Methodis Serierum et Fluxionum* , written in 1671, but he failed to get it published. It did not appear in print until John Colson produced an English translation in 1736 (nine years after Newton's death).

In fact this sort of behavior became the hallmark of Newton's scientific career. He often did not publish his important discoveries, at least not in a timely manner, and then he became very angry and aggressive when others made similar discoveries independently. Certainly Newton's behavior helped to fuel the rivalry between the Leibniz camp and the Newton camp over the discovery of calculus. This rivalry was so widespread and so intense that it adversely affected the development of British mathematics and science for a long time after Newton's death.

Cambridge University re-opened in 1667. On the strength of his recent scientific work, Newton was granted a major fellowship in July of 1668. Barrow played a key role in helping to bring Newton's work to the attention of influential British mathematicians and scientists. In 1669 Barrow resigned the Lucasian Professorship at Cambridge so that he could pursue matters of divinity; he was able to arrange that Isaac Newton be his successor in the Chair. Newton's first scientific activity as Lucasian Chair was a study of light. He correctly analyzed (contrary to well-founded beliefs that dated back to Aristotle) that white light was composed of a spectrum of color, and that the different colored light refracted at different angles. This led him to find fault with refracting telescopes and to construct his own reflecting telescope. After donating such a telescope to the Royal Society, Newton was elected a fellow—a great honor.

Shortly thereafter, Newton's first scientific publication appeared. It proposed a corpuscular (as opposed to a wave-based) theory of light. The overall opinion of the paper was positive, but a number of prominent scientists, including Hooke and Huygens, publicly criticized the work. Newton and Hooke continued to have bad relations until Hooke's death in 1703. Newton delayed the publication of his *Opticks* until 1704.

Newton was always conflicted about seeking fame and recognition while at the same time risking criticism. A consequence of his fisticuffs with Hooke and Huygens is that he withdrew from public discourse. Newton also had bitter disagreements with the English Jesuits in Liège over his theory of color; this led to a violent exchange of letters. In 1678 Newton appears to have suffered a nervous breakdown. His mother died in the following year and he withdrew further into his shell, mixing as little as possible with people for a number of years.

By 1666 Newton had formulated a version of his three laws of motion. These are so significant that we record them here in a simple, nonrigorous form:

First Law: A particle will stay at rest or continue at a constant velocity unless acted upon by an external unbalanced force.

Second Law: $F = ma$: The net force on an object is equal to the mass of the object multiplied by its acceleration.

Third Law: Every action has an equal and opposite reaction.

It was also around this time that Newton, studying Kepler's laws of planetary motion (discussed below), discovered his celebrated inverse square law for gravitational attraction. This says that the gravitational force exerted by body A on body B will be proportional to the masses of the two bodies and to the reciprocal of the square of the distance of the two bodies.[1] Typical of Newton, he did not publish this result until 1687 in his book *Philosophiae Naturalis Principia Mathematica*.

At this point we take a digression to put some of Newton's contributions to astronomy and cosmology into perspective. Tycho Brahe (1546–1601) was one of the great astronomers of the Renaissance. Through painstaking scientific procedure, he recorded reams and reams of data about the motions of the planets. His gifted student Johannes Kepler (1571–1630) was anxious to get his hands on Brahe's data, because he had ideas about formulating mathematical laws about the motions of the planets. But Brahe and Kepler were both strong-willed men. They could not see eye-to-eye on many things. And Brahe feared that Kepler would use his data to confirm the Copernican theory about the solar system (namely that the *sun*, not the earth, was the center of the system—a notion that ran counter to religious dogma). As a result, during Tycho Brahe's lifetime Kepler did not have access to Brahe's numbers.

But providence intervened in a strange way. Tycho Brahe had been given an island by his sponsor on which to build and run his observatory. As a result, Tycho was obliged to attend certain social functions—just to show his appreciation, and to report on his progress. At one such function, Tycho drank an excessive amount of beer, his bladder burst, and he died. Kepler was able to negotiate with Tycho Brahe's family to get the data that he so desperately needed. And thus the course of scientific history was forever altered.

Kepler did *not* use deductive cogitation, nor the axiomatic method, nor the strategy of mathematical proof to derive his three laws of planetary motion. Instead he simply stared at the hundreds of pages of planetary data that Brahe had provided, and he performed myriad calculations. At around this same time John Napier (1550–1617) was developing his theory of logarithms. These are terrific calculation tools, and would have simplified Kepler's task immensely. But Kepler could not understand the derivation of logarithms (see [KEP]), and so refused to use them. He did everything the hard way. Imagine what Kepler could have done with a computer!—but he probably would have refused to use one just because he didn't understand how the central processing unit (CPU) worked.

In any event, we tell here of Kepler and Napier because the situation is perhaps a harbinger of modern agonizing over the use of computers in mathematics. There are those who argue that the computer can enable us to see things—both calculationally and visually—that we could not see before. And there are those who say that all those calculations are good and well, but they do not constitute a mathematical proof. Nonetheless it seems that the first can inform the second, and a productive symbiosis can be created.

It is worthwhile at this juncture to enunciate Kepler's three very dramatic laws:

1. The orbit of each planet is in the shape of an ellipse. The sun is at one focus of that ellipse.

[1]From a modern perspective, this law is nearly obvious once one realizes that the gravitational potential on spheres centered about the body must be constant.

2. A line drawn from the center of the sun to the planet will, when the planet rotates, sweep out area at a constant rate.

3. The square of the time for one full orbit is proportional to the cube of the length of the major axis of the elliptical orbit.

It was about a century later that Edmond Halley (1656–1742), one of Isaac Newton's (1642–1727) few friends, was conversing with him about various scientific issues. Halley asked the great scientist what must be the shape of the orbits of the planets, given Newton's seminal inverse-square law for gravitational attraction. Without hesitation, Newton replied, "Of course it is an ellipse." Halley was shocked. "But can you prove this?" queried Halley. Newton said that he had indeed derived a proof, but then he had thrown the notes away. Halley was beside himself. This was the problem that he and his collaborators had studied for a great many years with no progress. And now the great Newton had solved the problem and then frivolously discarded the solution. Halley *insisted* that Newton reproduce the proof. Doing so required an enormous effort by Newton, and led in part to his writing of the celebrated *Principia*, published in 1687—perhaps the greatest scientific work ever written.

Newton had another nervous breakdown in 1693; after that he retired from research.[2] He kept his post at Cambridge until 1701, but he moved to London to assume the positions of Warden and then Master of the Royal Mint. It is worth noting that Newton's scientific life comprised about 28 years; his "post-scientific life" comprised 34 years. Newton was in fact a very effective head of the Mint. He led it through the difficult period of recoinage, and he implemented important measures to deal with counterfeiting. In 1703, Isaac Newton was elected President of the Royal Society. He was re-elected each year until his death. He was also knighted by Queen Anne—the first scientist to be so honored. His post-scientific years were not, however, uniformly happy. He was frequently consumed by the controversy with Leibniz over the invention of calculus. Newton used his position as President of the Royal Society to appoint an "impartial" committee to decide whether he or Leibniz was the inventor of the calculus. He (anonymously) wrote the official report of the committee which was published by the Royal Society, and he then wrote a review (again anonymously) which appeared in the Philosophical Transactions of the Royal Society.

A charming and frequently told story about this period of Newton's life concerns a famous old problem from mechanics. Called the *brachistochrone*, the question is what path in space, connecting two points, will allow a ball to fall most quickly from the upper point to the lower point—see Figure 8.1. Galileo incorrectly stated in 1638 in his *Two New Sciences* that this curve was an arc of a circle. Johann Bernoulli solved the problem (by reference to the previously analysed *tautochrone* curve) before posing it to readers of *Acta Eruditorum* in June, 1696. Five mathematicians responded with solutions: Isaac Newton, Jakob Bernoulli (Johann's brother), Gottfried Leibniz, Ehrenfried Walther von Tschirnhaus and Guillaume de l'Hôpital. Four of the solutions (excluding l'Hôpital's) were published in the May 1697 edition of the same publication.

[2]A modern theory is that in fact Newton suffered from Asperger's Syndrome. Asperger's syndrome is one of several autism spectrum disorders (ASD) characterized by difficulties in social interaction and by restricted, stereotyped patterns of behavior, interests and activities. AS is distinguished from the other ASDs in having no general delay in language or cognitive development. Although not mentioned in standard diagnostic criteria, motor clumsiness and atypical use of language are frequently reported.

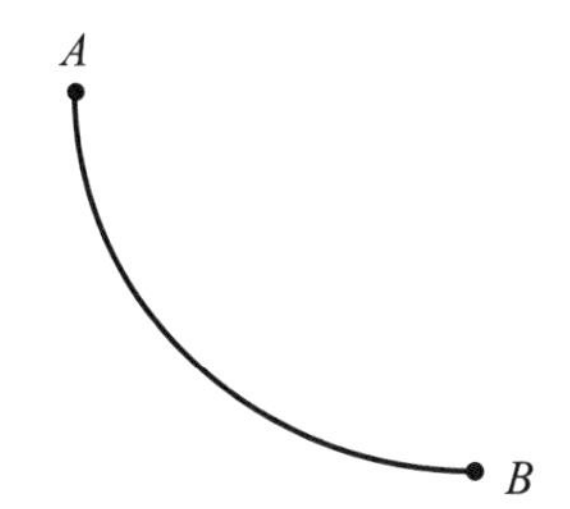

Figure 8.1. The brachistochrone.

It should be remembered that Newton had been retired from scientific life for several years when these events occurred. Newton came home from a long day at the Mint, read Bernoulli's problem in the scientific periodical, and just sat down and solved it. When Bernoulli saw Newton's solution to his problem (which solution was in fact contributed anonymously), Bernoulli is famously quoted to have said, "I can recognize the lion by the print of his paw."

The solution of the *brachistochrone* is a remarkable curve called the *cycloid*. A cycloid is generated by putting a dot on the edge of a disc and then rolling the disc down a linear path: the path of the dot is the cycloid. Examine Figure 8.2 to see the shape. An interesting side benefit of the solution of the *brachistochrone* problem is that it also solves the *tautochrone* problem (mentioned above). The *tautochrone* asks what shape path has the property that a ball rolling down it will reach the bottom in the same time no matter where it begins on the path. The answer, again is the cycloid.

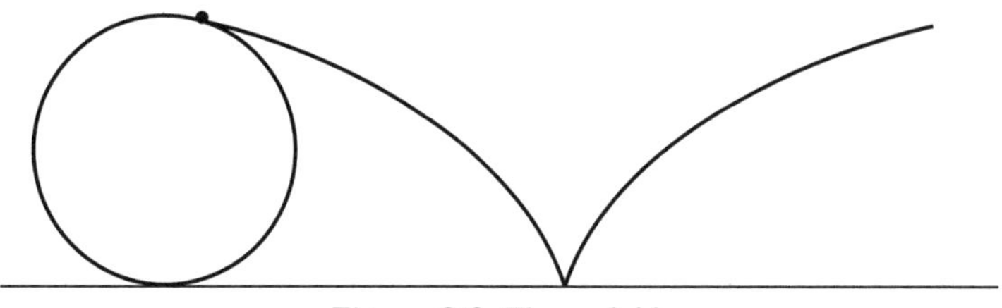

Figure 8.2. The cycloid.

The reader should not be confused by this discussion. Given two points A and B in space as in Figure 8.1, clearly the *shortest path* between them is a straight line. But that is not what is being asked here. For the *brachistochrone*, we seek that path of least time for the ball to travel from A to B. Again see Figure 8.1. For the *tautochrone*, we seek the path so that all balls, beginning anywhere on the path, will reach the bottom point B in the same time interval.

The solution of these problems is difficult and subtle; we cannot treat it here. But it is a fascinating part of mathematical history to know Newton's role in this story.

Further Reading

E. T. Bell, Newton after three centuries, *The American Mathematical Monthly* 49(1942), 553–575.

V. Frederick Rickey, Isaac Newton: man, myth, and mathematics, *The College Mathematics Journal* 18(1987), 362–389.

8.2 The Idea of the Integral

There are so many major scientific accomplishments to Newton's credit that we could fill several volumes with even a brief description of some of them. We shall concentrate here on the fundamental theorem of calculus. This is the central idea—linking the derivative and the integral—that makes calculus such a powerful tool in mathematics and science. A simplified version of the result was first published by James Gregory (1638–1675) around 1670. Isaac Barrow (1630–1677), who was Newton's teacher, formulated and proved a general version of the result—also around 1670. Isaac Newton realized the full power of this important idea and developed it into a powerful body of techniques. Gottfried Wilhelm von Leibniz (1646–1716), who was really the theoretician for the development of calculus, systematized the mathematical developments surrounding the fundamental theorem.

We have already discussed the concept of derivative in Chapter 7. We shall say more about it here. But first let us introduce and describe the integral. The motivation for the integral concept is the calculation of area. Consider the curve $y = x^2$ in Figure 8.3. We wish to calculate the shaded area shown in Figure 8.4—above the x-axis, below the curve, and between the lines $x = 1$ and $x = 2$. Of course we realize that this is a nonstandard shape. When we learn in school about areas, we see how to find the area of a rectangle and a triangle. These are fairly easy, as they are based on rectilinear reasoning: the area of a rectangle is the product of its length and its width. And a triangle is half of a rectangle. It is difficult to say anything about areas of more exotic shapes. We are taught that the area inside a circle of radius r is πr^2, but it is difficult to understand the mathematical justification for this formula.

The key idea to understanding the indicated area is to try to divide it up into simpler subregions. The method that we are about to describe harkens back to our discussion of Archimedes's analysis of the circle in Subsection 1.3.2.

We begin by dividing the interval $[1, 2]$ (the base of the shaded region) into small equal subintervals. We do this by inserting points $x_0, x_1, x_2, \ldots, x_{k-1}, x_k$ with

$$1 = x_0 < x_1 < x_2 < \cdots < x_{k-1} < x_k = 2$$

that are equally spaced. This collection of points is called a *partition*. When the points are equally spaced then we call this a *uniform partition*. Since these $k + 1$ points determine k intervals, we can be sure that the length of each small interval is $1/k$. Refer to Figure 8.5.

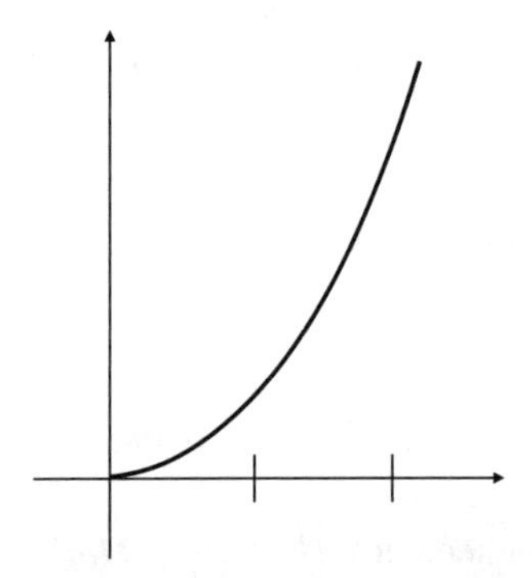

Figure 8.3. The curve $y = x^2$.

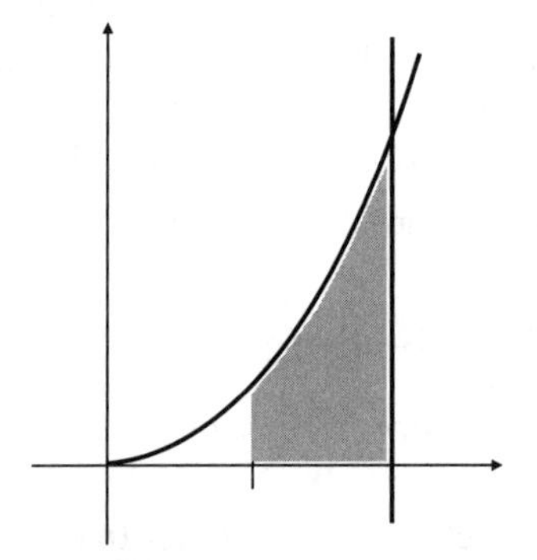

Figure 8.4. The area under the curve.

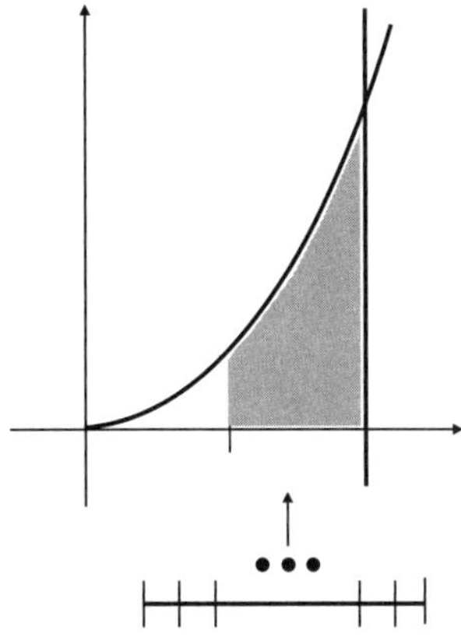

Figure 8.5. A partition of the interval [1, 2].

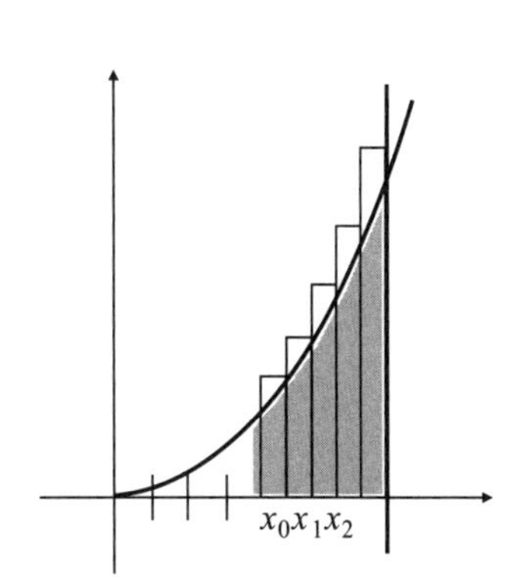

Figure 8.6. Approximating rectangles.

We call this collection of points a *uniform partition* of the interval [1, 2].

Now, over each of these subintervals of length $1/k$, we erect a rectangle that just touches the graph of $y = x^2$ on the right. Look at Figure 8.6. We stress that the union of all these rectangles does not literally *equal* the shaded region whose area we are trying to calculate. But it gives a *very good approximation*. And it is quite easy to calculate the area of each of these rectangles.

Look at the first rectangle on the left (Figure 8.7). It has base of length $1/k$ and height equal to $f(x_1)$. Thus the area of the first rectangle is

$$A_1 = f(x_1) \cdot (1/k) .$$

Likewise (Figure 8.8), the second rectangle has base equal to $1/k$ and height equal to $f(x_2)$. So it has area

$$A_2 = f(x_2) \cdot (1/k) .$$

Going on, the jth rectangle has base $1/k$ and height $f(x_j)$.

In sum, the total of all the areas of all the rectangles is equal to

$$\mathbf{A}^k = A_1 + A_2 + \cdots + A_k = \sum_{j=1}^{k} f(x_j) \cdot \frac{1}{k} .$$

This number $\mathbf{A}^k$ is an *approximation* to the true area of the shaded region. We call the quantity $\mathbf{A}^k$ a *Riemann sum*. If we make the number k larger, thus increasing the number

Figure 8.7. The area of the first rectangle.

Figure 8.8. The area of the second rectangle.

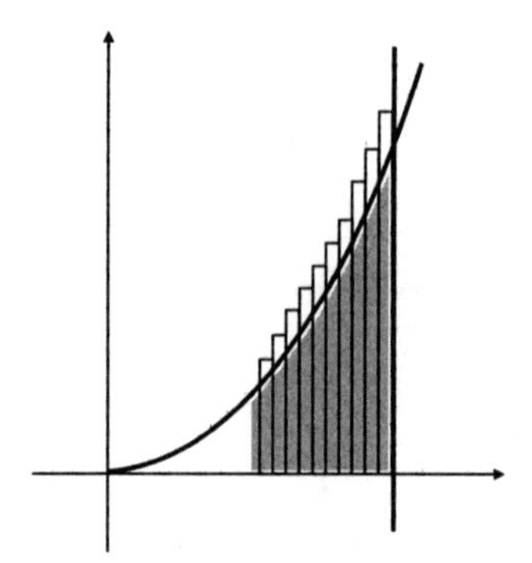

Figure 8.9. A finer partition.

of rectangles (and concomitantly making each rectangle narrower) then the resulting union of rectangles gives a better approximation to the true area of the shaded region (Figure 8.9). We *define* the area of the shaded region to be the "limit" of the approximating areas $\mathbf{A}^k$ as $k \to \infty$.

In fact there is some standard mathematical notation associated with this process, and this is a good time to learn it. The symbol aggregate

$$\int_1^2 x^2\,dx$$

stands for the area *under* the curve $y = x^2$, *above* the x-axis, and *between* the limits $x = 1$ and $x = 2$. More generally, the notation

$$\int_a^b f(x)\,dx$$

stands for (when f is positive) the area under the curve $y = f(x)$, above the x-axis, and between the limits $x = a$ and $x = b$. In case f is not necessarily positive, it is best to forget the pictorial interpretation and instead resort to the mathematics. We let

$$\int_a^b f(x)\,dx = \lim_{k\to\infty} \sum_{j=1}^{k} f(x_j) \cdot \frac{b-a}{k},$$

where $a = x_0 < x_1 < x_2 < \cdots < x_{k-1} < x_k = b$ is a uniform partition of $[a, b]$.

The expression $\int_a^b f(x)\,dx$ is called *the integral of the function f from a to b.*

Further Reading

Leonard Gillman, An axiomatic approach to the integral, *The American Mathematical Monthly* 100(1993), 16–25.

David M. Bressoud, How should we introduce integration?, *The College Mathematics Journal* 23(1992), 296–298.

8.3 Calculation of the Integral

Certainly the discussion in the last section has a compelling elegance, and helps us to understand the concept of area. But it would not be completely satisfying if we could not actually *calculate* some areas. Now we will learn how to actually calculate—as an explicit numerical quantity—the area under the curve $y = x^2$. As a preliminary step, we need to learn to calculate certain sums.

According to legend, Carl Friedrich Gauss (whom we shall discuss in detail in a later chapter) discovered at the age of 7 a formula for summing the numbers

$$1 + 2 + 3 + \cdots + 98 + 99 + 100 .$$

What Gauss realized is that it is propitious to write

$$S = 1 + 2 + 3 + \cdots + 98 + 99 + 100$$

and then to write the sum twice as

$$\begin{array}{ccccccccccccccc} S & = & 1 & + & 2 & + & 3 & + & \cdots & + & 98 & + & 99 & + & 100, \\ S & = & 100 & + & 99 & + & 98 & + & \cdots & + & 3 & + & 2 & + & 1. \end{array}$$

Now we notice that each column sums to 101. And there are 100 columns. The conclusion is that the total sum (which of course equals $2S$) is $100 \cdot 101 = 10100$. Thus

$$S = 1 + 2 + 3 + \cdots + 98 + 99 + 100 = 5050 .$$

Similar reasoning may be applied to any sum

$$T = 1 + 2 + 3 + \cdots + (N-2) + (N-1) + N .$$

We write

$$\begin{array}{ccccccccccccccc} T & = & 1 & + & 2 & + & 3 & + & \cdots & + & (N-2) & + & (N-1) & + & N, \\ T & = & N & + & (N-1) & + & (N-2) & + & \cdots & + & 3 & + & 2 & + & 1. \end{array}$$

Observe that each column sums to $(N + 1)$. And there are a total of N columns. Thus

$$2T = N \cdot (N + 1)$$

or

$$T = \frac{N \cdot (N + 1)}{2} .$$

In fact this formula was known to the Pythagoreans in ancient Greece. Nonetheless, it is common today to call it Gauss's formula.

EXAMPLE 8.1. We calculate that

$$1 + 2 + \cdots + 59 + 60 = \frac{60 \cdot 61}{2} = 1830 .$$

A slightly more interesting sum is

$$S = \sum_{j=12}^{39} j = 12 + 13 + \cdots + 38 + 39 .$$

We may treat this sum by writing it as

$$S = \left[\sum_{j=1}^{39} j \right] - \left[\sum_{j=1}^{11} j \right] = \frac{39 \cdot 40}{2} - \frac{11 \cdot 12}{2} = 780 - 66 = 714 .$$

Now we need to generalize our discussion of sums to sums of squares of successive integers. Particularly, we wish to sum

$$T = 1^2 + 2^2 + \cdots + (N-1)^2 + N^2 .$$

No trick as simple as Gauss's idea will work for this problem. Instead we do the following. Write

$$\begin{aligned}(j+1)^3 - j^3 &= \left[j^3 + 3j^2 + 3j + 1\right] - j^3 \\ &= 3j^2 + 2j + 1 .\end{aligned}$$

Summing from $j = 1$ to $j = N$ gives

$$\sum_{j=1}^{N}[(j+1)^3 - j^3] = 3\sum_{j=1}^{N} j^2 + 3\sum_{j=1}^{N} j + 3\sum_{j=1}^{N} 1 .$$

Now almost all the terms on the left-hand side cancel out (write out the first several summands to see this). The first term on the right is $3T$. The second term on the right is 3 times Gauss's sum. The last sum on the right is equal to N. As a result of these observations, we may conclude that

$$(N+1)^3 - 1^3 = 3T + 3 \cdot \frac{N(N+1)}{2} + 3N .$$

We may solve for T and thus find that

$$T = \frac{2N^3 + 3N^2 + N}{6} .$$

This formula was actually known to Archimedes, though he was likely not the first to discover it.

For You to Try Find a formula for

$$1^m + 2^m + 3^m + \cdots + (N-1)^m + N^m$$

by considering $(j+1)^{m+1} - j^{m+1}$ and imitating our argument for summing squares.

Now that we have these two summation formulas we can tackle the job of finding the area under the parabola:

EXAMPLE 8.2. Let us calculate

$$\int_1^2 x^2\, dx .$$

We fix a positive integer k and examine

$$\sum_{j=1}^{k}\left(1+\frac{j}{k}\right)^2 \cdot \frac{1}{k} = \sum_{j=1}^{k}\left[\frac{1}{k} + \frac{2j}{k^2} + \frac{j^2}{k^3}\right] = \frac{1}{k}\cdot\sum_{j=1}^{k} 1 + \frac{2}{k^2}\cdot\sum_{j=1}^{k} j + \frac{1}{k^3}\cdot\sum_{j=1}^{k} j^2 .$$

Now the first sum is trivially equal to k, the second can be handled with Gauss's formula, and the third can be treated with our formula for the sum of squares. The result is

$$\frac{1}{k}\cdot k + \frac{2}{k^2}\cdot\frac{k(k+1)}{2} + \frac{1}{k^3}\cdot\frac{2k^3+3k^2+k}{6} = 1 + (1+1/k) + \frac{2+3/k+1/k^2}{6} .$$

This is the approximation to the true area given by the sum of the areas of the rectangles. Our job now is to take the limit as $k \to \infty$. But it is plain that the limit as $k \to \infty$ is $7/3$. The area under the parabola, above the x-axis, and between $x = 1$ and $x = 2$ is therefore $7/3$.

For You to Try Calculate the integral

$$\int_2^4 5x^2 - x\,dx\,.$$

Further Reading

Richard Barshinger, Calculus II and Euler, *The American Mathematical Monthly* 101(1994), 244–249.

R. G. Douglas, One needs more than the algorithmic approach, *The College Mathematics Journal* 16(1985), 5–6.

8.4 The Fundamental Theorem of Calculus

We can see from the last example that calculating an integral from first principles can be a tedious chore. So it is of interest to have a technique that makes the analysis of integrals natural and straightforward. That is what the Fundamental Theorem of Calculus is all about.

Suppose now that we are given a positive, continuous function $y = f(x)$. See Figure 8.10. For simplicity we will take the domain of f to be the interval $[0, 1]$. For $0 < a \leq 1$, we define $\mathcal{A}(x)$ to be the area under the curve, above the x-axis, to the right of $x = 0$, and to the left of $x = a$. Refer to Figure 8.11.

Our goal is to differentiate the function $\mathcal{A}$. Referring back to Sections 7.2 and 7.3, we examine the Newton quotient

$$\frac{\mathcal{A}(a+h) - \mathcal{A}(a)}{h}\,.$$

Geometrically, this quantity is exhibited in Figure 8.12.

Now, following the philosophy that we used to set up the integral, we are going to approximate the area represented by the Newton quotient with a rectangle. See Figure 8.13.

For convenience this time we have let the approximating rectangle touch the curve on

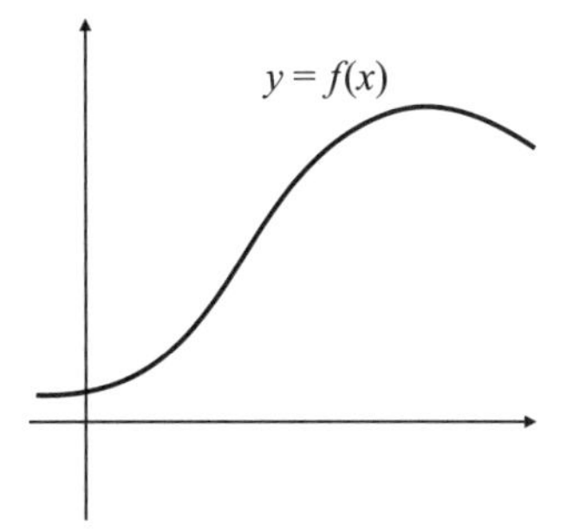

Figure 8.10. The function $y = f(x)$.

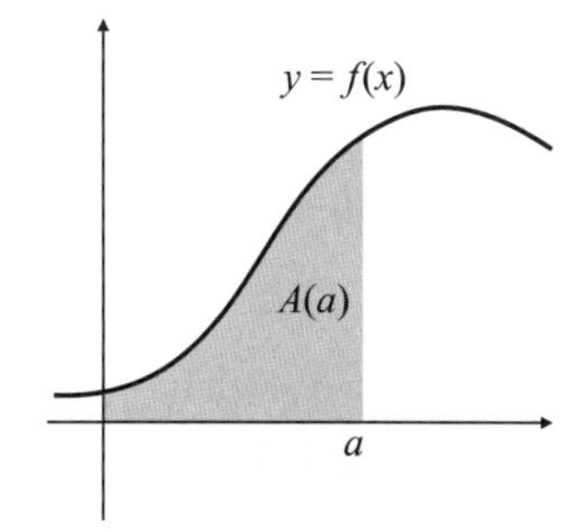

Figure 8.11. The area function $\mathcal{A}$.

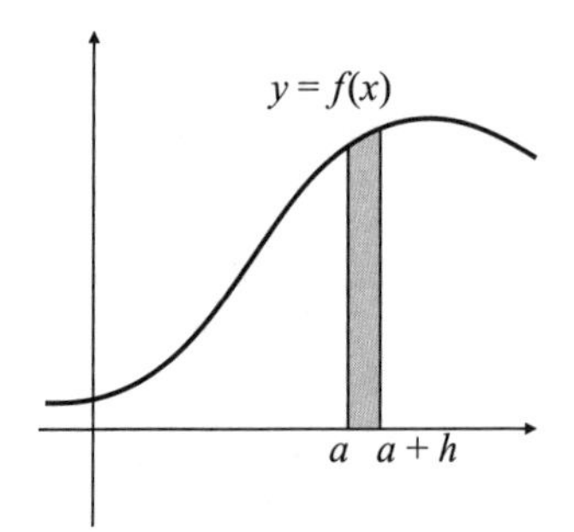

Figure 8.12. The Newton quotient for the area function.

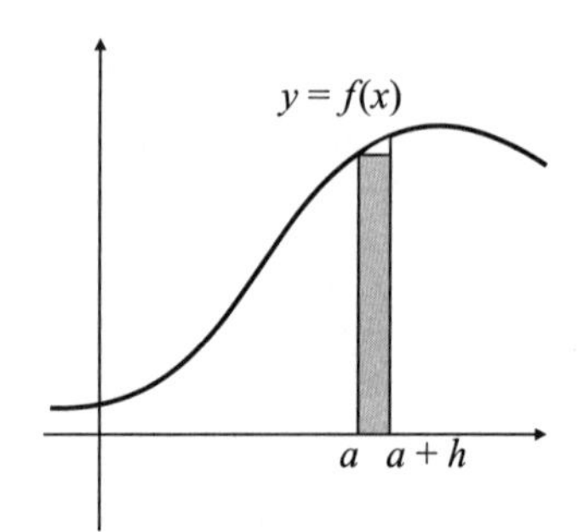

Figure 8.13. Approximation of the Newton quotient.

the left, so it has height $f(a)$. Then we see that the Newton quotient is

$$\frac{\mathcal{A}(a+h)-\mathcal{A}(a)}{h} \approx \frac{(\text{area of box})}{h} = \frac{f(a)\cdot h}{h} = f(a)\,.$$

In summary,

$$\lim_{h\to 0} \frac{\mathcal{A}(a+h)-\mathcal{A}(a)}{h} = f(a)\,.$$

This is a truly profound statement. It says that the derivative of the area function is the original function f. We summarize with a theorem:

Theorem 8.1. *Let f be a continuous function on the interval $[0, M]$. Define*

$$\mathcal{A}(a) = \int_0^a f(x)\,dx\,.$$

Then the derivative of $\mathcal{A}(a)$ is $f(a)$.

Let us use this information to re-do Example 8.2. We are given the function $f(x) = x^2$. Let us try to guess a function $F(x)$ whose derivative is $f(x)$. We might suppose that F is a power of x, and we might further guess that the power should be 1 greater than 2, or 3. So we guess that $F(x) = x^3$. That does not quite work, for $F'(x) = 3x^2$. This causes us to adjust our guess to $F(x) = x^3/3$. Then $F'(x) = x^2$ as desired. So we see that the area function for this particular f is

$$\mathcal{A}(a) = \int_0^a f(x)\,dx = \int_0^a x^2\,dx = \frac{a^3}{3}\,.$$

Thus the area between $x = 1$ and $x = 2$ is

$$\mathcal{A}(2) - \mathcal{A}(1) = \frac{2^3}{3} - \frac{1^3}{3} = \frac{7}{3}\,.$$

This is of course consistent with the answer that we obtained when we first worked Example 8.2.

The procedure that we have just performed to calculate $\int_1^2 x^2\,dx$ gives in fact a very general paradigm for calculating an integral $\int_a^b f(x)\,dx$:

1. First find an "antiderivative" $F(x)$ for $f(x)$. This is a function F such that $F' = f$.
2. Then the value of the integral is $F(b) - F(a)$.

EXAMPLE 8.3. Let us calculate

$$\int_0^3 x^3 - 4x + 1\, dx .$$

To do so, we first need to find an antiderivative for the integrand. Following the procedure that we used in the last example (which consisted mainly of educated guessing), we find that $F(x) = x^4/4 - 2x^2 + x$ (the reader should check this by differentiating F). Then

$$\int_0^3 x^3 - 4x + 1 = F(3) - F(0) = \left[\frac{3^4}{4} - 2 \cdot 3^2 + 3\right] - \left[\frac{0^4}{4} - 2 \cdot 0^2 + 0\right] = \frac{21}{4} .$$

For You to Try Calculate the integral

$$\int_1^3 x^4 - 5x^2 + 3x\, dx .$$

For You to Try Calculate the integral

$$\int_0^5 6x^7 - 3x^2 - 9\, dx .$$

Further Reading

W. R. Ransom, The fundamental theorem of the differential calculus, *The American Mathematical Monthly* 62(1955), 361–363.

Jack Wagner, Barrow's fundamental theorem, *The College Mathematics Journal* 32(2001), 58–59.

8.5 Some Preliminary Calculations

We need to know some specific and fairly sophisticated derivatives in order to perform some of the calculations in the next section. We isolate the necessary techniques here.

Lemma 8.1. *We have the limit*

$$\lim_{h \to 0} \frac{\sin h}{h} = 1 .$$

Proof. Examine Figure 8.14, which exhibits the angle h (measured in radians, of course) and the trigonmetric triangles that are relevant to this angle. We see that the length of $\overline{OB}$ is $1 - \epsilon$. That is the base of a right triangle with hypotenuse 1. Hence the height of the triangle (segment $\overline{AB}$) has, by the Pythagorean theorem, length $\sqrt{2\epsilon - \epsilon^2}$.

Now clearly the arc of the circle extending from A to C (this is the arc whose length is the radian measure of the angle h) has length less than the combined lengths of $\overline{AB}$ and $\overline{BC}$—just look at the rectangle that these two segments determine in the figure. The length of $\overline{BC}$ is just ϵ. Also note that segment $\overline{AB}$ is shorter than this arc. So we have the following facts:

1. Certainly

$$\frac{\sin h}{h} = \frac{|\overline{AB}|}{h} < 1 .$$

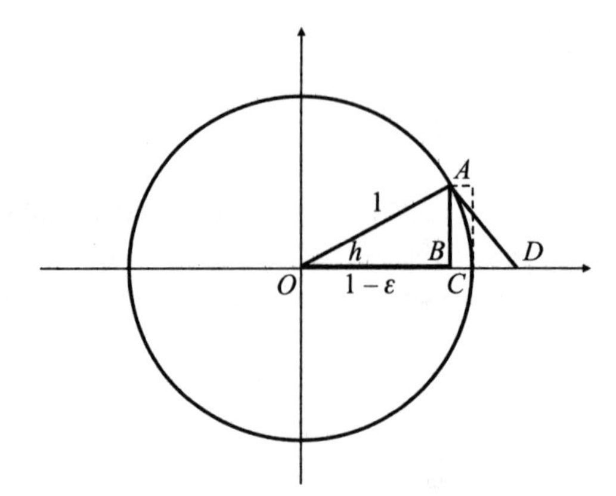

Figure 8.14. The limit of $(\sin h)/h$.

2. In the reverse direction,

$$\frac{\sin h}{h} \geq \frac{\sqrt{2\epsilon - \epsilon^2}}{|\overline{BC}| + |\overline{AB}|} = \frac{\sqrt{2\epsilon - \epsilon^2}}{\epsilon + \sqrt{2\epsilon - \epsilon^2}}.$$

Putting these two facts together yields

$$\frac{\sqrt{2\epsilon - \epsilon^2}}{\epsilon + \sqrt{2\epsilon - \epsilon^2}} \leq \frac{\sin h}{h} < 1.$$

As $h \to 0$ then $\epsilon \to 0$, hence the expression on the left tends to 1. We conclude that

$$\lim_{h\to 0} \frac{\sin h}{h} = 1.$$

□

Lemma 8.2. *We have the limit*

$$\lim_{h\to 0} \frac{\cos h - 1}{h} = 0.$$

Proof. We set $h = 2s$ and write

$$\frac{\cos h - 1}{h} = \frac{\cos 2s - 1}{2s} = \frac{\cos^2 s - \sin^2 s - 1}{2s} = \frac{-2\sin^2 s}{2s} = -(\sin s) \cdot \left(\frac{\sin s}{s}\right).$$

Thus we may calculate

$$\begin{aligned}
\lim_{h\to 0} \frac{\cos h - 1}{h} &= \lim_{s\to 0} \frac{\cos 2s - 1}{2s} \\
&= \lim_{s\to 0} \left[(-\sin s) \cdot \left(\frac{\sin s}{s}\right)\right] \\
&= \lim_{s\to 0} (-\sin s) \cdot \lim_{s\to 0} \left(\frac{\sin s}{s}\right) \\
&= \lim_{s\to 0} (-\sin s) \cdot 1 \\
&= 0.
\end{aligned}$$

□

For You to Try Calculate the limit

$$\lim_{h\to 0} \frac{[\tan h]^2}{h}.$$

For You to Try Calculate the limit

$$\lim_{h\to 0} \frac{\cos^2 h - 1}{h^2}.$$

Now we wish to differentiate the sine function.

Proposition 8.1. *We have*

$$(\sin x)' = \cos x.$$

Proof. We calculate

$$\begin{aligned}\lim_{h\to 0} \frac{\sin(x+h) - \sin x}{h} &= \lim_{h\to 0} \frac{\sin x \cos h + \cos x \sin h}{h} \\ &= \sin x \cdot \lim_{h\to 0} \frac{\cos h - 1}{h} + \cos x \cdot \lim_{h\to 0} \frac{\sin h}{h}.\end{aligned}$$

The two lemmas tell us how to evaluate the two limits, and we obtain the desired result. □

Proposition 8.2. *We have*

$$(\cos x)' = -\sin x.$$

Proof. We calculate

$$\begin{aligned}\lim_{h\to 0} \frac{\cos(x+h) - \cos x}{h} &= \lim_{h\to 0} \frac{\cos x \cos h - \sin x \sin h - \cos x}{h} \\ &= \lim_{h\to 0} \cos x \cdot \frac{\cos h - 1}{h} - \sin x \cdot \frac{\sin h}{h} \\ &= -\sin x.\end{aligned}$$

That completes the proof. □

We conclude this section with some remarks about change of variable in an integral. Suppose we are considering an integral of the form

$$\int_a^b f(x)\,dx.$$

Let us examine the sum that approximates this integral:

$$\sum_{j=1}^{k} f(j/k) \cdot 1/k. \qquad (*)$$

Now each factor $1/k$ may be written (in view of the fact that the partition is $a = x_0 < x_1 < x_2 < \cdots < x_{k-1} < x_k = b$) as

$$\frac{1}{k} = x_j - x_{j-1}.$$

Now let $\varphi : [c, d] \to [a, b]$ be a monotone increasing or monotone decreasing, continuously differentiable, function. We may think of each point x_j as the image under φ of some t_j. So we have

$$\frac{1}{k} = \varphi(t_j) - \varphi(t_{j-1}) .$$

By the definition of the derivative, this last line may be rewritten as

$$\frac{1}{k} \approx \varphi'(t_j) \cdot (t_j - t_{j-1}) .$$

So we may write the sum $(*)$ as

$$\sum_{j=1}^{k} f(\varphi(t_j))\varphi'(t_j) \cdot [t_j - t_{j-1}] .$$

This leads to the identity of integrals

$$\int_a^b f(x)\, dx = \int_c^d f(\varphi(t)) \cdot \varphi'(t)\, dt .$$

We will get good use out of this "change of variables" formula in the next section.

For You to Try Perform the change of variable $x = t^4$ in the integral

$$\int_2^4 x^3 + x\, dx .$$

For You to Try Perform the change of variable $x = \cos t$ in the integral

$$\int_0^1 x\sqrt{1 - x^2}\, dx .$$

Further Reading

A. M. Bruckner, Creating differentiability and destroying derivatives, *The American Mathematical Monthly* 85(1978), 554–562.

M. R. Spiegel, On the derivatives of trigonometric functions, *The American Mathematical Monthly* 63(1956), 118–120.

P. K. Subramanian, Successive differentiation and Leibniz's theorem, *The College Mathematics Journal* 35(2004), 274–282.

8.6 Some Examples

Here we provide some examples to illustrate the power of the integral, and of the Fundamental Theorem of Calculus.

EXAMPLE 8.4. Let us calculate the area of a circle of radius 1. We think of the circle as centered at the origin, and given by the equation

$$x^2 + y^2 = 1 .$$

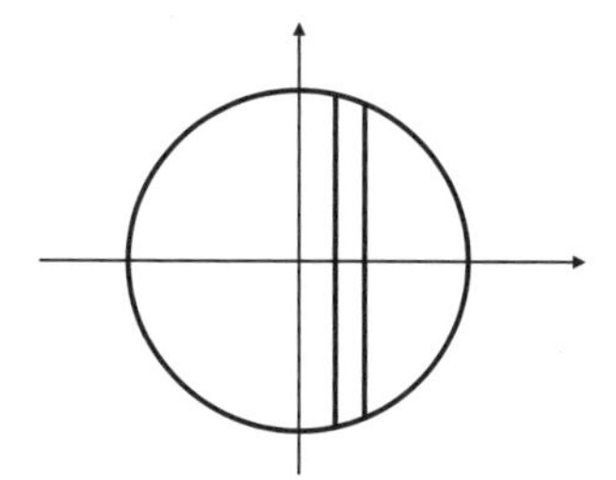

Figure 8.15. The area inside a circle.

Refer to Figure 8.15 as we reason along. We think of the x-axis, between $x = -1$ and $x = 1$, as broken up into $2k$ subintervals of equal lengths $1/k$. Above each such subinterval we erect a rectangle of height equal to the abscissa of the circle at the right-hand endpoint. Again refer to the figure. Adding up the areas of these rectangles, we obtain

$$\sum_{j=-k+1}^{k} 2\sqrt{1-(j/k)^2} \cdot (1/k)$$

as the area of the aggregate of the rectangles. But of course this is a summation approximation for an integral. We conclude that the area A of the circle is given by

$$A = \int_{-1}^{1} 2\sqrt{1-x^2}\, dx .$$

It is our job then to calculate this integral.

We utilize the change of variables $x = \sin t$. Then the integral becomes

$$\int_{-\pi/2}^{\pi/2} 2\sqrt{1-\sin^2 t} \cos t\, dt = \int_{-\pi/2}^{\pi/2} 2(\cos t) \cdot (\cos t) dt = \int_{-\pi/2}^{\pi/2} 2\cos^2 t\, dt .$$

Now it is useful to notice that

$$\cos 2t = \cos^2 t - \sin^2 t = 2\cos^2 t - 1 ,$$

hence

$$2\cos^2 t = 1 + \cos 2t .$$

Thus we may rewrite our integral as

$$\int_{-\pi/2}^{\pi/2} 1 + \cos 2t\, dt .$$

This integral is easy enough to evaluate. It equals

$$\left[t + \frac{\sin 2t}{2} \right]_{-\pi/2}^{\pi/2} = \frac{\pi}{2} - \frac{-\pi}{2} = \pi .$$

We see then that the area inside the circle is π. This result is consistent with the formula for the area inside a circle that you learned in school. □

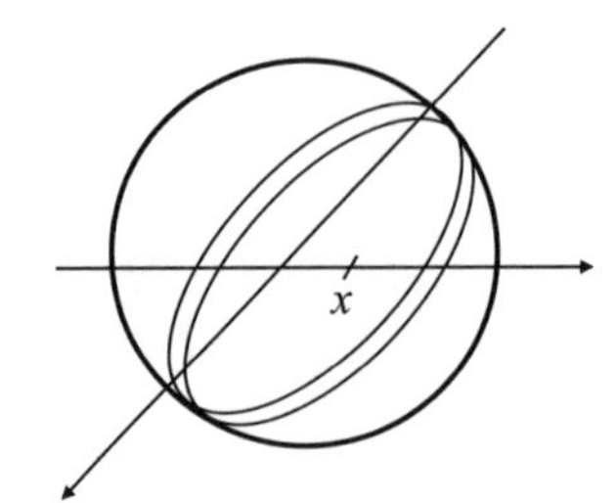

Figure 8.16. The volume of the sphere.

EXAMPLE 8.5. Let us calculate the volume inside a sphere of radius 1. Examining Figure 8.16, we think of the sphere as centered at the origin. We examine slices that are perpendicular to the x-y plane and parallel to the y axis as shown in the figure. There are $2k$ slices of equal thickness $1/k$.

The slice at position j/k, $-k < j \leq k$, is a disc of radius $\sqrt{1-(j/k)^2}$ and thickness $1/k$. So it has volume

$$V_j = \pi \cdot \left[\sqrt{1-\left(\frac{j}{k}\right)^2}\,\right]^2 \cdot \frac{1}{k}.$$

The sum of the volumes of these slices gives an approximation to the volume inside the sphere:

$$V \approx \sum_{j=-k+1}^{k} \pi \cdot \left[\sqrt{1-\left(\frac{j}{k}\right)^2}\,\right]^2 \cdot \frac{1}{k}.$$

As $k \to \infty$, this approximation tends to an integral. We find that

$$V = \int_{-1}^{1} \pi \left[\sqrt{1-x^2}\right]^2 dx .$$

This is an easy integral to evaluate:

$$\begin{aligned} V &= \int_{-1}^{1} \pi \left[\sqrt{1-x^2}\right]^2 dx \\ &= \int_{-1}^{1} \pi \left[1-x^2\right] dx \\ &= \pi \left[x - \frac{x^3}{3}\right]_{-1}^{1} . \end{aligned}$$

The notation on the right side of the last line indicates that the function $x - x^3/3$ is to be evaluated at 1 and -1 and the results subtracted.

So we find that

$$V = \pi \cdot \left[1 - \frac{1}{3}\right] - \pi \cdot \left[-1 + \frac{1}{3}\right] = \frac{4\pi}{3}.$$

This is in fact the familiar formula for the volume inside a sphere of radius 1.

For You to Try Imitate the method of the last example to calculate the volume inside a sphere of radius r.

For You to Try Use the method developed here to determine the volume inside the paraboloid

$$\mathcal{P} = \{(x, y, z) : 0 < z < 1 - x^2 - y^2\}.$$

We now give a quick review of the logarithm function so that the next example will make sense.

First, an *exponential function* is one of the form

$$E_a(x) = a^x$$

for some positive base a. It is convenient for us to take $a > 1$, and we show a typical graph in Figure 8.17.

Plainly the exponential function is increasing, so it is one-to-one. Thus it has an inverse function which we graph in Figure 8.18.

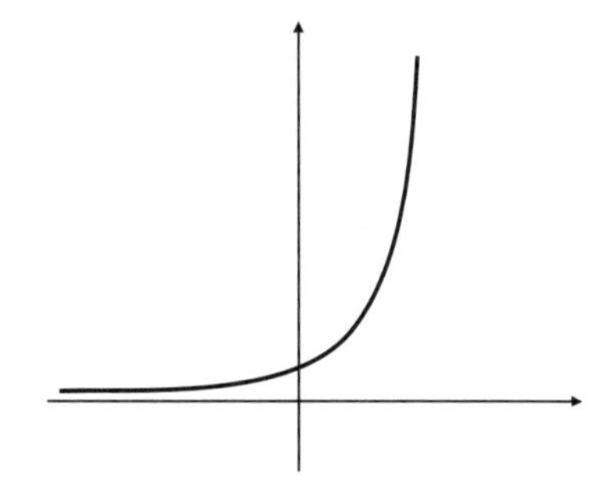

Figure 8.17. An exponential function.

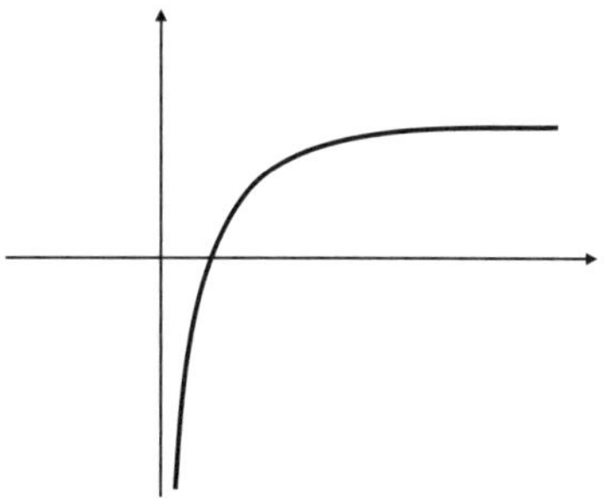

Figure 8.18. The inverse of the exponential function (or log function).

We call this function a *logarithmic function*. Of course there are many different exponential functions (depending on a) and, correspondingly, there are many different logarithmic functions. The particular logarithmic function whose graph has slope 1 at $x = 1$ is special; we call it the *natural logarithm* and denote it by $\ln x$. We can now calculate the derivative of $\ln x$ at any point:

$$(\ln x)' = \lim_{h \to 0} \frac{\ln(x+h) - \ln(x)}{h} = \lim_{h \to 0} \frac{\ln\left(\frac{x+h}{x}\right)}{h},$$

where we have used the standard fact that the difference of logs is the log of the quotient. This last equals

$$\lim_{h \to 0} \frac{\ln\left(1 + \frac{h}{x}\right)}{h} = \frac{1}{x} \cdot \lim_{h \to 0} \frac{\ln\left(1 + \frac{h}{x}\right)}{\frac{h}{x}}.$$

Since the function ln has slope 1 at $x = 1$, this last limit is equal to 1, and the entire expression equals $1/x$. We conclude then that

$$(\ln x)' = \frac{1}{x}.$$

This is a very important formula which we shall now use decisively to solve a problem.

EXAMPLE 8.6. It begins to snow at a steady rate (in terms of depth of the snow) some time before noon. A snow plow begins plowing at a steady rate (in terms of volume of snow) at precisely noon. The plow clears twice as much area in the first hour as it does in the second hour. When did it start snowing? [Refer back to Exercise 10 in Chapter 2, where we had a preliminary look at this problem.]

At first it may seem that there is insufficient information to answer the question. Everything is too vague. But a little analysis proves otherwise.

- Let A be the length of time before noon when it started snowing.
- Let t be the time that the snow plow has been plowing, with $t = 0$ denoting noon (the time when the plow begins its work).
- Let $v(t)$ denote the speed of the snow plow.
- Let $\ell(t)$ be the linear feet cleared by the snowplow at time t.
- Let $D(t)$ be the depth of the snow at time t.

Let us declare that the snow plow clears 2 units of snow in the first hour after noon and 1 unit of snow in the second hour. It doesn't really matter what the unit is, so we leave it unspecified. Certainly we know that $\ell(0) = 0$, $\ell(1) = 2$, and $\ell(2) = 3$.

The depth of the snow is a linear function of time. Clearly the speed at which the snow plow clears snow is inversely proportional to the depth of the snow. So we may say that there are constants of proportionality k' and k such that

$$v(t) = \frac{k'}{D(t)} = \frac{k}{t + A} .$$

Observe that we take into account the unit of time A that it was snowing before the plow began its work.

Certainly $d\ell/dt = v(t)$ so we may write

$$\frac{d\ell}{dt} = \frac{k}{t + A} ,$$

with initial conditions $\ell(0) = 0$, $\ell(1) = 2$, and $\ell(2) = 3$. Let us apply the integral from 0 to time T to both sides of this last equation. Thus

$$\int_0^T \frac{d\ell}{dt}\, dt = \int_0^T \frac{k}{t + A}\, dt .$$

We apply the Fundamental Theorem of Calculus to the left-hand side and the information about the derivative of the logarithm to the right-hand side. The result is

$$\ell(T) = k \cdot \ln(T + A) + C ,$$

where C is our abbreviation for $-k \cdot \ln(0 + A)$.

Applying our boundary conditions gives

$$\begin{aligned} 0 &= \ell(0) = k \cdot \ln A + C , \\ 2 &= \ell(1) = k \cdot \ln(1 + A) + C , \\ 3 &= \ell(2) = k \cdot \ln(2 + A) + C . \end{aligned}$$

Conveniently, we have three equations in three unknowns.

The first of these equations tells us that

$$C = -k \cdot \ln A\,.$$

Substituting this information into the other two equations gives

$$2 = k \cdot \ln(1 + A) - k \ln A$$

and

$$3 = k \cdot \ln(2 + A) - k \ln A\,.$$

As a result,

$$k = \frac{2}{\ln((1 + A)/A)}$$

and

$$k = \frac{3}{\ln((2 + A)/A)}\,.$$

We may equate the two expressions for k and do some algebra to find that

$$\frac{2}{\ln((1 + A)/A)} = \frac{3}{\ln((2 + A)/A)}$$

or

$$2 \cdot \ln\left(\frac{2 + A}{A}\right) = 3 \cdot \ln\left(\frac{1 + A}{A}\right).$$

This simplifies to

$$\ln\left(\frac{1 + A}{A}\right)^3 - \ln\left(\frac{2 + A}{A}\right)^2 = 0\,.$$

Hence

$$\ln\left(\frac{(1 + A)^3}{A(2 + A)^2}\right) = 0\,.$$

Now the exponential a^x takes the value 1 only at 0. Hence the logarithm function ln takes the value 0 only at 1. We may conclude from the last line then that

$$\frac{(1 + A)^3}{A(2 + A)^2} = 1\,.$$

This multiplies out to

$$A^3 + 3A^2 + 3A + 1 = A(4 + 4A + A^2)\,.$$

We finally see that we must solve the quadratic equation

$$A^2 + A - 1 = 0\,.$$

Using the quadratic formula, we find that the only positive solution to this last equation is

$$A = \frac{-1 + \sqrt{5}}{2} \approx 0.618\,.$$

Thus the snow begins to fall about 37 minutes before noon, or at approximately 11:23 am.

If an object moves down the real line with variable velocity $v(t)$, then we may analyze its position as follows. Say that the object's motion is parameterized by time t, and that time ranges over $0 \le t \le 1$. We may divide the unit interval into k equal pieces, with endpoints

$$0 = x_0 < x_1 < x_2 < \cdots < x_{k-1} < x_k = 1\,,$$

and approximate the motion (with position denoted by $p(t)$) on the jth piece by

$$p(j/k) - p((j-1)/k) \approx v(j/k) \cdot \frac{1}{k}\,.$$

Now sum both sides from $j = 1$ to $j = k$. We obtain

$$\sum_{j=1}^{k} p(j/k) - p((j-1)/k) \approx \sum_{j=1}^{k} v(j/k) \cdot \frac{1}{k}\,.$$

The left-hand side collapses to $p(1) - p(0)$. The right-hand side is of course the approximating sum for an integral. We deduce that

$$p(1) - p(0) = \int_0^1 v(t)\,dt\,.$$

Of course there is nothing special about the unit interval in this analysis. We only used it to make the calculations come out neatly. We have the following general fact:

Theorem 8.2. *Let an object move down the real line with velocity v over the time interval $a \le t \le b$. Then the motion of the object, from start to finish, is given by the position function p with*

$$p(b) - p(a) = \int_a^b v(t)\,dt\,.$$

This result reiterates something that we already know from our discussion of the derivative in Section 7.3: Namely, if p is the position of an object in motion, then the velocity of the object is given by the derivative $p'(t)$. The Fundamental Theorem of Calculus allows us to translate back and forth between the language of the theorem and the language of this paragraph.

A similar analysis shows that velocity is the integral of acceleration. Or, put in other words, acceleration is the derivative of velocity.

We now illustrate these ideas with an example.

EXAMPLE 8.7. Imagine that an object is dropped from a window and hits the ground 4 seconds later. What is the height of the window?

The key fact is that an object falling near the Earth's surface accelerates at a rate of about $-32\text{ft.}/\text{sec.}^2$. Thus we begin with the equation

$$a(t) = -32\,.$$

Now we integrate both sides from 0 to T, where $t = 0$ is the time when the object is released from the window. The result is

$$v(T) - v(0) = -32T\,.$$

Of course the initial velocity of a falling body is 0, so $v(0) = 0$. Thus we may write

$$v(T) = -32T \, .$$

We integrate both sides again to find that

$$p(T) - p(0) = -16T^2 \, .$$

Now our aim is to find $p(0)$. What we know is that $p(4) = 0$ (i.e., the object hits the ground after 4 seconds). We may plug this information into the last equation to obtain

$$0 - p(0) = -16 \cdot 4^2 \, .$$

As a result, $p(0) = 256$. We conclude that the height of the window is 256 feet.

EXAMPLE 8.8. An archer who is $6'6''$ tall shoots an arrow into the air, from a height of 5 feet, at an angle of 45°. The arrow strikes the ground 300 feet from the archer. What is the greatest height that the arrow achieves?

We must keep track of both the forward motion and the vertical motion of the arrow, and we do so with two functions $p(t)$ and $h(t)$. We imagine the archer standing at the origin in the x-y plane, as shown in Figure 8.19.

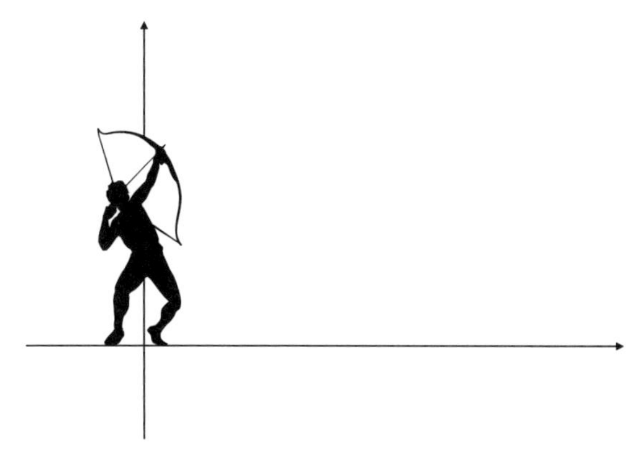

Figure 8.19. The archer shooting the arrow.

We know that

$$h''(t) = -32 \, .$$

Reasoning as in the last example, we find that

$$h'(t) = -32t + C \, .$$

Here C is the initial vertical speed of the arrow. Now, since the arrow is launched at a 45° angle, the initial vertical speed and the initial horizontal speed are the same. Also the forward motion is constant velocity. So we may conclude that $p(t) = Ct$. Refer to Figure 8.20.

Integrating one more time yields

$$h(t) = -16t^2 + Ct + E \, ,$$

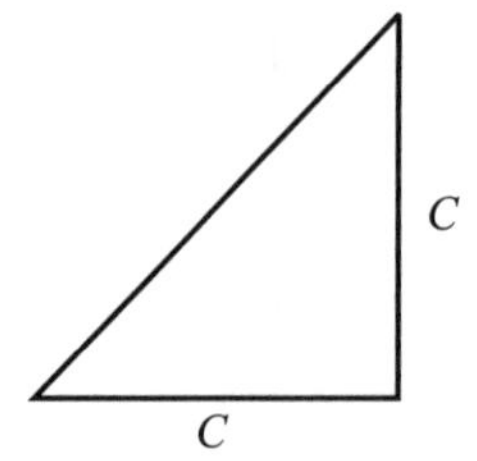

Figure 8.20. Vertical and horizontal components of velocity.

where E is the initial height. But that is known to be 5. Hence

$$h(t) = -16t^2 + Ct + 5.$$

Now we know, up to the unknown constant C, both the height function and the horizontal position function. The arrow's motion ceases when the arrow is 300 feet from the archer or when

$$Ct = p(t) = 300.$$

We conclude that this occurs at time $t = 300/C$. But, at that same time, the height is 0. Hence

$$0 = h(300/C) = -16\left(\frac{300}{C}\right)^2 + C \cdot \frac{300}{C} + 5.$$

We may solve this equation for C using the quadratic formula and find that $C \approx 68.71$.

Thus

$$h(t) = -16t^2 + 68.71t + 5.$$

The arrow reaches its greatest height when $h'(t) = 0$ or

$$0 = -32t + 68.71.$$

We see that this occurs at time $t = 2.147$. At that time the height is

$$h(2.147) = -16 \cdot (2.147)^2 + 68.71 \cdot 2.147 + 5 = 78.77.$$

The maximum height of the arrow is 78.77 feet.

For You to Try An object is dropped from a window with initial downward velocity of 10 feet per second. It strikes the ground 3 seconds later. How high is the window?

Further Reading

David Garber and Boaz Tsaban, A mechanical derivation of the area of the sphere, *The American Mathematical Monthly* 108(2001), 10–15.

Pam Littleton and David A. Sanchez, Dipsticks for cylindrical storage tanks—exact and approximate, *The College Mathematics Journal* 32(2001), 352–358.

Exercises

1. Calculate these integrals:

   (a) $\int e^{x^2} \cdot 2x \, dx$

   (b) $\int \frac{\ln x}{x} \, dx$

   (c) $\int [\sin(e^x)] \cdot e^x \, dx$

   (d) $\int \frac{[\sin(\ln x)] \cdot [\cos(\ln x)]}{x} \, dx$

2. The function $\varphi(x) = e^{-x^2}$ is very important in probability and statistics. It is known as the *Gaussian distribution*. It is useful to have a function Φ such that $\Phi' = \varphi$. But it is in fact impossible to write down an explicit formula for such a Φ. Explain instead why Φ exists, and write a non-explicit formula for it.

3. It is *not* true that

$$\int f(x) \cdot g(x) \, dx = \left[\int f(x) \, dx\right] \cdot \left[\int g(x) \, dx\right].$$

   Give an example to explain why this is so.

   But there is a substitute fact (known as *integration by parts*):

$$\int f(x) g'(x) \, dx = f(x) g(x) - \int g(x) f'(x) \, dx. \qquad (\dagger)$$

   Verify that this formula is correct for the functions $f(x) = x^2$ and $g(x) = x^3$. Use the formula ($\dagger$) to evaluate $\int x \cdot \sin x \, dx$.

4. Use the integration by parts formula from Exercise 3 to evaluate $\int \ln x \, dx$. Do so by letting $f(x) = \ln x$ and $g(x) = 1$.

5. An archer shoots an arrow into the air with an initial velocity of 75 feet per second. The initial height of the shot is 5 feet. The arrow strikes the ground precisely 200 feet from the archer. At what angle did the archer shoot the arrow?

6. An object is dropped from a 100-foot-high window and strikes the ground with a velocity of 100 feet per second. From what height is the object dropped?

7. On the planet Zork a falling object, near the surface of the planet, has acceleration 55 feet per second squared. If an object is thrown straight up with initial velocity 100 feet per second, then how high will it rise? After how many seconds will it strike the surface of the planet?

8. Calculate the integral

$$\int \frac{\cos x}{\sin x} \, dx.$$

9. **Project:** Read the description of the cycloid curve in the text. Write down parametric equations for a cycloid curve. Details of this construction may be found in [BLK].

10. Project: Calculate the derivative of the function

$$F(x) = \int_{x^2}^{x^3} \sin(\ln t)\, dt\,.$$

See [KRA1], [BLK] for further discussion of this type of problem.

11. Project: We know from our experience with polar coordinates in the plane that the relationship between a Cartesian integral and the corresponding polar integral is given by

$$dx\, dy = r\, dr\, d\theta\,.$$

This is best explained because the Jacobian matrix of the coordinate change

$$x = r\cos\theta\,,$$
$$y = r\sin\theta$$

is given by

$$\mathcal{J} = \begin{pmatrix} \dfrac{\partial x}{\partial r} & \dfrac{\partial x}{\partial \theta} \\ \dfrac{\partial y}{\partial r} & \dfrac{\partial y}{\partial \theta} \end{pmatrix} = \begin{pmatrix} \cos\theta & -r\sin\theta \\ \sin\theta & r\cos\theta \end{pmatrix}.$$

Notice that

$$\det \mathcal{J} = r\,.$$

If $f(x, y)$ is a given function of Cartesian coordinates, then we let $\widetilde{f}(r, \theta)$ be the corresponding function in polar coordinates. Then the general change of variables formula tells us that

$$\iint f(x, y)\, dxdy = \iint \widetilde{f}(r, \theta) \det \mathcal{J}\, drd\theta = \iint \widetilde{f}(r, \theta) r\, drd\theta\,.$$

In general, if $\Phi : U \to V$ is a change of coordinates, with $\Phi(x, y) = (\varphi_1(x, y), \varphi_2(x, y))$, then

$$\iint_V f(\widetilde{x}, \widetilde{y})\, d\widetilde{x} d\widetilde{y} = \iint_U f \circ \Phi(x, y) \det \mathcal{J}(x, y)\, dxdy\,, \qquad (\star)$$

where

$$\mathcal{J} = \begin{pmatrix} \dfrac{\partial \varphi_1}{\partial x} & \dfrac{\partial \varphi_1}{\partial y} \\ \dfrac{\partial \varphi_2}{\partial x} & \dfrac{\partial \varphi_2}{\partial y} \end{pmatrix}.$$

We call $\mathcal{J}$ the Jacobian matrix of the mapping Φ.

Explain why the transformation formula $(\star)$ is true by first considering the case when Φ is linear and then treating the general case by approximating a smooth map Φ by a linear map. The book [KRA1] has a thoroughgoing treatment of this topic.

The Complex Numbers and the Fundamental Theorem of Algebra

9.1 A New Number System

What is remarkable about the discussion we are about to provide is that we are going to *construct* the complex numbers. We shall *not* say, "Let us pretend that the number -1 has a square root and we'll build a number system around it." That is the sort of thinking that can lead to contradictions and paradoxes, and is best avoided. We will instead construct our new number system with tools that we have at hand. Such a constructivist approach gives our mathematics a solid foundation that we can rely on, and that we can be certain will not lead to later conundrums.

9.2 Progenitors of the Complex Number System

The complex numbers evolved over a period of several centuries. They did not spring at once from the mind of any particular individual. This number system arose from a need to have solutions to all polynomials. While a polynomial like

$$p(x) = x^2 - 5x + 6$$

has roots $x = 2$ (that is to say, $p(2) = 2^2 - 5 \cdot 2 + 6 = 0$) and $x = 3$ (that is to say, $p(3) = 3^2 - 5 \cdot 3 + 6 = 0$), the polynomial

$$q(x) = x^2 + x + 1$$

has no evident real roots. In fact it requires a larger number system—the complex numbers—in which to find roots of the polynomial q.

We shall say a few words here about some of the people who contributed to the development of the complex numbers.

Further Reading

R. B. McClenon, A contribution of Leibniz to the history of complex numbers, *The American Mathematical Monthly* 30(1923), 369-374.

Garret Sobczyk, The hyperbolic number plane, *The College Mathematics Journal* 26(1995), 268–280.

9.2.1 Cardano

We have treated the life of Girolamo Cardano (1501 C.E.–1576 C.E.) elsewhere in this book. Cardano did not understand the complex numbers very well. But he was the first to use them to solve polynomial equations. For example, Cardano would have understood what it meant to say that $1+i$ and $1-i$ are roots of the polynomial $p(x) = x^2-2x+2$. We should certainly mention that Rafael Bombelli (1526–1572) was one of the first to calculate effectively with the complex numbers.

9.2.2 Euler

Leonhard Euler (1707 C.E.–1783 C.E.) was born in Basel, but the family moved to Riehen when he was only one year old. Euler's father Paul attended lectures of Jacob Bernoulli and lived in Bernoulli's house when he was a student. He was good friends with Johann Bernoulli (Jacob's brother). But in fact the father became a Protestant minister. Paul's strong mathematical background served the young Euler in good stead, for he was able to provide some mathematical training for his young son. Euler led a rich and varied life that is chronicled in [FUE] and [THE].

Leonhard Euler was sent to the University of Basel in 1720, at the age of fourteen. His father expected him to enter the ministry. The level of education in Basel was very poor, and there was no mathematics. So Euler engaged in general studies. He did study mathematics on his own, and he took some private lessons. It was thus a matter of great good fortune that Johann Bernoulli discovered Euler's talents. Euler's remarks on the matter were

> ...I soon found an opportunity to be introduced to a famous professor Johann Bernoulli.... True, he was very busy and so refused flatly to give me private lessons; but he gave me much more valuable advice to start reading more difficult mathematical books on my own and to study them as diligently as I could; if I came across some obstacle or difficulty, I was given permission to visit him freely every Sunday afternoon and he kindly explained to me everything I could not understand...

In 1723 Euler wrote a thesis comparing the philosophical ideas of Descartes and Newton. He thus earned his Masters Degree. He then engaged in exclusive studies of theology. In fact Euler found the study of theology, Hebrew, and Greek to be unsatisfying. With Johann Bernoulli's help, he obtained his father's permission to switch his studies to mathematics. By 1726 Euler had completed his studies, and actually published a paper. His second article, in 1727, won second prize in a contest for new ideas in shipbuilding.

Euler needed an academic post, and the position of Nicolaus Bernoulli in St. Petersburg happened to open up at the time. Euler was lucky enough to secure the position, but he deferred his acceptance because he was also applying for a physics position in Basel. He failed to obtain the latter post, and so found himself in St. Petersburg in May, 1727. He joined the St. Petersburg Academy of Sciences two years after it was founded by Catherine I, wife of Peter the Great. Daniel Bernouli and Jakob Hermann arranged for Euler to be appointed to the mathematical-physical division of the Academy rather than to the physiology post that he originally had been offered. This certainly suited his talents nicely.

St. Petersburg offered Euler quite a number of brilliant and stimulating colleagues, including

- Jakob Hermann (geometry);
- Daniel Bernoulli (geometry, applied mathematics);
- Christian Goldbach (analysis, number theory);
- F. Maier (trigonometry);
- Joseph Nicholas Delisle (astronomy and geography).

Euler began his time in Russia by serving as a medical lieutenant in the Russian Navy. He actually assumed his Professorship in 1730. He was then a full member of the Academy, and was thus able to relinquish his Navy post.

Daniel Bernoulli held the Senior Chair in Mathematics in St. Petersburg. But he was unhappy in Russia and departed in 1733. At that time Euler assumed the Senior Chair. The resulting financial enhancement enabled Euler to marry Katharina Gsell, daughter of a painter from the St. Petersburg Gymnasium. Leonhard and Katharina had a total of 13 children, although only 5 of them survived infancy. Euler liked to observe that he made a number of his most notable discoveries while holding a baby in his arms with others playing at his feet.

Leonhard Euler began having health problems in 1735. He had a severe fever which threatened his life. In 1740 he lost an eye, possibly because of eyestrain brought on by cartographic work. Euler won the grand prize of the Paris Academy both in 1738 and 1740. As a result, his reputation was at the highest level in those days. At the same time, political conditions for foreigners in Russia were becoming quite uncomfortable. As a result Euler accepted a position at the Academy of Science in Berlin. In fact Euler was Director of Mathematics for the new Academy. In a letter to a friend, Euler indicated that the King was his special benefactor, and he had complete freedom to spend his professional time as he wished. He received salaries both from Russia and from Germany. He was able to spend some of his funds to help equip his former Academy in St. Petersburg.

Euler spent twenty-five years at the Berlin Academy. During that time he wrote 380 scientific papers and several books. Among these latter were:

- a book on the calculus of variations;
- a book on the calculation of planetary orbits;
- a book on artillery and ballistics;
- a book on analysis;
- a book on shipbuilding and navigation;
- a book on the motion of the moon;
- a book on differential calculus;
- a book containing a popular exposition of scientific ideas.

In 1759 Leonhard Euler became the President of the entire Berlin Academy. He served in that position for four years, but was rather alarmed when Frederick the Great then planned to appoint d'Alembert to that post (Euler and D'Alembert had had professional disagreements). d'Alambert declined the offer, but Euler decided that it was time for him to leave Berlin. In 1766, he returned to a position in St. Petersburg, much to the chagrin of Frederick. Soon after, Euler became almost entirely blind; also his house was destroyed by fire. He was able to save only himself and his mathematical manuscripts. In spite of the loss of his sight, Euler continued his work on optics, algebra, and lunar motion. Remarkably, he produced almost half of his total scientific output *after* going blind.

Of course, without sight, Euler required help in order to do his work. His son Johann Albrecht Euler was Chair of Physics at the Academy in St. Petersburg. Another son, Christoph Euler, had a military career. The other members of the academy, including W. L. Krafft, A. J. Lexell, and N. Fuss, were generous with their time and assistance. Fuss was in fact Euler's grandson-in-law; he became the great man's formal assistant in 1776. It should be stressed that Fuss's work was not merely clerical; he was in many ways a scientific consultant and collaborator.

Euler died of a brain hemorrhage on September 18, 1783. He had a full day of scientific activity, including vigorous discussions of the newly-discovered planet Uranus. But he was struck down, and lost consciousness, at 5:00 PM with the cry "I am dying." He expired at 11:00 P.M.

Leonhard Euler was one of the most prolific scientists of all time. The St. Petersburg academy continued to publish Euler's unpublished manuscripts for 50 years after he died. He had an impact on almost all parts of modern mathematics, and many parts of engineering, astronomy, and physics as well.

Of particular interest to us are Euler's contributions to complex analysis. He published his theory of logarithms of complex numbers in 1751. He investigated analytic functions of a complex variable in several different contexts, including the study of orthogonal trajectories and cartography. He discovered the important Cauchy-Riemann equations in 1777 (although it seems that he was anticipated by d'Alembert in 1752). It may be noted here that Euler's theory of the complex logarithm interacts very nicely with Argand's geometric theory of complex numbers (see below). As is often the case in mathematics, different streams of thought flow together and create new synergies.

9.2.3 Argand

Jean-Robert Argand (1768–1822) was an accountant and bookkeeper in Paris. His interest in mathematics was strictly as an amateur. Very little is known about his childhood or his education. His parents were named Jacques and Eves. The quiet life of Jean-Robert Argand is told in [FEH].

Argand had two children: a boy who lived his adult life in Paris and a girl who later married and moved to Stuttgart, Germany.

Argand is remembered for providing us with geometric interpretations of the complex numbers. As we shall see below, we can think of a complex number as an ordered pair of real numbers. Thus we can associate to the complex number a point in the plane. This is now known as the *Argand plane*, and the resulting picture is called an *Argand diagram*. Perhaps more interesting, and certainly more profound, is the fact that multiplication by the complex number i can be interpreted as rotation through an angle of $+90°$. We see this because if $z = x + iy$ then $iz = -y + ix$. Certainly the vector $\langle -y, x\rangle$ is perpendicular to $\langle x, y\rangle$. And a moment's thought shows that in fact iz is a $90°$ rotation of z in the *counterclockwise* direction (just try a specific example, such as $z = 1 + 1i$).

Since Argand was not a university scholar, he was not plugged into the regular academic system of publication and accreditation. In fact it is through an interesting sequence of accidents that we now associate Argand's name with this collection of ideas.

It is notable that the first publication of the geometric interpretation of the complex numbers was authored by Caspar Wessel. In fact Wessel notes the concept in a (unpublished) work of 1787 but it appeared in published form, under Wessel's byline, in a paper of 1797. That paper actually appeared in print in 1799. The paper received scant attention from the mathematical community.

Indeed it was not until 1895, when S. C. Juel drew attention to the work and Sophus Lie actually republished it that Wessel began to receive some credit for these ideas. Like Argand himself, Wessel was not an academic mathematician. He was in fact a surveyor. So it is not surprising that he was not a part of the flow of scholarly discourse.

Argand in fact published his own ideas in a small book—published at his own expense!—in 1806. Such publications generally are not widely noticed; by contrast, Wessel's work was published by the Royal Danish Academy. To make matters worse, Argand's name did not appear on his own book, so that even those few who noticed the work had no idea to whom to attribute it.

As luck would have it, a copy of Argand's work was sent to the noted mathematician Adrien-Marie Legendre. He, in turn, sent it to François Français. Still, neither man knew the identity of the author of this privately published volume. When François Français died, his brother Jacques came to be in charge of his papers. Finding Argand's book, Jacques took a great interest in geometric representations of the complex numbers. In 1813 he published a tract describing these ideas. He could easily have claimed them to be his own, but he did not. In fact he announced in the work that the ideas came from the work of an unknown mathematician and he asked that that mathematician come forward and claim credit.

Jacques Français's article appeared in Gergonne's journal *Annales de mathématiques*, and Argand read it. He responded, acknowledged that he was the author of the ideas, and submitted to that same journal a revised and updated version of his ideas. The upshot of these publishing events was a public row, and there is nothing like clamourous dissension to gain real publicity for a set of ideas.

For the mathematician F. J. Servois claimed that complex numbers should not be viewed geometrically—the only correct way, in his view, to think about the complex number system is algebraically. Argand and Français disagreed. In the end, the geometric viewpoint won out, and has proved to be a valuable source of ideas and powerful tools in modern mathematics.

Although Argand is certainly best known, and best remembered, for his contributions to the geometric theory of complex analysis, he in fact published a number of other works. He was the first to formulate a version of the fundamental theorem of algebra for general polynomials (i.e., polynomials with complex coefficients). His proof, although it contained a few defects, was very close to a modern proof of this important theorem. He published several papers amplifying his theory of the geometric interpretation of complex numbers, and he published several commentaries on the work of other mathematicians. His final publication, in 1816, was about combinatorics and counting.

Argand is remembered as a gifted amateur mathematician who made significant contributions that received only belated recognition. His work was significant and timely and has had lasting value.

9.2.4 Cauchy

Augustin Louis Cauchy (1789 C.E.–1857 C.E.) was born in Paris during a tumultuous period of French history. His father feared for the family's safety because of the political events connected with the French revolution, so he moved the family to Arcueil. In this new location life was hard. The family often did not have adequate food. Cauchy developed into a seminal mathematician; the story of his life is recounted in [BEL].

We shall treat the details of Cauchy's life in Chapter 10. Suffice it for now to say that Cauchy had a profound influence over the development of complex analysis. The Cauchy-Riemann equations, the Cauchy integral theorem, and the Cauchy integral formula are all named after him. These are among the most central and far-reaching ideas in the subject. In modern treatments, all the key ideas of complex analysis flow from the Cauchy integral formula.

Cauchy led a chaotic and unsatisfying personal life. But his influence over modern mathematics continues to be profound.

9.2.5 Riemann

Bernhard Riemann (1826 C.E.–1866 C.E.) was born into a poor family with a Lutheran minister father. He was tormented by disease and poverty all his life, and he died at the young age of forty. We treat his life in greater detail in Chapter 15.

Even so, Riemann achieved a great many mathematical triumphs during his short time on earth. He discovered the Cauchy-Riemann equations, created Riemann surfaces, and developed the Riemann mapping theorem. Much of the geometric theory of complex analysis is due to Riemann. The Riemann zeta function arises from considerations of complex analysis, but has become of seminal importance in number theory. The distribution of the zeros of the Riemann zeta function contains profound information about the distribution of the prime numbers. Riemannian geometry was created by Riemann as part of his oral examinations for Gauss; but in fact this idea of geometry now plays a major role in modern research on complex variables. It also had a profound influence on Einstein as he developed his general theory of relativity. We shall say more on this matter in Chapter 15.

Riemann certainly left his mark on complex analysis. But he also studied geometry, partial differential equations, calculus, Abelian functions, Fourier series, and many other parts of mathematics. His contributions are still the basis for much modern research.

Further Reading

E. T. Bell, Successive generalizations in the theory of numbers. *The American Mathematical Monthly* 34(1927), 55–75.

Yves Nievergelt, How (not) to solve quadratic equations, *The College Mathematics Journal* 34(2003), 90–104.

9.3 Complex Number Basics

The tale of the genesis of complex numbers is a long and elaborate one. In recent years entire volumes are being prepared to lay out this history. And the story has many heros, including Euler, Argand, De Moivre, Hamilton, Riemann, Cauchy, and Weierstrass. The idea of a number system that contains a square root for -1 was counterintuitive, and the mathematical sophistication needed to "just construct" a number system did not exist 300 years ago. It took some real courage to come up with these ideas. The presentation here will take advantage of modern ideas, and will be quite elegant and sophisticated.

We are already familiar with the real numbers $\mathbb{R}$. Just to review, these are all numbers given by decimal expansions. These include the whole numbers or integers (denoted by $\mathbb{Z}$), the fractions or rational numbers (denoted by $\mathbb{Q}$), and the irrational numbers. An integer has a decimal expansion with no nonzero digits to the right of the decimal point. Examples of integers are 2.0, -6.0, 15.0. A rational number has just finitely many nonzero digits to the right of the decimal point, or else finitely many digits that repeat infinitely often. Examples of rational numbers are

$$\begin{aligned}\frac{3}{4} &= 0.75\,,\\ \frac{9}{10} &= 0.9\,,\\ \frac{1}{3} &= 0.3333\overline{3}\ldots\,,\\ \frac{125}{999} &= .125125\overline{125}\ldots\,,\end{aligned}$$

where the overbar indicates that the designated string is repeated infinitely often.

An irrational number has a decimal expansion that goes on indefinitely and never repeats. These are the most subtle numbers in the system, and they are difficult to identify (although, in our chapter on Pythagoras, we learned for instance that $\sqrt{2}$ is irrational). The decimal expansion for $\sqrt{2}$ is

$$\sqrt{2} = 1.414213562\ldots\,,$$

where the dots indicate that the string of integers in the decimal expansion goes on indefinitely *but there is no repetition* (because $\sqrt{2}$ is irrational).

Another irrational number is π. Its decimal expansion is

$$\pi = 3.141592654\ldots\,,$$

where again the dots indicate that the decimal expansion goes on indefinitely *but there is no repetition* (because the number π is irrational).

All of these numbers taken together constitute the real number system $\mathbb{R}$. We typically *picture* the real number system as a number line (Figure 9.1).

Now we will begin our construction of the complex number system. We will create a new number system $\mathbb{C}$ consisting of all *ordered pairs* of real numbers. Thus an element

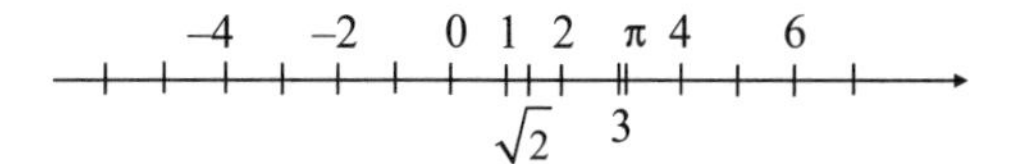

Figure 9.1. The real number line.

of $\mathbb{C}$ is a pair (x, y) of real numbers. As examples, $(3, -2)$, $(-6, 1.74)$, and $(\pi, -\sqrt{2})$ are complex numbers. Now, in order to have a useful system of numbers, we need to know the arithmetic operations on $\mathbb{C}$.

If (a, b) and (c, d) are complex numbers, then we define

$$(a, b) + (c, d) = (a + c, b + d).$$

As an example,

$$(-3, 6) + (2, 4) = (-3 + 2, 6 + 4) = (-1, 10).$$

We define subtraction similarly:

$$(a, b) - (c, d) = (a - c, b - d).$$

For instance,

$$(-3, 6) - (2, 4) = (-3 - 2, 6 - 4) = (-5, 2).$$

Observe that the number $0 = (0, 0)$ is the *additive identity* in our new number system. This means that if you add this 0 to any complex number then that number is reproduced:

$$(a, b) + (0, 0) = (a + 0, b + 0) = (a, b).$$

Now multiplication is more complicated. It would be a *mistake* to define

$$(a, b) \cdot (c, d) = (ac, bd). \tag{$*$}$$

Why is this a mistake? It seems so obvious that we should do multiplication like this. But with the definition $(*)$, it would hold that $(1, 0) \cdot (0, 1) = (0, 0)$. And we do not want the product of two nonzero numbers to be zero. So definition $(*)$ will not do.

Thus our definition of multiplication will be non-obvious. But, as you will see, it will get the job done in a very nice way. If (a, b) and (c, d) are complex numbers, then we set

$$(a, b) \cdot (c, d) = (ac - bd, ad + bc). \tag{$**$}$$

Let us look at a couple of examples to be sure we understand the idea. First,

$$(-3, 2) \cdot (4, 6) = ((-3) \cdot 4 - 2 \cdot 6, (-3) \cdot 6 + 2 \cdot 4) = (-24, -10).$$

Second,

$$(2, 8) \cdot (1, -9) = (2 \cdot 1 - 8 \cdot (-9), 2 \cdot (-9) + 8 \cdot 1) = (74, -10).$$

Now the justification for the rather exotic[1] definition in $(**)$ of multiplication is that it gives us the results that we want. First of all, we want to have a complex number that plays the role of "1". This is the *multiplicative identity*. If you multiply any number by 1, then you should get that number back again. The complex number that will play this role for us is $(1, 0)$. Let us see why. Let (a, b) be any other complex number. Then, according to $(**)$,

$$(1, 0) \cdot (a, b) = (1 \cdot a - 0 \cdot b, 1 \cdot b + 0 \cdot a) = (a, b).$$

[1] Certainly Leonhard Euler (1707 C.E.–1783 C.E.) knew how to multiply complex numbers. But it was William Rowan Hamilton (1805 C.E.–1865 C.E.) who came up with the algebraic formalism that we are using here.

So we see explicitly that multiplication by $1 = (1, 0)$ *reproduces* any complex number. For a specific example, take the complex number $(-3, 7)$. Then

$$(1, 0) \cdot (-3, 7) = (1 \cdot -3 - 0 \cdot 7, 1 \cdot 7 + 0 \cdot -3) = (-3, 7)\,.$$

You can check for yourself that $(-3, 7) \cdot (1, 0) = (-3, 7)$ as well.

It is also the case—and this is a very important property of the complex numbers—that any nonzero complex number has a multiplicative inverse. This means that, given a nonzero complex number (a, b), we can find another complex number whose product with (a, b) is the unit $(1, 0)$. In fact—and again you may find this a bit surprising—the multiplicative inverse of (a, b) is

$$\left(\frac{a}{a^2 + b^2}, \frac{-b}{a^2 + b^2}\right).$$

Let us test this out using rule $(**)$:

$$\begin{aligned}
&(a, b) \cdot \left(\frac{a}{a^2 + b^2}, \frac{-b}{a^2 + b^2}\right) \\
&= \left(a \cdot \frac{a}{a^2 + b^2} - b \cdot \frac{-b}{a^2 + b^2}, a \cdot \frac{-b}{a^2 + b^2} + b \cdot \frac{a}{a^2 + b^2}\right) \\
&= \left(\frac{a^2 + b^2}{a^2 + b^2}, 0\right) = (1, 0)\,.
\end{aligned}$$

For a concrete example of multiplicative inverse, consider the complex number $(-2, 1)$. According to what we have just said, its multiplicative inverse should be $(-2/5, -1/5)$. Let us test this assertion:

$$\begin{aligned}
(-2, 1) \cdot \left(\frac{-2}{5}, \frac{-1}{5}\right) &= \left(-2 \cdot \frac{-2}{5} - 1 \cdot \frac{-1}{5}, -2 \cdot \frac{-1}{5} + 1 \cdot \frac{-2}{5}\right) \\
&= \left(\frac{5}{5}, 0\right) = (1, 0)\,.
\end{aligned}$$

Finally, we can divide complex numbers. Let us divide (a, b) by (c, d):

$$\frac{(a, b)}{(c, d)} = (a, b) \cdot \frac{1}{(c, d)}\,.$$

Of course we know what the multiplicative inverse of (c, d) is, and we use that information now. Thus

$$\begin{aligned}
\frac{(a, b)}{(c, d)} &= (a, b) \cdot \frac{1}{(c, d)} \\
&= (a, b) \cdot \left(\frac{c}{c^2 + d^2}, \frac{-d}{c^2 + d^2}\right) \\
&= \left(a \cdot \frac{c}{c^2 + d^2} - b \cdot \frac{-d}{c^2 + d^2}, a \cdot \frac{-d}{c^2 + d^2} + b \cdot \frac{c}{c^2 + d^2}\right) \\
&= \left(\frac{ac + bd}{c^2 + d^2}, \frac{-ad + bc}{c^2 + d^2}\right).
\end{aligned}$$

It always helps to look at a concrete example: The quotient

$$\begin{aligned}\frac{(2,-5)}{(1,6)} &= (2,-5)\cdot\frac{1}{(1,6)}\\ &= (2,-5)\cdot\left(\frac{1}{37},\frac{-6}{37}\right)\\ &= \left(2\cdot\frac{1}{37}-(-5)\cdot\frac{-6}{37}, 2\cdot\frac{-6}{37}+(-5)\cdot\frac{1}{37}\right)\\ &= \left(\frac{-28}{37},\frac{-17}{37}\right).\end{aligned}$$

And the wonderful thing about mathematics is that we can check our work: If

$$\frac{(2,-5)}{(1,6)} = \left(\frac{-28}{37},\frac{-17}{37}\right)$$

then it should be the case that

$$(1,6)\cdot\left(\frac{-28}{37},\frac{-17}{37}\right) = (2,-5).$$

Let us try it and see:

$$\begin{aligned}(1,6)\cdot\left(\frac{-28}{37},\frac{-17}{37}\right) &= \left(1\cdot\frac{-28}{37}-6\cdot\frac{-17}{37}, 1\cdot\frac{-17}{37}+6\cdot\frac{-28}{37}\right)\\ &= \left(\frac{-28+102}{37},\frac{-185}{37}\right) = (2,-5).\end{aligned}$$

So now we have a working number system: We can add, subtract, multiply, and divide. In the next section we shall return to Gauss's Fundamental Theorem of Algebra, and get an idea of why it is actually true that every nonconstant polynomial has a root.

Before we turn to that task, let us note a special feature—for some purposes the most important feature—of the complex number system. *In the complex numbers, the number -1 has a square root.* Now our number 1, the unity, is $(1,0)$. And therefore -1 is $(-1,0)$. Its square root is in fact $(0, 1)$. Let us verify this claim:

$$(0,1)\cdot(0,1) = (0\cdot 0 - 1\cdot 1, 0\cdot 1 + 1\cdot 0) = (-1,0),$$

as claimed.

In practice, when we use the complex numbers, we use the simpler symbol 1 to denote $(1,0)$ and the special symbol i to denote $(0, 1)$. This means that we can write any complex number (a,b) as

$$(a,b) = a\cdot(1,0) + b\cdot(0,1) = a\cdot 1 + b\cdot i = a + bi.$$

If you consult mathematics books[2] about the complex numbers, this is how you will find them written. Observe that, in this new notation, $i\cdot i = -1$.

[2]It is worth noting that electrical engineers use the letter j to denote the square root of -1. This rarely gives rise to any confusion, since one can tell from context whether one is dealing with mathematicians or engineers.

Just for practice, let us add two complex numbers using our new notation:

$$(3 - 9i) + (4 + 6i) = (3 + 4) + i((-9) + 6) = 7 - 3i\,.$$

Now let us multiply two complex numbers using the new notation:

$$\begin{aligned}(3 - 9i) \cdot (4 + 6i) &= 3 \cdot 4 + 3 \cdot 6i - 9i \cdot 4 - 9i \cdot 6i \\ &= 12 + 18i - 36i - 54i^2 \\ &= 12 - 18i - 54 \cdot (-1) \\ &= 66 - 18i\,.\end{aligned}$$

For You to Try Verify that the complex numbers $7 - i$ and $7 + i$ are both roots of the polynomial equation $p(x) = x^3 - 15x^2 + 64x - 50$.

For You to Try Calculate the reciprocal of the complex number $z = 4 - 5i$ using the $x + iy$ notation for complex numbers.

Further Reading

D. H. Potts, Axioms for complex numbers, *The American Mathematical Monthly* 70(1963), 974–975.

Richard C. Weimer, Can the complex numbers be ordered?, *The Two-Year College Mathematics Journal* 7(1976), 10–12.

9.4 The Fundamental Theorem of Algebra

Gauss's theorem (the celebrated *Fundamental Theorem of Algebra*) is that *any polynomial* with real coefficients

$$p(z) = a_0 + a_1 z + a_2 z^2 + \cdots + a_{k-1} z^{k-1} + a_k z^k\,,$$

$a_j \in \mathbb{R}$, of degree at least 1 has a complex root. To get an idea of why this is true, let us return to our original definition of complex numbers in the last section: ordered pairs (x, y) of real numbers. It is natural to picture the complex numbers in a plane. [This is called, for historical reasons, an *Argand diagram*.] See Figure 9.2. Now we will think of the polynomial function p as mapping the complex plane to itself: plug in a complex number

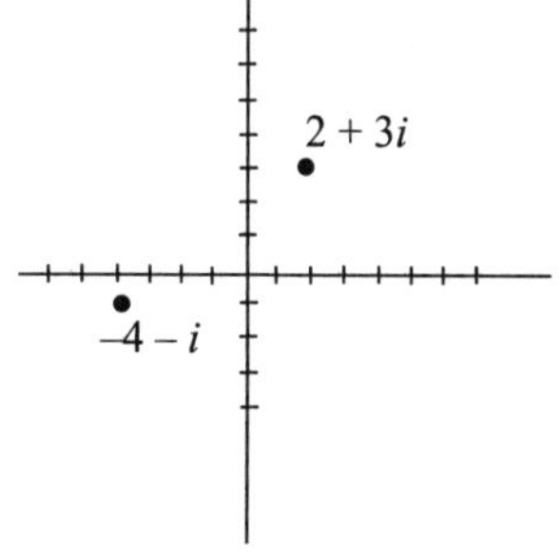

Figure 9.2. An Argand diagram.

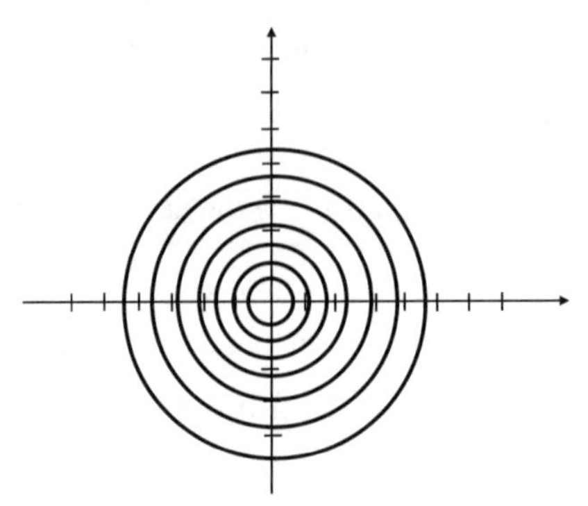

Figure 9.3. The complex plane as a union of circles.

z and $p(z)$ is a new complex number. We may suppose that the constant term a_0 of the polynomial is nonzero—for otherwise 0 itself would be a root of the polynomial and there would be nothing more to prove.

Now we think of the complex plane as a union of circles centered at the origin—Figure 9.3. If we consider the action of p on a *very large* circle—with some huge radius R— then of course the term of the polynomial that is most significant is the top degree term $a_k z^k$. This point perhaps merits a moment's thought. For specificity, look at the particular polynomial

$$p(z) = z^3 - 4z + 5 .$$

If we plug a number like 100 into this polynomial, then the lead term is 10^6. The other two terms—-400 and 5—are miniscule by comparison. The same thing happens if we plug in a large complex number like $50 + 50i$. Then the lead term is $-25000 + 25000i$. The other two terms—$-200 - 200i$ and 5—are miniscule by comparison.

Thus the "lead" term dwarfs all the other terms when it is applied to a complex number on that circle of the huge radius R (Figure 9.4). And what it does to that huge circle is it wraps it around itself k times. The *image* of the circle of radius R under that top-order monomial is another huge circle of radius $|a_k|R^k$. See Figure 9.5.

Let us instead now consider the action of the polynomial p on a very tiny circle, with some very small (much less than 1) radius r, centered at the origin (Figure 9.6). Now larger

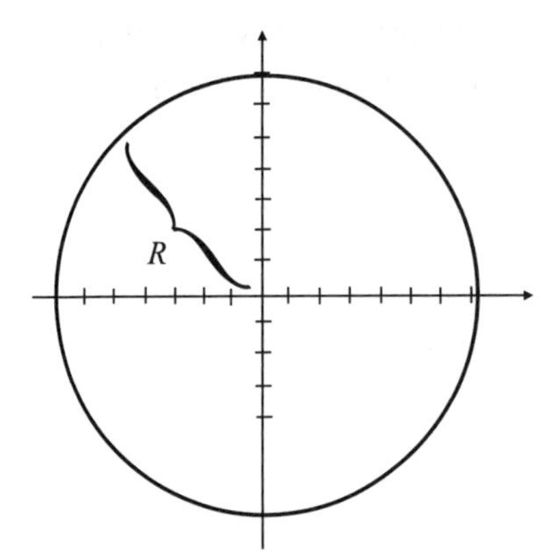

Figure 9.4. The image of a large circle.

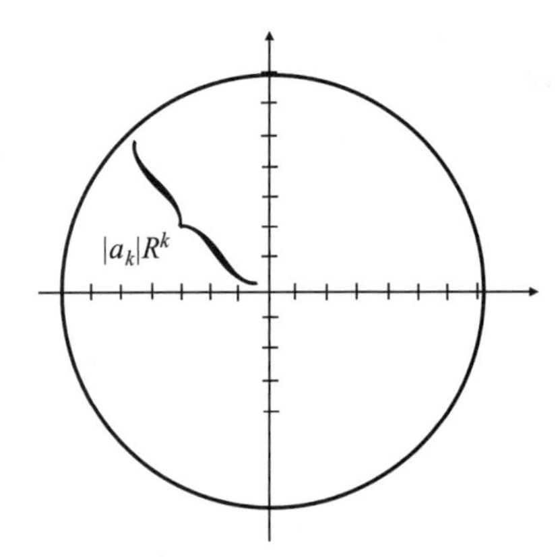

Figure 9.5. More on the large circle.

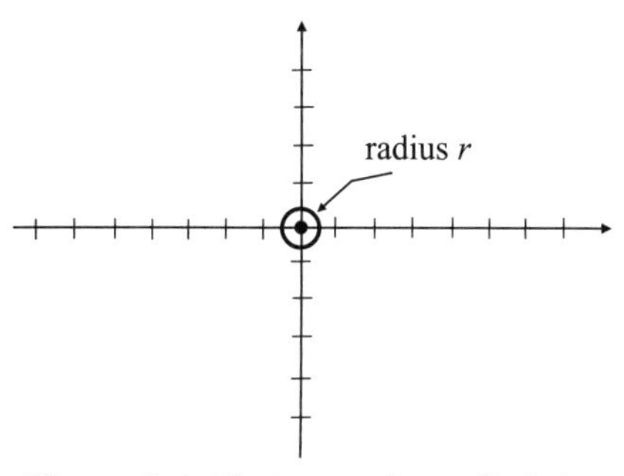

Figure 9.6. The image of a small circle.

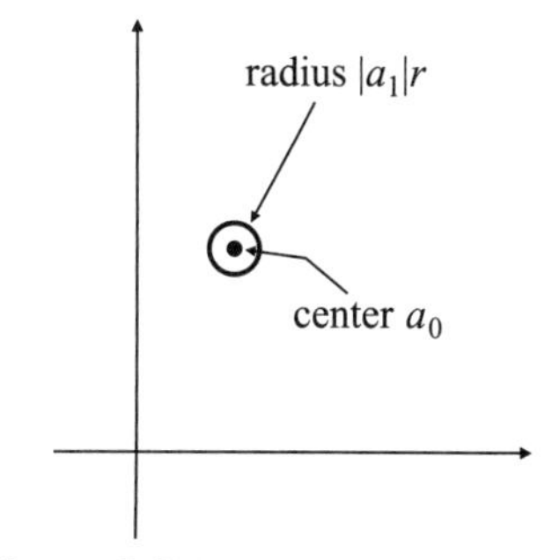

Figure 9.7. More on the small circle.

powers of the variable, lying on this little circle, will make it even smaller. Again let us look at the specific polynomial

$$p(z) = z^3 - 4z + 5 .$$

If we plug in the number $z = 0.01$ then the lead term yields 10^{-6}. By comparison, the constant term 5 (by comparison) is quite large. Likewise, if we plug in a small complex number like $0.01 + 0.01i$, then the lead term is $-0.00002 + 0.00002i$ while the constant term 5 (by comparison) is quite large.

So the most significant terms in the action of p on this very small circle centered at the origin are the zero-degree term a_0 and the first-degree term $a_1 z$. The image of the little circle centered at the origin under $a_0 + a_1 z$ is another circle of radius $|a_1|r$—*centered at* a_0. See Figure 9.7.

Now the important thing to notice, as you examine Figures 9.4, 9.5, 9.6, and 9.7, is that the one circle of radius R is mapped to a circle that surrounds the origin, and the other circle of radius r is mapped to a circle that does *not* surround the origin. All the circles in between—as the radius ranges from R to r—will have images that vary continuously between the two images that we have just described. See Figure 9.8. And therefore, as the figure illustrates, one of those images *will have to pass through the origin*. But *that means that the polynomial p maps some point to the origin*. Which means that there is a root.

The last paragraph is the nub of the argument, and bears repeating. As circles in the domain of the polynomial vary continuously, so will their images vary continuously. We have identified some circles in the domain whose images surround the origin. And we

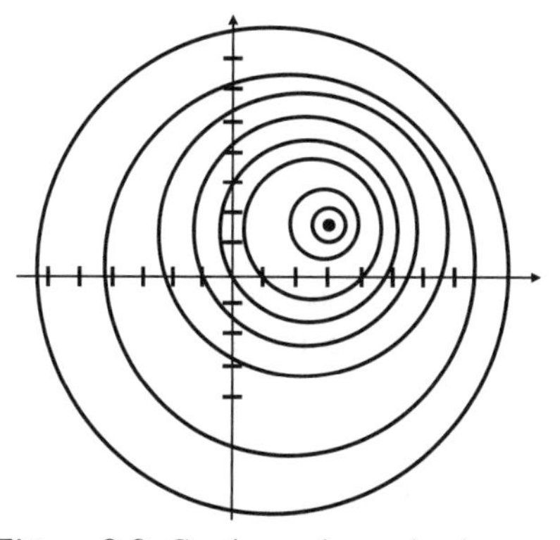

Figure 9.8. Continuously varying images.

have identified some other circles whose images do not surround the origin. By continuous variation of the images, some image of some circle must *pass through the origin*. But that means that the polynomial takes the value 0. We have thus found a root.

It should be stressed that the argument we have just presented is rather abstract. It demonstrates the *existence* of a root, but gives no hint as to how to *find* that root. We shall discuss the latter issue below.

EXAMPLE 9.1. Find all roots of the polynomial $z^2 + 2z + 5$.

Solution. We use the quadratic formula with $a = 1, b = 2$, and $c = 5$. The result is

$$z = \frac{-2 \pm \sqrt{2^2 - 4 \cdot 1 \cdot 5}}{2 \cdot 1} = \frac{-2 \pm \sqrt{-16}}{2} = \frac{-2 \pm 4i}{2} = -1 \pm 2i\,.$$

Thus the roots of the polynomial are $-1 + 2i$ and $-1 - 2i$.

For You to Try Find all roots of the polynomial $z^2 + 6z + 10$.

Now let us return to our general discussion of the fundamental theorem of algebra. We have shown that the polynomial

$$p(z) = a_0 + a_1 z + a_2 z^2 + \cdots + a_{k-1} z^{k-1} + a_k z^k$$

of degree $k \geq 1$ must have a root, even though we cannot say how to find it. Let us call this root r_1. Now we divide the polynomial p by $(z - r_1)$. Of course there will be a quotient q_1, and there will be a remainder. And the remainder will have to be of degree lower than the degree of $(z - r_1)$, otherwise we could keep on dividing. In other words, the remainder is some constant c_1. Thus

$$p(z) = (z - r_1) \cdot q_1(z) + c_1\,.$$

Now let $z = r_1$. Thus

$$0 = p(r_1) = (r_1 - r_1) \cdot q_1(r_1) + c_1$$

or

$$0 = c_1\,.$$

We find that the remainder (the constant) c_1 is 0. Thus

$$p(z) = (z - r_1) \cdot q_1(z)\,.$$

In other words, $(z - r_1)$ divides evenly into p.

Now if the quotient q_1 is not constant (i.e., is a polynomial of degree at least 1), then we may apply the Fundamental Theorem of Algebra to q_1. And find a root r_2. And, by the reasoning we just used, $(z - r_2)$ will evenly divide q_1 with some quotient q_2. Thus we may now write

$$p(z) = (z - r_1) \cdot (z - r_2) \cdot q_2(z)\,.$$

We may continue this reasoning, producing additional roots, until the degree of the polynomial is exhausted (i.e., the degree of the quotient q_k is zero, so q_k is a constant a_k). The result is

$$p(z) = a_k (z - r_1) \cdot (z - r_2) \cdots (z - r_k)\,,$$

where a_k is the leading coefficient of the polynomial p. This calculation produces such an important result that we now enunciate it formally:

Theorem 9.1. *Let*

$$p(z) = a_0 + a_1 z + a_2 z^2 + \cdots + a_{k-1} z^{k-1} + a_k z^k$$

be a polynomial. Then there are k complex roots $r_1, r_2, \ldots, r_k$ of this polynomial. [Some of the roots may be repeated.] The polynomial may be factored in terms of these roots as:

$$p(z) = a_k (z - r_1) \cdot (z - r_2) \cdots (z - r_k) .$$

We call this the factorization of p into linear factors.

This is the standard factorization of a polynomial. Note that the roots r_j are, in general, complex. And they may not be distinct. For example, the polynomial $p(z) = z^3 + 3z^2 - 4$ factors as $p(z) = (z - 1)(z + 2)(z + 2)$.

Further Reading

Sharon Barrs, James Braselton, and Lorraine Braselton, A rational root theorem for imaginary roots, *The College Mathematics Journal* 34 (2003), 380–382.

S. Stein, The fundamental theorem of algebra, *The American Mathematical Monthly* 61 (1954), 109.

9.5 Finding the Roots of a Polynomial

Once the degree of a polynomial exceeds 4, it is quite difficult to find the roots. There are many algebra and computer techniques for finding (or at least approximately finding) these roots. In the present section we will present one elegant little trick which is often useful.

Suppose that p is a polynomial with integer coefficients. And suppose that we want to see whether p has any roots that are rational numbers. Let us guess a solution of the form $r = c/d$. We may as well suppose that the fraction c/d is reduced to lowest terms, so that c and d have no common prime factors. In order to keep things simple, let us suppose that our polynomial has degree 2 (the same reasoning will apply to a polynomial of any degree). So the polynomial has the form

$$p(z) = \alpha z^2 + \beta z + \gamma , \qquad (\star)$$

where α, β, and γ are integers. Let us substitute our guess into equation $(\star)$. Thus

$$0 = p(r) = \alpha \left(\frac{c}{d}\right)^2 + \beta \left(\frac{c}{d}\right) + \gamma .$$

Multiplying through by d^2 yields

$$0 = \alpha c^2 + \beta c d + \gamma d^2 .$$

Let us rearrange this equation as

$$\gamma d^2 = -\alpha c^2 - \beta c d = -c(\alpha c + \beta d) .$$

We notice that c divides the right-hand side, so c must divide the left-hand side. But c certainly does not divide d. So c must divide γ.

Reasoning similarly, let us now rearrange the equation as

$$\alpha c^2 = -\gamma d^2 - \beta c d = -d(\gamma d + \beta c)\,.$$

We notice that d divides the right-hand side, so d must divide the left-hand side. But d certainly does not divide c. So d must divide α.

We have discovered this algorithm (we only reasoned for second-degree polynomials, but the argument works for polynomials of any degree):

If

$$p(z) = a_0 + a_1 z + a_2 z^2 + \cdots + a_{k-1} z^{k-1} + a_k z^k$$

is a polynomial with integer coefficients, then any rational root $r = c/d$ of p, reduced to lowest terms, must satisfy

- The numerator c divides a_0 evenly.
- The denominator d divides a_k evenly.

EXAMPLE 9.2. Find the rational roots of

$$p(z) = 2 - 7z + 6z^2\,.$$

Solution. The factors of the constant term are $\pm 1, \pm 2$. The factors of the top-order (second-degree) terms are $\pm 1, \pm 2, \pm 3$. Therefore the possible rational roots of p are

$$\pm\frac{1}{1}, \pm\frac{1}{2}, \pm\frac{1}{3}, \pm\frac{2}{1}, \pm\frac{2}{2}, \pm\frac{2}{3}.$$

It is a simple matter to plug these values into the polynomial and see whether any of them gives the value 0. We find that $1/2$ and $2/3$ are roots of the polynomial p.

For You to Try Use the method just developed to find all the rational roots of the polynomial

$$p(z) = 1 - 6z - 24z^2 + 64z^3\,.$$

Further Reading

John W. Pratt, Finding how many roots a polynomial has in $(0, 1)$ or $(0, \infty)$, *The American Mathematical Monthly* 86(1979), 630–637.

Don Redmond, Finding rational roots of polynomials, *The College Mathematics Journal* 20(1989), 139–141.

Exercises

1. Find all roots (three of them!) of the polynomial $p(z) = z^3 - 5z - 2z^2 + 6$.

2. Find all roots (three of them!) of the polynomial $q(z) = z^3 - 3z^2 + z - 3$. [*Hint:* Some of the roots will be complex numbers.

3. Consider the polynomial

$$q(z) = z^3 - 6z^2 + 13z - 20\,.$$

Given that $1 - 2i$ is a root (verify this claim!), find all three roots of q. Notice that two of them are complex. Further notice that one of these complex roots is the conjugate of the other in the sense that one has the form $a + ib$ with $a, b \in \mathbb{R}$ while the other has the form $a - ib$.

4. Refer to Exercise 3. Suppose that

$$p(z) = a_0 + a_1 z + a_2 z^2 + \cdots + a_k z^k$$

is a polynomial with all the coefficients a_j real numbers. Assume that the complex number $\alpha = a + ib$ is a root of p. Then verify that $\overline{\alpha} = a - ib$ will also be a root of p. We call $\overline{\alpha}$ the *conjugate* of α.

5. Let $\beta = 1 + i$. Let $\beta' = \beta^2$ and $\beta'' = \beta^3$. Find a polynomial p of degree 3 such that β, β', β'' are all roots of p.

6. The polynomial $q(z) = z^5 - z^4 - 13z^3 + 23z^2 - 14z + 24$ factors as a cubic polynomial times a quadratic polynomial. Find that factorization. Now find all the roots of q that you can. Discuss this problem in class.

7. The roots of the polynomial $z^3 - 1$ will be the "cube roots of unity". Find those three numbers, and verify directly that the cube of each one is equal to 1.

8. Find all rational roots of the polynomial

$$p(z) = 32z^3 + 20z^2 - 11z + 1\,.$$

9. Write the complex number $1/[3 - 7i]$ in the form $a + ib$.

10. Write the complex number

$$\alpha = \frac{3 - 2i}{4 + 7i}$$

in the form $a + ib$.

11. Find all the (complex) roots of the polynomial

$$p(z) = z^3 + (1 - i)z^2 + (2i - 1)z + (i + 3)\,.$$

[*Hint:* One of the roots is $z = i$.]

12. The polynomial

$$q(z) = z^3 + (1 - 3i)z^2 + (-8 - 2i)z + (2 + 6i)$$

has the complex number $z = i$ as a double root. Explain what this statement means, and verify that it is true.

13. What are all the complex roots of the polynomial equation $z^5 + 1$?

14. Give a justification for the statement

> If α is a complex number, then there is another complex number β such that $\beta^2 = \alpha$.

Can you suggest a method for finding β once α is given? Discuss this problem in class. Can you enlist the computer to help? Is β unique? How many such β are there?

15. **Project:** Find all cube roots of the complex number $\alpha = 1 + i$. Sketch these three numbers on an Argand diagram. The books [GRK], [KRA4] give background on this idea.

16. **Project:** Let p be a polynomial of degree k. Let $\alpha_1, \alpha_2, \ldots, \alpha_k$ be the complex roots of p (counting multiplicities). Let Γ be the smallest convex polygon in the complex plane that contains these roots. Then Γ will contain the roots of the derivative p'. Explain why this is so. See [GRK] for the provenance of this result.

17. **Project:** Suppose that a polynomial p has all its roots on the real line. Then what can you say about the coefficients of this polynomial? See [GRK].

10

Carl Friedrich Gauss: The Prince of Mathematics

10.1 Gauss the Man

Carl Friedrich Gauss (1777–1855) is widely held to have been the greatest mathematician who ever lived. Often called "the prince of mathematics," he—along with Archimedes and Isaac Newton—played a dominant role in the development of the subject. Gauss's genius was recognized quite early in his life. Gauss exercised a phenomenal influence on mathematics throughout his entire life, and he continues to do so today. A detailed account of his life appears in [BUH] and [DUN].

According to legend, when Gauss was 7 years old his teacher Büttner assigned the class to sum up the integers from 1 to 100. Thinking this would take the kids all afternoon, the teacher applied himself to other tasks. But young Gauss realized that one could write these 100 numbers as

$$\begin{array}{ccccccccccccccc} S & = & 1 & + & 2 & + & 3 & + & \cdots & + & 98 & + & 99 & + & 100\,, \\ S & = & 100 & + & 99 & + & 98 & + & \cdots & + & 3 & + & 2 & + & 1\,. \end{array}$$

Here we are calling the desired sum S. Now each column sums to 101, and there are 100 columns. So we see that

$$2S = 100 \cdot 101 = 10100\,,$$

hence

$$S = 5050\,.$$

This is quite a sophisticated argument for a 7-year-old, and attracted quite a lot of attention.

In 1788 Gauss began his education at the Gymnasium, where he learned High German and Latin. After receiving a stipend from the Duke of Brunswick-Wolfenbüttel, Gauss entered Brunswick Collegium Carolinum in 1792. At the academy Gauss independently discovered Bode's law, the binomial theorem, and the arithmetic-geometric mean, as well as the law of quadratic reciprocity and the prime number theorem. It may be noted that Gauss elicited the prime number theorem by staring for many hours at tables of prime numbers. After considerable thought, he came to the conclusion that, if $\pi(n)$ denotes the number of primes less than or equal to n, then

$$\pi(n) \approx \frac{n}{\ln n}.$$

Gauss was never able to prove this result, but the idea fascinated him all his life. It was not until 1896 that Hadamard and de la Vallée Poussin independently proved the sharp result that

$$\lim_{n\to\infty} \frac{\pi(n)}{n/\ln n} = 1.$$

Quadratic reciprocity is also a very profound result in number theory. It considers the solvability of certain quadratic diophantine equations modulo various primes. Gauss referred to this theorem as "the golden theorem", and he produced eight different proofs (six that he published, and two that were found posthumously) of it during his lifetime. We shall say more about quadratic reciprocity in the discussion that follows.

In 1795 Gauss left Brunswick to study at the University of Göttingen. Gauss's teacher there was Kästner. Gauss left Göttingen in 1798 without a diploma, but by this time he had made one of his most important discoveries—the construction of a regular 17-gon by ruler and compass.

Gauss returned to Brunswick where he received a degree in 1799. After the Duke of Brunswick had agreed to continue Gauss's stipend, he requested that Gauss submit a doctoral dissertation to the University of Helmstedt. He already knew Pfaff, who was chosen to be his advisor. Gauss's dissertation was a treatment of the fundamental theorem of algebra. This is the result that any nonconstant polynomial has a complex root. Gauss found four different proofs of the fundamental theorem in his lifetime—the first ever in his doctoral dissertation. We provide a discussion and proof of this theorem in Section 9.4.

It should be understood that much of the mathematics of the preceding 350 years had concentrated on solving polynomial equations. But the methods (such as we discuss for the Arabs in Chapter 4) concerned concrete methods for finding specific solutions of specific equations. Gauss's treatment of the fundamental theorem of algebra was profoundly original, for it treated *all* polynomials in an abstract manner. It did *not* show how to find the roots of a given polynomial; it merely showed that those roots exist.

Since Gauss had a stipend from the Duke, he did not need to find a job and he did not need to spend his time teaching. He was able to concentrate all his attentions on research, and that he did. In 1801 he published *Disquisitiones Arithmeticae*, certainly one of the most important and profound works on number theory that has ever been written. Even to this day people study Gauss's book, which contained other results as well—such as his construction of the 17-gon.

In 1801 Gauss developed an interest in astronomy, and produced results about the orbital positions of Ceres. For these data, Gauss developed and used the method of least squares—an important statistical technique that bears Gauss's name today. Gauss also studied the orbital positions of Pallas. Gauss married Johanna Ostoff on 9 October, 1805. In 1807 Gauss left Brunswick to take up the position of director of the Göttingen observatory. This astronomical work was to take much of Gauss's time for the remainder of his life.

Gauss arrived in Göttingen in late 1807. In 1808 his father died, and a year later Gauss's wife Johanna died after giving birth to their second son, who was to die soon after her.

Gauss was married for a second time the next year, to Minna the best friend of Johanna, and they had three children.

In 1809 Gauss published his second book, *Theoria Motus Corporum Coelestium in Sectionibus Conicis Solem Ambientium*. This was a two-volume treatise on the motions of the planets. Gauss's contributions to theoretical astronomy stopped after 1817, although he went on making observations until the age of 70.

Gauss spent most of his time during this period developing a new observatory, completed in 1816, but he still found the time to work on other projects. His publications during this time include *Disquisitiones Generales Circa Seriem Infinitam*, a rigorous treatment of series and an introduction to hypergeometric functions, *Methodus Nova Integralium Valores per Approximationem Inveniendi*, an essay on approximate integration, *Bestimmung der Genauigkeit der Beobachtungen*, a treatment of statistical estimators, and *Theoria Attractionis Corporum Sphaeroidicorum Ellipticorum Homogeneorum Methodus Nova Tractata*. The latter work was inspired by geodesic problems (i.e., mapping the surface of the earth). In fact, Gauss found himself more and more interested in geodesy in the 1820s. In 1818 Gauss carried out a geodesic survey of the state of Hanover.

In 1822 Gauss won the Copenhagen University Prize with *Theoria Attractionis Corporum Sphaeroidicorum Ellipticorum Homogeneorum Methodus Nova Tractata* together with the idea of mapping one surface onto another so that the two are similar in their smallest parts. This paper was published in 1825 and led to the much later publication of Untersuchungen über Gegenstände der Höheren Geodäsie (1843 and 1846).

From the early 1800s Gauss had an interest in the question of the possible existence of a non-Euclidean geometry. He discussed this topic at length with Farkas Bolyai. In a book review in 1816 he discussed proofs which deduced the axiom of parallels from the other Euclidean axioms, suggesting that he believed in the existence of non-Euclidean geometry, although he was rather vague. Gauss confided in Schumacher, telling him that he believed his reputation would suffer if he admitted in public that he believed in the existence of such a geometry.

In 1831 Farkas Bolyai sent to Gauss his son János Bolyai's work on the subject. Gauss replied

> If I begin with the statement that I dare not praise such a work, you will of course be startled for a moment: but I cannot do otherwise; to praise it would amount to praising myself; for the entire content of the work, the path which your son has taken, the results to which he is led, coincide almost exactly with my own meditations which have occupied my mind for from thirty to thirty-five years. On this account I find myself surprised to the extreme.
>
> My intention was, in regard to my own work, of which very little up to the present has been published, not to allow it to become known during my lifetime. Most people have not the insight to understand our conclusions and I have encountered only a few who received with any particular interest what I communicated to them. In order to understand these things, one must first have a keen perception of what is needed, and upon this point the majority are quite confused. On the other hand, it was my plan to put all down on paper eventually, so that at least it would not finally perish with me.
>
> So I am greatly surprised to be spared this effort, and am overjoyed that it happens to be the son of my old friend who outstrips me in such a remarkable way.

This seemingly well-intentioned statement from Gauss had a devastating effect on young Bolyai, and the novice was tormented by it for the rest of his life. Nonetheless, Bolyai and Lobachevsky are credited today with the creation of non-Euclidean geometry.

Gauss had a major interest in differential geometry, and published many papers on the subject. *Disquisitiones Generales Circa Superficies Curva* (1828) was his most renowned work in this field. In fact, this paper rose from his interest in geodesy, but it also contained such fundamental geometrical ideas as Gaussian curvature.

The period 1817–1832 was a particularly distressing time for Gauss. He took in his sick mother in 1817, who stayed until her death in 1839, while he was arguing with his wife and her family about whether they should go to Berlin. He had been offered a position at Berlin University and Minna and her family were keen to move there. Gauss, however, never liked change and decided to stay in Göttingen. In 1831 Gauss's second wife died after a long illness.

In 1831, Wilhelm Weber arrived as Professor of Physics in Göttingen. Gauss had worked on physics before 1831, publishing *Über ein Neues Allgemeines Grundgesetz der Mechanik*, which contained the principle of least constraint, and *Principia Generalia Theoriae Figurae Fluidorum in Statu Aequilibrii*, which discussed forces of attraction. These papers were based on Gauss's potential theory, which proved of great importance in his work on physics.

In 1832, Gauss and Weber began investigating the theory of terrestrial magnetism after Alexander von Humboldt attempted to obtain Gauss's assistance in making a grid of magnetic observation points around the Earth. Gauss was excited by this prospect and by 1840 he had written three important papers on the subject. These papers all dealt with the current theories on terrestrial magnetism, including Poisson's ideas, absolute measure for magnetic force, and an empirical definition of terrestrial magnetism.

Gauss and Weber achieved much in their six years together. They discovered Kirchhoff's laws, as well as building a primitive telegraph device which could send messages over a distance of 5,000 feet (this anticipated Samuel Morse by a year or two). This occupation produced many concrete results. The Magnetischer Verein and its journal were founded, and the atlas of geomagnetism was published, while Gauss and Weber's own journal in which their results were published ran from 1836 to 1841.

In 1837, Weber was forced to leave Göttingen when he became involved in a political dispute. From this time, Gauss's activity gradually decreased. He still produced letters in response to fellow scientists' discoveries—usually remarking that he had known the methods for years but had never felt the need to publish.

Gauss spent the years from 1845 to 1851 updating the Göttingen University widow's fund. This work gave him experience in financial matters, which he later exploited to make his fortune through shrewd investments in bonds.

Two of Gauss's last doctoral students were Moritz Cantor and Richard Dedekind. Both of them, especially Dedekind, went on to become distinguished mathematicians.

Gauss presented his golden jubilee lecture in 1849, fifty years after his diploma had been granted by Helmstedt University. It was appropriately a variation on his dissertation of 1799.

From 1850 onwards Gauss's work was again nearly all of a practical nature although he did approve Riemann's probationary lecture and doctoral thesis (in which Riemann invented what we now call Riemannian geometry—see Chapter 15). His last known scientific exchange was with Gerling. He discussed a modified Foucault pendulum in 1854. He was also able to attend the opening of the new railway link between Hanover and Göttingen, but this proved to be his last outing. His health deteriorated slowly, and Gauss died in his sleep early in the morning of 23 February, 1855.

Further Reading

William Dunham, Euler and the fundamental theorem of algebra, *The College Mathematics Journal* 22(1991), 282–293.

George Bruce Halsted, Gauss and the non-Euclidean geometry, *The American Mathematical Monthly* 7(1900), 247–252.

10.2 The Binomial Theorem

One of Gauss's earliest interests in mathematics—even when he was a teenager—was in the binomial theorem. This result is part of the bedrock of the subject, and it comes up in all aspects of analysis and algebra.

In our discussion of this result, we shall use some important notation that comes up frequently in counting arguments. We will revisit these ideas in Section 14.3. The fundamental question here is how to choose k objects from among n, $0 \le k \le n$.

10.2.1 A Few Words About Permutations

Suppose that we are given k objects. In how many different orders may we put this collection of objects?

Take the simple instance of $k = 3$. Call the objects A, B, C. Then the possible orders for these objects are

$$
\begin{array}{ccc}
A & B & C \\
A & C & B \\
B & A & C \\
B & C & A \\
C & A & B \\
C & B & A
\end{array}
$$

There are no other possibilities. Just six orderings.

For a general positive integer k, we can reason as follows:

- In choosing an object for the first position, there are k choices.
- In choosing an object for the second position, there are $k - 1$ choices (as we have already selected on object for the first position).
- In choosing an object for the third position, there are $k - 2$ choices (as we have already selected two objects for the first two positions).
- ...and so forth.

Thus we see that the number of possible orderings is

$$k \cdot (k-1) \cdot (k-2) \cdot \cdots \cdot 3 \cdot 2 \cdot 1 \,.$$

This expression is so common in mathematics that we give it the name "k factorial", and we write it $k!$. Thus

$$k! = k \cdot (k-1) \cdot (k-2) \cdot \cdots \cdot 3 \cdot 2 \cdot 1 \,.$$

We have determined that there are $k!$ different orderings, or *permutations*, of k objects.

For You to Try You have seven marbles, two of them black and five of them white. How many different orderings are there of these marbles so that the first two are black?

10.2.2 Choosing and the Binomial Coefficients

Suppose that you have n objects and you are going to choose k of them (where $0 \le k \le n$). In how many different ways can you do this?

Just to illustrate the idea, take $n = 3$ and $k = 2$—see Figure 10.1. For convenience we have labeled the objects A, B, C. The different ways that we can choose 2 from among the three are

$$\{A, B\} \qquad \{A, C\} \qquad \{B, C\} \,.$$

There are no other possibilities.

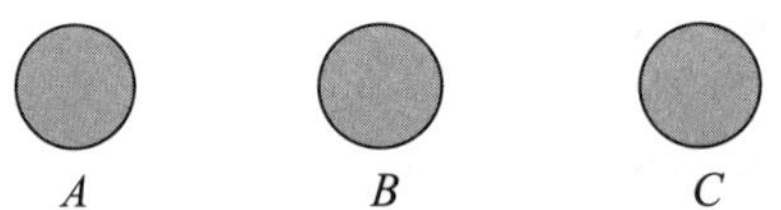

Figure 10.1. Three objects from which to choose two.

It is of interest to try to analyze this problem in general. So imagine now that we have n objects as shown in Figure 10.2. We are going to select k of them, with $0 \le k \le n$. We may note in advance that there is only one way to select 0 objects from among n—you just do not select any! So we may assume that k is at least 1.

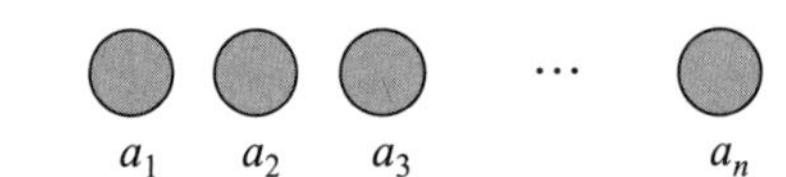

Figure 10.2. Selecting k objects from among n.

We select the first object. There are n different ways to do this—either you select the first one a_1, or the second one a_2, or ..., or the last one a_n.

Now let us think about selecting the second object. One object has already been selected, so there are $n-1$ objects left. And we may select any one of them with no restriction. So there are $n-1$ ways to select the second object. See Figure 10.3.

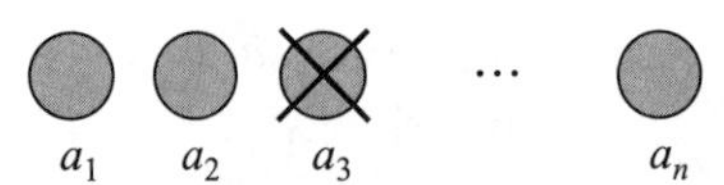

Figure 10.3. One object selected and $n-1$ remaining.

Now the pattern emerges. When we go to select the third object, there are $n-2$ choices remaining. So there are $n-2$ ways to select the third object.

And on it goes. There are $n-3$ ways to select the fourth object, $n-4$ ways to select the fifth object, ..., $n-k+1$ ways to select the kth object.

Altogether then we have

$$n \cdot (n-1) \cdot (n-2) \cdot \cdots \cdot (n-k+1)$$

ways to select k objects from among n total. But we have overlooked an important fact. We could have selected those k objects in any of $k!$ different orders. So in fact the number of ways to select k objects from among n is

$$\frac{n \cdot (n-1) \cdot (n-2) \cdot \cdots \cdot (n-k+1)}{k!}.$$

This is a very common expression in mathematics, and we give it the name n *choose* k. We denote this quantity by $\binom{n}{k}$. In fact it is customary to write it as

$$\binom{n}{k} = \frac{n!}{(n-k)!k!}.$$

EXAMPLE 10.1. In how many different ways can we choose two objects from among five?

Solution. The answer is immediate:

$$\binom{5}{2} = \frac{5!}{(5-2)!2!} = \frac{120}{6 \cdot 2} = 10.$$

In fact if the five objects are a, b, c, d, e, then the ten possible choices of two objects are:

$$\{a,b\}, \{a,c\}, \{a,d\}, \{a,e\}, \{b,c\},$$

$$\{b,d\}, \{b,e\}, \{c,d\}, \{c,e\}, \{d,e\}.$$

□

EXAMPLE 10.2. How many different 5-card poker hands are there in a standard 52-card deck of cards?

Solution. The answer is

$$\binom{52}{5} = \frac{52!}{(52-5)!5!} = 2{,}598{,}960.$$

□

EXAMPLE 10.3. In a 5-card poker hand, what are the chances of having 3-of-a-kind?

Solution. For the three matching cards, the first card that we select can be anything. So there are 52 possibilities. But the next one must match it. So there are only three choices for the second card (remember that there are four of each kind of card in the deck). And for the third card there are then just two possibilities. The other two cards can be completely random, so there are 49 and 48 possibilities for those two cards. Thus the total number of ways to have a hand with 3-of-a-kind is

$$\frac{52 \cdot 3 \cdot 2 \cdot 49 \cdot 48}{6} = 122304\,.$$

Note the little surprise: We have divided by 6 since we must divide out all the different possible orders of the same set of three matching cards; the number of permutations of three objects is $3! = 6$.

Now the *odds* of getting 3-of-a-kind is the ratio

$$\frac{122304}{2598960} = 0.047058824$$

(using the count of all possible hands from the previous example) or slightly less than 1 in 20. □

For You to Try What are the odds of getting two pairs in a standard 5-card poker hand?

For You to Try What are the odds of getting four-of-a-kind in a standard 5-card poker hand?

For You to Try You have a pot of beads. The beads are all identical in size and shape, but come in two different colors. You wish to make a beaded necklace consisting of ten beads. How many different necklaces could you make? Of course *the order* of the beads will be important: Beads in the order black-white-black-white do *not* give the same necklace as black-black-white-white. Note also that two necklaces are *equivalent*, and count as just one necklace, if a rotation of one gives the other. [After you have solved this problem, try replacing "ten" with n and "two" with k and solve it again.]

Now we turn to the binomial theorem. In its simplest form, the binomial theorem says this:

Theorem. *Let n be a positive integer and x an unknown. Then*

$$(x+1)^n = x^n + \binom{n}{1}x^{n-1} + \binom{n}{2}x^{n-2} + \cdots + \binom{n}{n-2}x^2 + \binom{n}{n-1}x + 1\,.$$

Note here the systematic use of the "choose" notation that we introduced above. This is one of many mathematical contexts in which "choose" is a useful and convenient idea.

EXAMPLE 10.4. We calculate that

$$(x+1)^2 = x^2 + \binom{2}{1}x + 1 = x^2 + 2x + 1\,.$$

EXAMPLE 10.5. We calculate that

$$(x+1)^3 = x^3 + \binom{3}{1}x^2 + \binom{3}{2}x + 1 = x^3 + 3x^2 + 3x + 1\,.$$

We mention for the record, and this matter will be discussed in some detail below, that a more general (but by no means the most general) form of the binomial theorem is

$$(a+b)^n = a^n + \binom{n}{1}a^{n-1}b + \binom{n}{2}a^{n-2}b^2 + \binom{n}{3}a^{n-3}b^3 + \cdots$$
$$+\binom{n}{n-3}a^3b^{n-3} + \binom{n}{n-2}a^2b^{n-2} + \binom{n}{n-1}ab^{n-1} + \binom{n}{n}b^n.$$

Notice that if we set $b = 1$ then we get the more elementary form of the theorem that we discussed above.

10.2.3 Pascal's Triangle

The idea of what we now call *Pascal's triangle* actually goes back to Yanghui in about the twelfth century in China (and the Chinese call the object *Yanghui's triangle*). But it was Blaise Pascal (1623–1662) who really developed the concept and showed its importance and context in modern mathematics and probability theory.

Pascal's triangle is a triangle formed according to the following rule (see Figure 10.4).

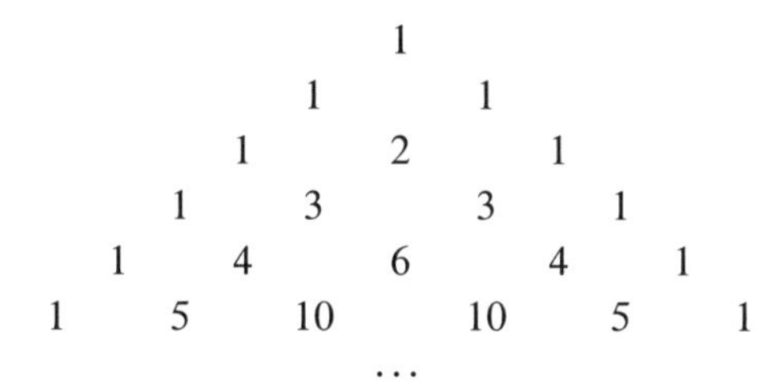

Figure 10.4, Pascal's Triangle

The rule for forming Pascal's triangle is this:

- A 1 goes at the top vertex.
- Each term in each subsequent row is formed by adding together the two numbers that are to the upper left and upper right of the given term.

It is convenient in our discussion to refer to the very top row of Pascal's triangle (with a single digit 1 in it) as the 0th row. The next row is the first row. And so forth. So the 0th row has one element, the 1st row has two elements, the 2nd row has three elements, and so forth.

Thus, in the first row, the leftmost term has nothing to its upper left and a 1 to its upper right. The sum of these is 1. So the leftmost term in the second row is 1. Likewise the rightmost term in the second row is 1.

For the second row, the leftmost term has nothing to its upper left and a 1 to its upper right. So this new term is 1. Likewise the rightmost term in the third row is 1. But the middle term in the third row has a 1 to its upper left and a 1 to its upper right. Therefore this middle term equals $1 + 1 = 2$. That is what we see in Pascal's triangle.

For the third row, we see as usual that the leftmost term and the rightmost terms are both 1 (in fact this property holds in all rows). But the second term in the fourth row has a 1 to its upper left and a 2 to its upper right. Therefore this second term is equal to 3. Likewise the third term is equal to 3.

The rest of Pascal's triangle is calculated similarly. The triangle is obviously symmetric from left to right, about a vertical axis through the upper vertex. The kth row has $k + 1$ terms. What is the significance of these numbers?

One obvious significance is the relation of the triangle to the binomial theorem. Consider the quantity

$$(a+b)^n = a^n + \binom{n}{1}a^{n-1}b + \binom{n}{2}a^{n-2}b^2 + \binom{n}{3}a^{n-3}b^3 + \cdots$$
$$+ \binom{n}{n-3}a^3b^{n-3} + \binom{n}{n-2}a^2b^{n-2} + \binom{n}{n-1}ab^{n-1} + \binom{n}{n}b^n .$$

Now we will examine this important formula in the first several specific instances:

n = 0: $(a+b)^0 = 1$

n = 1: $(a+b)^1 = a+b$

n = 2: $(a+b)^2 = a^2 + 2ab + b^2$

n = 3: $(a+b)^3 = a^3 + 3a^2b + 3ab^2 + b^3$

n = 4: $(a+b)^4 = a^4 + 4a^3b + 6a^2b^2 + 4ab^3 + b^4$

n = 5: $(a+b)^5 = a^5 + 5a^4b + 10a^3b^2 + 10a^2b^3 + 5ab^4 + b^5$

We see that the coefficients that occur for **n = 0** are just the same as the zeroeth row of Pascal's triangle: namely, a single digit 1. The coefficients that occur for **n = 1** are just the same as the first row of Pascal's triangle: namely, 1 and 1. The coefficients that occur for **n = 2** are just the same as the second row of Pascal's triangle: namely 1, 2, 1. And so forth. Of course if we think about how the binomial expression $(a + b)^n$ is multiplied out, then we see that the coefficients are formed by the very same rule that forms Pascal's triangle. And that explains why the rows of Pascal's triangle give the binomial coefficients.

Another remarkable fact is that the sum of the numbers in the nth row of Pascal's triangle is 2^n. For example, in the third row, $1+3+3+1 = 8 = 2^3$. This is again a fundamental property of the binomial coefficients that can be verified directly as follows. Note that

$$\begin{aligned}
2^n &= (1+1)^n \\
&= 1^n + \binom{n}{1}1^{n-1}\cdot 1 + \binom{n}{2}1^{n-2}\cdot 1^2 + \binom{n}{3}1^{n-3}\cdot 1^3 + \cdots \\
&\quad + \binom{n}{n-3}1^3\cdot 1^{n-3} + \binom{n}{n-2}1^2\cdot 1^{n-2} + \binom{n}{n-1}1\cdot 1^{n-1} + \binom{n}{n}1^n \\
&= \binom{n}{0} + \binom{n}{1} + \binom{n}{2} + \cdots + \binom{n}{n-2} + \binom{n}{n-1} + \binom{n}{n}.
\end{aligned}$$

A pleasing interpretation of the rows of Pascal's triangle can be given in terms of coin tosses:

- If we toss a coin once, then there are two possible outcomes: 1 heads and 1 tails. This information is tabulated in row 1 of Pascal's triangle.
- If we toss a coin twice, then there are three possible outcomes: two heads (which can occur just one way), a head and a tail (which can occur two ways—heads-tails or tails-heads), and two tails (which can occur just one way). This information is tabulated by 1—2—1 in the second row of Pascal's triangle.
- If we toss a coin three times, then there are four possible outcomes: three heads (which can occur just one way), two heads and a tail (which can occur three ways—heads-heads-tails, heads-tails-heads, or tails-heads-heads), two tails and a head (which can occur three ways—tails-tails-heads, tails-heads-tails, or heads-tails-tails), and three tails (which can occur just one way). This information is tabulated by 1—3—3—1 in the third row of Pascal's triangle.
- And so forth.

Pascal's triangle is also a useful mnemonic for carrying information about the choose function. For this purpose we number the rows 0, 1, 2, etc. as usual. We also number the *terms* in each row 0 ,1, 2, etc. Now suppose we wish to know how many different ways we can choose three objects from among five (in fact this came up in an example in the last section—the answer was 10). Simply go to row 5 of the triangle, term 3, and we see the answer to be 10. Or if we want to know how many different ways to choose five items from among five (the answer is obviously 1), we go to row 5 and look at the 5th term (remembering to count from 0).

Of course Pascal's triangle is not magic—it is mathematics. And the mathematical explanation behind everything we have said here is the fundamental formula

$$\binom{n}{k} = \binom{n-1}{k-1} + \binom{n-1}{k}.$$

You may test this formula by hand, or on your calculator. It is easy to confirm rigorously using elementary algebra. And it simply says (if we think of the kth element in the nth row of Pascal's triangle as a_{nk}) that $a_{nk} = a_{(n-1)(k-1)} + a_{(n-1)k}$. This just says that the element a_{nk} is formed by adding the two elements directly in the row above it.

For You to Try Prove by induction that the terms of the kth row of Pascal's triangle add up to 2^k.

10.2.4 The Binomial Theorem in the Nonintegral Case

Gauss realized that the binomial theorem is true even when the exponent n is *not* an integer (he was anticipated here by Isaac Newton). He calculated the formula in this case as well—it turns out then to have infinitely many terms (it is what we call a *series*)—and determined when and how it makes sense.

As an example, let us consider the expression $(1+x)^{1/2}$. This is just the square root of $(1+x)$. If we postulate that

$$(1+x)^{1/2} = a_0 + a_1 x + a_2 x^2 + \cdots,$$

then we may calculate that

$$\begin{aligned} 1+x &= \left[(1+x)^{1/2}\right]^2 \\ &= \left[a_0 + a_1x + a_2x^2 + a_3x^3 + \cdots\right]^2 \\ &= a_0^2 + [2a_0a_1]x + [2a_0a_2 + a_1^2]x^2 + [2a_3a_0 + 2a_1a_2]x^3 + \cdots . \end{aligned}$$

Identifying powers of x on either side, we obtain the recursion

$$1 = a_0^2 ,$$

$$1 = 2a_0a_1 ,$$

$$0 = 2a_0a_2 + a_1^2 ,$$

$$0 = 2a_0a_3 + 2a_1a_2 ,$$

and so forth.

We select the solution $a_0 = 1$ of the first of these equations. Then the second equation yields $a_1 = 1/2$, the third equation yields $a_2 = -1/8$, the fourth equation yields $a_3 = 1/16$, and so forth. So we find that

$$(1+x)^{1/2} = 1 + \frac{1}{2}x - \frac{1}{8}x^2 + \frac{1}{16}x^3 + - \cdots .$$

Notice that the expression on the right has infinitely many terms.

It is of interest to know how to generalize the "choose" function so that we may generate these coefficients automatically. And then to know for which values of x the formula is valid. In fact Isaac Newton treated the binomial expansion for negative exponents. But it was Gauss who formulated the result for *arbitrary complex exponents*; we discuss Gauss's theorem below.

Theorem. *Let r be any complex number. Then*

$$(x+y)^r = \sum_{k=0}^{\infty} \binom{r}{k} x^{r-k} y^k .$$

Here

$$\binom{r}{k} \equiv \frac{1}{k!} \prod_{n=0}^{k-1} (r-n) = \frac{r(r-1)(r-2)\cdot \cdots \cdot (r-(k-1))}{k!} .$$

Note that the definition of $\binom{r}{k}$ used in this theorem is consistent with what we have said before about the "choose" function. For, when r and k are integers,

$$\binom{r}{k} = \frac{r!}{(r-k)!k!} = \frac{r(r-1)(r-2)\cdot \cdots \cdot (r-(k-1))}{k!} .$$

It is enlightening to apply the theorem to the case $(1+x)^{1/2}$, which we have already treated in a rather *ad hoc* fashion above. So now $r = 1/2$. If we take x to be 1 and y to be x, then we find that

$$\begin{aligned}(1+x)^{1/2} &= \sum_{k=0}^{\infty} \binom{1/2}{k} 1^{1/2-k} x^k \\ &= \sum_{k=0}^{\infty} \binom{1/2}{k} x^k .\end{aligned}$$

We may calculate that

$$\binom{1/2}{0} = 1 ,$$

$$\binom{1/2}{1} = \frac{1/2}{1} = \frac{1}{2} ,$$

$$\binom{1/2}{2} = \frac{(1/2)(-1/2)}{2} = -\frac{1}{8} ,$$

$$\binom{1/2}{3} = \frac{(1/2)(-1/2)(-3/2)}{3!} = \frac{1}{16} .$$

Thus, according to the theorem, the binomial expansion of $(1+x)^{1/2}$ is

$$(1+x)^{1/2} = 1 + \frac{1}{2}x - \frac{1}{8}x^2 + \frac{1}{16}x^3 + - \cdots ,$$

and that is consistent with the calculation that we did before.

But the theorem also contains some surprises. We could instead take $y = 1$ and $x = x$. Then we find that

$$\begin{aligned}(x+1)^{1/2} &= \sum_{k=0}^{\infty} \binom{1/2}{k} x^{1/2-k} 1^k \\ &= \sum_{k=0}^{\infty} \binom{1/2}{k} x^{1/2-k} .\end{aligned}$$

Of course the coefficients are calculated just as before, but we find now that the specified expansion is

$$(x+1)^{1/2} = x^{1/2} + \frac{1}{2}x^{-1/2} - \frac{1}{8}x^{-3/2} + \frac{1}{16}x^{-5/2} + - \cdots .$$

This is an expansion *not* in integer powers of x but instead in negative fractional powers of x. Most likely such an expansion is less useful than the one we derived (twice) above, but it is certainly of mathematical interest.

10.2.5 Proof of the Binomial Theorem

We should say a few words about why the binomial expansion is true. We shall in fact describe two ways to understand this result.

First Verification of the Binomial Theorem. Let us examine the basic form of the binomial theorem that we considered at first:

$$(x+1)^n = x^n + \binom{n}{1}x^{n-1} + \binom{n}{2}x^{n-2} + \cdots + \binom{n}{n-2}x^2 + \binom{n}{n-1}x + 1 .$$

Consider the left-hand side of this equation. It is

$$(x+1)^n = \underbrace{(x+1)\cdot(x+1)\cdot(x+1)\cdot\cdots\cdot(x+1)}_{n \text{ times}} .$$

The only way that the term x^j can arise from this product is if x is selected j times from among the n factors. This can occur in precisely $\binom{n}{j}$ ways. So the correct coefficient in front of x^j is $\binom{n}{j}$. And that is the proof of the binomial theorem. □

Second Verification of the Binomial Theorem. This proof is by induction. We shall treat induction in some detail in Section 19.2. Here we use the method in an *ad hoc* fashion.

First we note the following useful fact about the "choose" function:

$$\binom{n}{j} + \binom{n}{j-1} = \binom{n+1}{j} .$$

This is true because

$$\begin{aligned}
\binom{n}{j} + \binom{n}{j-1} &= \frac{n!}{(n-j)!j!} + \frac{n!}{(n-(j-1))!(j-1)!} \\
&= \frac{n!}{(n-j)!j!} + \frac{n!j/(n-(j-1))}{(n-j)!j!} \\
&= \frac{n![1+j/(n-(j-1))]}{(n-j)!j!} \\
&= \frac{n!\,[1+j/(n-(j-1))]((n+1)-j)}{((n+1)-j)!j!} \\
&= \frac{n![n+1-j+j]}{((n+1)-j)!j!} \\
&= \frac{(n+1)!}{((n+1)-j)!j!} \\
&= \binom{n+1}{j} .
\end{aligned}$$

Now we certainly know that the basic binomial theorem is true for $n = 1$ and $n = 2$. Suppose now that it is known to be true for some n. Thus

$$(x+1)^n = x^n + \binom{n}{1}x^{n-1} + \binom{n}{2}x^{n-2} + \cdots + \binom{n}{n-2}x^2 + \binom{n}{n-1}x + 1 .$$

Now multiply both sides by $(x+1)$. We obtain

$$(x+1)^{n+1} = x^{n+1} + \left[\binom{n}{1} + 1\right] x^n + \left[\binom{n}{2} + \binom{n}{1}\right] x^{n-1} + \left[\binom{n}{3} + \binom{n}{2}\right] x^{n-2}$$
$$+ \cdots + \left[\binom{n}{n-1} + \binom{n}{n-2}\right] x^2 + \left[\binom{n}{n} + \binom{n}{n-1}\right] x + 1 .$$

Using the previous identity for the choose function, we see that this simplifies to

$$(x+1)^{n+1} = x^{n+1} + \binom{n+1}{1} x^n + \binom{n+1}{2} x^{n-1} + \binom{n+1}{3} x^{n-2}$$
$$+ \cdots + \binom{n+1}{n-1} x^2 + \binom{n+1}{n} x + 1 ,$$

which is the binomial theorem for the index $n+1$. That completes our inductive argument, and the result is proved. □

Of course we still have not explained the binominal theorem in the most general case, for a complex exponent. And we haven't discussed for *which* values of x the infinite series representation is valid. It requires a good dose of calculus to give these matters a thorough and rigorous treatment. But we can give some indications here of why things work the way they do.

Begin by re-examining our expansion for $(x+1)^{1/2}$. We have

$$(1+x)^{1/2} = 1 + \frac{1}{2}x - \frac{1}{8}x^2 + \frac{1}{16}x^3 + - \cdots .$$

If we want to derive an expansion for $(x+1)^{3/2}$, we need only cube both sides:

$$(x+1)^{3/2} = \left[1 + \frac{1}{2}x - \frac{1}{8}x^2 + \frac{1}{16}x^3 + - \cdots\right]^3$$

or

$$(x+1)^{3/2} = 1 + \frac{3}{2}x + \frac{3}{8}x^2 - \frac{1}{16}x^3 + - \cdots .$$

Note that this expansion is consistent with that prescribed in our theorem on the general binomial expansion. You should check this for yourself.

Of course similar reasoning may be used to find an expansion for $(x+1)^{m/2}$ for any positive integer m.

In like manner, we could find an expansion for $(x+1)^{1/3}$. For we simply postulate that $(x+1)^{1/3} = a_0 + a_1 x + a_2 x^2 + \cdots$ and then write

$$x + 1 = \left[(x+1)^{1/3}\right]^3 = \left[a_0 + a_1 x + a_2 x^2 + \cdots\right]^3 .$$

We write out the right-hand side, find the recursions for the coefficients a_j, and then successively solve for the a_j. The results will be the same as in the general binomial expansion theorem.

Finally we need to comment on negative exponents. Suppose that we wish to find an expansion for $(x+1)^{-1}$. We follow our earlier procedure and set $(x+1)^{-1} = a_0 + a_1 x + a_2 x^2 + \cdots$. Then we write

$$\begin{aligned} 1 &= (x+1) \cdot (x+1)^{-1} = (x+1) \cdot [a_0 + a_1 x + a_2 x^2 + \cdots] \\ &= a_0 + [a_0 + a_1]x + [a_1 + a_2]x^2 + [a_2 + a_3]x^3 + \cdots . \end{aligned}$$

Identifying powers of x on both sides of this equation, we find that

$$1 = a_0 ,$$

$$0 = a_0 + a_1 ,$$

$$0 = a_1 + a_2 ,$$

$$0 = a_2 + a_3 ,$$

and so forth. We may solve these equations in turn to find that

$$a_0 = 1 , \; a_1 = -1 , \; a_2 = 1 , \; a_3 = -1 ,$$

and so forth. Thus

$$(x+1)^{-1} = 1 - x + x^2 - x^3 + - \cdots .$$

Of course an expansion for any $(x+1)^{-m}$ for any positive integer m may be obtained from this last equation just by taking powers. And now an expansion for $(x+1)^{-m/n}$, any positive integers m and n, may be had by imitating our argument for $(x+1)^{1/2}$ and $(x+1)^{1/3}$.

In summary, we now see why the generalized binomial theorem is true for $(x+1)^r$ with r any rational exponent—positive or negative. The argument for r irrational is more complicated, and we cannot present it here.

The question of for which values of x these expansions are valid is subtle, and requires deep ideas from the theory of series. We cannot discuss it. Suffice it to say that, if the exponent r is any complex number, and if $|x| < 1$, then the binomial representation of $(x+1)^r$ is valid: The infinite sum converges, and it converges back to $(x+1)^r$.

Further Reading

J. L. Brenner and R. C. Lyndon, Proof of the fundamental theorem of algebra, *The American Mathematical Monthly* 88(1981), 253–256.

Michael D. Hirschhorn, The fundamental theorem of algebra, *The College Mathematics Journal* 29(1998), 276–277.

10.3 The Chinese Remainder Theorem

One of the hallmarks of Gauss's treatise *Disquisitiones Arithmeticae* was the systematic use of modular arithmetic. A particularly elegant and powerful result in modular arithmetic is the Chinese Remainder Theorem. Although this is a very old idea, it is one to which

Gauss contributed (in a more generalized form for polynomials). We shall discuss it here to give an indication of the kinds of questions that Gauss studied.

The Chinese Remainder Theorem was first treated by Sun Zi in a third-century book of mathematics. It was later republished in a 1247 book by Qin Jiushao.

A formal statement of the theorem is as follows:

Theorem 10.1. *Suppose that positive integers $m_1, m_2, \ldots m_k$ are given, and that they are relatively prime (i.e., have no common prime factors). Then, for any integers a_j, $0 \le a_j \le m_j - 1$, and for $j = 1, \ldots k$, there is an integer x such that*

$$\begin{aligned} x &= a_1 \bmod m_1 \,, \\ x &= a_2 \bmod m_2 \,, \\ &\cdots \\ x &= a_k \bmod m_k \,. \end{aligned} \tag{$\star$}$$

The solution x is unique modulo $M = m_1 \cdot m_2 \cdot \cdots \cdot m_k$.

We should like to say a few words about the proof of this theorem, and then to discuss some of its uses.

For $i = 1, 2, \ldots, k$, let

$$n_i = \prod_{j \neq i} m_j \,.$$

Then of course m_i and n_i are relatively prime. Thus there exist integers s_i and t_i (see Section 13.3 below) such that

$$s_i m_i + t_i n_i = 1 \,. \tag{$*$}$$

Set $u_i = t_i n_i$. Then

- $u_i = 1 - s_i m_i = 1 \bmod m_i$ by $(*)$.
- $u_i = 0 \bmod m_j$ for $j \neq i$ because n_i divides u_i.

Now we claim that

$$x \equiv a_1 u_1 + a_2 u_2 + \cdots + a_k u_k$$

does the job. For

$$\begin{aligned} x \bmod m_1 &= \sum_j (a_j u_j) \bmod m_1 \\ &= \sum_j (a_j \bmod m_1) \cdot (u_j \bmod m_1) \\ &= (a_1 \bmod m_1) \cdot 1 \\ &= a_1 \,. \end{aligned}$$

A similar calculation shows that $x \bmod m_j = a_j$ for every $j = 1, \ldots, k$.

Note that if x' is another solution of the system $(\star)$, then $x - x' = 0 \bmod m_j$ for every j. Hence, letting $M = \prod_j m_j$, $x - x' = 0 \bmod M$.

Further Reading

George Mackiw, Computing in abstract algebra, *The College Mathematics Journal* 27(1996), 136–142.

Oystein Ore, The general Chinese remainder theorem, *The American Mathematical Monthly* 59(1952), 365–370.

10.4 A Constructive Means for Finding the Solution x

In fact our proof of the Chinese Remainder Theorem gives a constructive means for finding x once the m_j and a_j are given. Let us illustrate this idea with an example.

EXAMPLE 10.6. Consider the problem of finding an integer x such that

$$x = 2 \bmod 3,$$
$$x = 3 \bmod 4,$$
$$x = 1 \bmod 5.$$

Using the Euclidean algorithm for 3 and $4 \cdot 5 = 20$, we find that $(-13) \cdot 3 + 2 \cdot 20 = 1$. Hence $s_1 = -13$ and $t_1 = 2$. We conclude then that $u_1 = 2 \cdot 20 = 40$.

Using the Euclidean algorithm for 4 and $3 \cdot 5 = 15$, we obtain that $(-11) \cdot 4 + 3 \cdot 15 = 1$. Hence $s_2 = -11$ and $t_2 = 3$. We conclude then that $u_2 = 3 \cdot 15 = 45$.

Finally, using the Euclidean algorithm for 5 and $3 \cdot 4 = 12$, we see that $5 \cdot 5 + (-2) \cdot 12 = 1$. So $s_3 = 5$ and $t_3 = -2$. We conclude then that $u_3 = (-2) \cdot 12 = -24$.

A solution x is therefore

$$x = 2 \cdot 40 + 3 \cdot 45 + 1 \cdot (-24) = 191.$$

All other solutions are congruent to 191 modulo 60 (note that $M = 3 \cdot 4 \cdot 5 = 60$), which means that they are all congruent to 11 modulo 60.

It is worth noting that the number x may be quite large even when the initial data m_j and a_j are rather small.

Further Reading

Bo Green, A project for discovery, extension, and generalization in abstract algebra, *The College Mathematics Journal* 31(2000), 329–332.

C. J. Mozzochi, A simple proof of the Chinese remainder theorem, *The American Mathematical Monthly* 74(1967), 998.

10.5 Quadratic Reciprocity and the Gaussian Integers

Certainly one of the triumphs of Gauss's forays into number theory (or "higher arithmetic", as it was sometimes called) was the celebrated quadratic reciprocity law. This was part and parcel of Gauss's investigations of modular arithmetic, and certainly was one of the big topics in his book *Disquisitiones Arithmeticae*.

We shall not attempt a proof of quadratic reciprocity, but shall introduce it and discuss it. It shall serve for us as an entree to the Gaussian integers, which we shall develop in greater detail in the next section.

Let r and m be fixed integers, and assume that r is not divisible by m. Is there an integer n such that

$$n^2 = r \bmod m\,? \qquad (*)$$

If such n exists, we call r a *quadratic residue* of m. If not, then we call r a *quadratic nonresidue* of m.

Supposing that r *is* a quadratic residue of m, then there is an n whose square, when divided by m, leaves remainder r. That is what equation $(*)$ says.

Let us illustrate these ideas by asking whether 13 is a quadratic residue of 17. If so, then one must be able to solve the congruence

$$n^2 = 13 \bmod 17\,.$$

Simply plugging in $n = 1, 2, \ldots$, we can determine that $n = 8, 25, 42, 59, \ldots$ are in fact solutions. For example,

$$8^2 = 64 = 3 \cdot 17 + 13 = 13 \bmod 17\,,$$

$$25^2 = 625 = 36 \cdot 17 + 13 = 13 \bmod 17\,,$$

and so forth. Thus we see that 13 *is indeed* a quadratic residue of 17.

By contrast, 5 is a quadratic nonresidue of 17. In fact calculations indicate that the only odd quadratic residues for 17 are 1, 9, 13, 15, etc. .

Naturally the question we wish to ask is "Given an integer m, what are its quadratic residues?" It is not difficult to see that one may restrict attention to the case when r and m are both (distinct) prime numbers.

Now the spirit of quadratic reciprocity is this:

> The congruence
>
> $$n^2 = r \bmod m$$
>
> and the congruence
>
> $$n^2 = m \bmod r$$
>
> (where r and m are distinct primes) are either both simultaneously solvable or both simultaneously unsolvable *unless* both of m and r have residue 3 modulo 4. In the special case that both *do* have residue 3 modulo 4, one of the equations is solvable and the other is not.

The quadratic reciprocity law was conceived by both Euler and Legendre, but they were unable to prove it. Gauss proved the result when he was just nineteen years old. Gauss was fascinated by quadratic reciprocity during his entire lifetime (and he also studied a generalization called *biquadratic reciprocity*), and produced six distinct proofs of the result.

We may test the law of quadratic reciprocity against instances that we have already examined. First consider the case $r = 13$, $m = 17$ which we considered at the start of our discussion. Of course

$$13 \bmod 4 = 1 \bmod 4$$

and

$$17 = 1 \bmod 4\,.$$

According to the law of quadratic reciprocity, the equations

$$n^2 = 13 \bmod 17$$

and

$$n^2 = 17 \bmod 13$$

should either both be solvable or both not. But we already know that the first is solvable. Hence so is the second. In fact

$$11^2 = 121 = 4 \bmod 13 = 17 \bmod 13\,.$$

Now let us look at $r = 5$, $m = 17$. Of course

$$5 = 1 \bmod 4$$

and

$$17 = 1 \bmod 4\,.$$

According to quadratic reciprocity, either

$$n^2 = 5 \bmod 17$$

and

$$n^2 = 17 \bmod 5$$

are both solvable or both not. Since we already know that the first of these is nonsolvable, we may conclude that neither is solvable.

Finally examine $r = 11$ and $m = 19$. We see that

$$11 = 3 \bmod 4$$

and

$$19 = 3 \bmod 4\,.$$

Thus the two equations

$$n^2 = 19 \bmod 11$$

and

$$n^2 = 11 \bmod 19$$

must have the property that one is solvable and the other not. Indeed, the first of these has no solutions while the second has the solutions $7, 26, 45, \ldots$.

Further Reading

Reinhard C. Laubenbacher and David J. Pengelley, Eisenstein's misunderstood geometric proof of the quadratic reciprocity theorem, *The College Mathematics Journal* 25(1994), 29–34.

Emma Lehmer, Rational reciprocity laws, *The American Mathematical Monthly* 85(1978), 467–472.

10.6 The Gaussian Integers

Gauss developed the technique of the Gaussian integers to further his studies of quadratic residues. This was a remarkable achievement for several reasons. First, the complex numbers (which, as we shall see, are the basis for the Gaussian integers) were not at all well understood in Gauss's day. So he had to master all of basic complex analysis so that he could proceed. Second, the complex-number based Gaussian integers were used as a tool to study questions of quadratic residues, even though those questions have nothing whatever to do with complex analysis. This was an astonishingly powerful and original thing for Gauss to do.

So what are the Gaussian integers? These are all numbers of the form $a + ib$, where a and b are ordinary rational integers and i is the square root of -1 from complex analysis. The Gaussian integers are commonly denoted by $\mathbb{Z}[i]$. Clearly the Gaussian integers are closed under ordinary addition and multiplication of numbers, so they form a ring (note that $1 = 1 + i0$ is the multiplicative identity and $0 = 0 + i0$ is the additive identity for this ring). If $x = a + ib$ is a Gaussian integer, then we define

$$d(x) = a^2 + b^2 .$$

This is not a norm in the usual sense—it is more like the square of a norm. But $d(x)$ will prove to be a useful quantity in our studies. Note in particular that if $x = a + ib$ and $x' = a' + ib'$ are Gaussian integers, then

$$\begin{aligned} d(x \cdot x') &= d([aa' - bb'] + i[ab' + a'b]) \\ &= (aa' - bb')^2 + (ab' + a'b)^2 \\ &= a^2a'^2 + b^2b'^2 + a^2b'^2 + a'^2b^2 \\ &= (a^2 + b^2) \cdot (a'^2 + b'^2) \\ &= d(x) \cdot d(x') . \end{aligned}$$

Thus d respects multiplication in the Gaussian integers.

Our goal is to prove the following result of Euler using the Gaussian integers:

Theorem. *Let p be an odd prime. Then $p = 1 \bmod 4$ if and only if $p = a^2 + b^2$ for some integers a and b. This latter expression is unique up to permutation of a and b or multiplication of either term by -1.*

In fact this result was first enunciated in a letter of Fermat to Mersenne, but (characteristically) Fermat provided no proof.

We have already noted that $\mathbb{Z}[i]$ forms a ring. Thanks to Emmy Noether and others, a basic artifact of the study of rings is the idea of an *ideal*. An ideal $\mathcal{I}$ in the commutative ring R is a subset that is closed under addition and such that if $r \in R$ and $i \in \mathcal{I}$ then $r \cdot i = i \cdot r \in \mathcal{I}$.

Now a basic fact about the Gaussian integers, of which we will make good use below, is this: the Gaussian integers form a Principal Ideal Domain (or PID). This means that any ideal in $\mathbb{Z}[i]$ has just one generator. We shall not prove this result, but note that if $\mathcal{I}$ is an ideal in $\mathbb{Z}[i]$, then the nonzero element x of $\mathcal{I}$ with $d(x)$ as small as possible will be the generator that we seek.

We note a couple of other simple facts about $\mathbb{Z}[i]$ before we proceed with a proof of the theorem. The only elements of $\mathbb{Z}[i]$ that have multiplicative inverses are $1, -1, i, -i$. We call these the *units*. They form a group under multiplication, and we call this the *multiplicative group*. Notice that if u is a unit, then $d(u) = 1$.

We shall prove Euler's theorem with a sequence of lemmas.

Lemma. *The ring of Gaussian integers forms a Unique Factorization Domain (UFD).*

Proof. In fact it is a standard result in algebra that any Principal Ideal Domain (PID) is a Unique Factorization Domain (UFD). This means that every element can be uniquely factored into irreducible factors. The Fundamental Theorem of Arithmetic—that every positive integer can be uniquely factored into prime factors—is the primordial example of such a result.

We shall prove now that every PID is a UFD. So let R be a ring which is a PID. We claim that every non-unit in R can be factored as a product of irreducibles.

If a is irreducible, then there is nothing to prove (it is already factored). If a is *not* irreducible, then $a = a_1 b_1$, with a_1, b_1 non-units. If now a_1, b_1 are both irreducible, then we are done. If, for instance, a_1 is not irreducible, write $a_1 = a_2 b_2$.

We keep going to express a as a product of irreducibles. This is all fine as long as the process stops. Why does it stop?

Set $a = a_0$. We have the increasing chain of ideals

$$a_0 R < a_1 R < a_2 R < \cdots .$$

The factorization process terminates if the chain does, and vice versa. Let $\mathcal{I}$ be the union of the ideals $a_j R$. Then $\mathcal{I}$ itself is an ideal. Since R is a PID, we know that $\mathcal{I} = bR$, for some element $b \in R$. Of course b must be in one of the ideals $a_j R$ (since $b \in \mathcal{I}$). So $b = a_j r$ for some $r \in R$. But of course $a_j \in \mathcal{I}$. Thus a_j divides b and b divides a_j. We conclude that $a_i R = a_j R$ for $i \geq j$. So the chain of ideals stabilizes.

Uniqueness follows familiar lines, and we omit the details. □

Lemma. *Let m be a Gaussian integer. If $d(m)$ is a prime (in the usual sense) then m is prime in $\mathbb{Z}[i]$.*

Proof. Seeking a contradiction, we suppose that $m = ab$, with a, b non-units in $\mathbb{Z}[i]$. Of course then $d(m) = d(a) \cdot d(b)$. But $d(m)$ is prime. Hence one of $d(a)$ and $d(b)$ is prime and the other is a unit (1 or -1). Since d takes only nonnegative values, we must conclude that one of $d(a)$ or $d(b)$ is 1. But that means that one of a or b is a unit in $\mathbb{Z}[i]$, and that is impossible. □

Now suppose that p is an odd prime and that $p = 1 \bmod 4$.

Lemma. *There exists an integer c such that $c^2 = -1 \bmod p$.*

Proof. Consider the multiplicative group for the field $\mathbb{Z}/p$. This consists of the $p - 1$ elements $1, 2, \ldots, p - 1$. This is a cyclic group, and there must be an element t of order 4 in the group since 4 divides $p - 1$ (the order of the group). Thus t^2 has order 2.

Now the polynomial $x^2 - 1$ has roots 1 and -1 in $\mathbb{Z}/p$, and only -1 actually has order 2. Thus we conclude that $t^2 = -1$. Now let c be an integer whose equivalence class modulo p is t. Then $c^2 \bmod p = t^2 \bmod p = -1 \bmod p$. That is what we wished to prove. □

We now know that p divides $1 + c^2$ in $\mathbb{Z}$. Hence certainly p divides $1 + c^2$ in $\mathbb{Z}[i]$ (which is a larger ring). Let us write

$$1 + c^2 = (1 + ic) \cdot (1 - ic).$$

Now we cannot have that p divides either $1 + ic$ or $1 - ic$ for then p (which is of course real) must divide 1, and that is impossible. So p is not a prime element of $\mathbb{Z}[i]$, hence is *not* irreducible.

We thus can write

$$p = (a + bi) \cdot (e + fi)$$

as a product of non-units. We calculate that

$$d(p) = d(a + bi) \cdot d(e + fi)$$

hence

$$p^2 = (a^2 + b^2) \cdot (e^2 + f^2).$$

The conclusion, then, is that either $a^2 + b^2 = p$ or $e^2 + f^2 = p$ or else one of these numbers is 1. But clearly neither is a unit. The only possible conclusion is

$$a^2 + b^2 = p$$

and

$$e^2 + f^2 = p.$$

Hence $p = a^2 + b^2$ as required. That proves our theorem.

For the uniqueness, suppose that $p = a^2 + b^2$ and $p = a'^2 + b'^2$. Then

$$d(a + ib) = d(a - ib) = d(a' + ib') = d(a' - ib') = p.$$

Hence $a + ib$, $a - ib$, $a' + ib'$, and $a' - ib'$ are all *prime* Gaussian integers. But we know that

$$(a + ib) \cdot (a - ib) = (a' + ib') \cdot (a' - ib').$$

By uniqueness of factorization in $\mathbb{Z}[i]$, either

$$a + ib = (a' + ib')u$$

or

$$a + ib = (a' - ib')u,$$

where u is a unit. Of course that means that u is one of $1, -1, i, -i$. But this just says that $a + ib$ and $a' + ib'$ differ by sign changes or swapping (as claimed).

Finally, let us consider the converse assertion. Suppose that $p = a^2 + b^2$. Since p is odd, one of a or b must be odd. And the other therefore must be even. The square of an even integer is 0 modulo 4, and the square of an odd integer is 1 modulo 4. Hence the residue of p modulo 4 is 1. That proves the result.

Further Reading

Walter Rudin, Unique factorization of Gaussian integers, *The American Mathematical Monthly* 68(1961), 907–908.

Robert G. Stein, Exploring the Gaussian integers, *The Two-Year College Mathematics Journal* 7(1976), 4–10.

Exercises

1. Suppose that you are playing poker with a standard 52-card deck of cards. You are dealt five cards. What is the probability that you hold a pair (i.e., two cards with the same face value, of different suits)?

2. Suppose that you are playing poker with a standard 52-card deck of cards. You are dealt five cards. What is the probability that you hold a straight (i.e., five cards in sequence, possibly of different suits)?

3. A peasant takes her cabbages to market. She does not know how many cabbages she has. But she knows that if she divides them up into baskets each holding seven cabbages then there are three left over. If instead she divides them up into baskets each holding five, then there are two left over. Finally, if she divides them up into baskets each holding eleven, then there are ten left over. The woman is sure that there are fewer than 100 cabbages. How many are there?

4. In the expression $(a+b)^{50}$, what is the coefficient of $a^{43}b^7$? What is the coefficient of $a^{16}b^{34}$?

5. What is the first row of Pascal's triangle in which the number 50 occurs? Why?

6. How many Gaussian integers are there on the circle $x^2 + y^2 = 25$? What is the circle, centered at the origin, of least radius that has four Gaussian integers on it? How about 8 Gaussian integers?

7. What is the least positive integer that has residue 2 modulo 3, 5, 7, and 9?

8. How many different permutations are there of the numbers

$$1\ 2\ 3\ 4\ 5?$$

   How many different permutations are there of these numbers so that the even numbers 2 and 4 appear in the first two positions?

9. How many different permutations are there of the numbers

$$1\ 2\ 3\ 4\ 5$$

   so that the even numbers and the odd numbers alternate in the list?

10. **Project:** Not every Gaussian integer has a square root which is a Gaussian integer. The number $2i$ does have such a square root. Explain why. The number $1+i$ does not. Explain why. Can you devise a simple test that will tell when a Gaussian integer has a Gaussian integer square root? See [VAN] for these basic ideas.

11. **Project:** A jar has 100 white marbles and 100 black marbles. You select five marbles at random (with your eyes closed). What is the probability that at least two of them are black? What is the probability that all of them are black? What is the probability that at least one of them is black? See [KRA5] for a discussion of this type of problem.

12. **Project:** Use the generalized binomial theorem to understand the value of $(1 + i)^{\sqrt{2}}$ to three decimal places of accuracy. The book [KRA1] discusses this idea.

11

Sophie Germain and the Attack on Fermat's Last Problem

11.1 Birth of an Inspired and Unlikely Child

Marie-Sophie Germain (1776–1831) was one of the most profound and remarkable figures in all of modern mathematics. Born in Paris the year of the American revolution, she was a young teenager during the time of the French revolution. It is possible that some of the revolutionary spirit got into her blood, for she was an independent thinker from an early age. Sophie's childhood home was a meeting place for liberal reform aficionados, so she was exposed to political and philosophical discussions at an early age. Sophie Germain's life was not a long one, but she was one of the most remarkable woman mathematicians who ever lived. Details on her life appear in [BIE].

Sophie Germain was born into a prosperous mercantile family, but her mental sophistication was in no way tainted by ordinary middle-class values. Because her mother was named Marie, and so was her sister, Marie-Sophie became known as "Sophie".

For her safety, Sophie's parents kept her at home, and away from school, during the most violent times of the French revolution. She diverted herself, and fought the tedium of being home alone, by reading the books in her father's library. She was particularly fascinated by the story of Archimedes. Of particular interest to the young girl was the account of Archimedes's untimely demise.

According to legend (and, as mentioned in Chapter 4 of this book), when invading Roman troops marauded Syracuse, where Archimedes was living, he contented himself with his mathematics. Marcellus (268 B.C.E.–208 B.C.E.), the general who commanded the conquering troops, commanded that the great scientist Archimedes should be protected. Archimedes's first intimation that the city had been sacked was the shadow of a Roman soldier falling across his diagram in the soil. One version of the story is that the heathen stepped on Archimedes's diagram, causing the mild-mannered scholar to become angry and exclaim, "Don't disturb my circles!" Enraged, the soldier drew his sword and slew Archimedes.

Sophie Germain was fascinated with the notion that a person could be so absorbed in anything that he would ignore a soldier and then die as a result. She concluded that mathematics must be quite a worthwhile subject, and she determined to study it.

Sophie began, as was appropriate, by studying geometry. She also learned Latin and Greek so that she could read the classical texts. Unfortunately, Sophie's rather mundane parents did not approve of her scholarly plan. It was commonly held by the *bourgeoisie* of the time that advanced education was inappropriate for young women (although a certain amount of sophistication in upper class women was tolerated—just because it lent spice to their social conversation). This was not just a matter of amused disapprobation. Sophie Germain's parents were frankly mortified that she would have a mathematical bent.

But Sophie was determined. She quietly studied her mathematics at night, after her parents had retired. When her parents discovered the subterfuge, they were outraged. They took away all Sophie's clothing, took away her lamps, and turned off the heat. Now, they figured, all she could do was bundle up in her bedclothes and sleep. But not Sophie Germain.

Sophie smuggled candles into her room, wrapped herself in quilts, and studied her books under the cover of some bedding. In this way she was able to learn her beloved mathematics. Eventually Sophie's parents realized that this is what she truly wanted to do. They declared her to be "incurable", and they reluctantly bestowed their blessing. With the largesse of her parents, Sophie Germain spent the time of the Reign of Terror studying differential calculus—and all without the aid of a tutor! Sophie Germain never married, and certainly never obtained a professional position. Her father supported her financially throughout her life.

In 1794 the very distinguished École Polytechnique was founded in Paris. Designed to educate the best minds in the country to become "mathematicians and scientists in the service of France," this institution remains to this day one of the finest universities in the world. Women were not allowed to enroll in the Polytechnique. But, through her indomitable determination, Sophie Germain managed to obtain copies of the notes from the courses. She studied them assiduously, and she was particularly enthralled with the teachings of Joseph-Louis Lagrange (1736–1813). She learned the name of a former student of Lagrange (M. LeBlanc), and at the end of the term submitted a paper on analysis to Lagrange using LeBlanc's name. Professor Lagrange was so impressed by the work that he demanded to meet the student who had written it. You can imagine his surprise to learn that said student was a young woman!

It is to Lagrange's credit that he did not let the common social prejudices affect his judgment. He not only approved of Sophie Germain's work, but he agreed to become her mentor. This development had significance on several levels. Not only did Lagrange share with Sophie his scientific erudition, but he was able to gain access for her to the top circle of scientists and scholars of the day. Until this point, Sophie had been inhibited in her scholastic growth not only because she was a woman, but also because her social status (middle class) denied her access to the most sophisticated scholarly circles. It still must be said that Sophie Germain's education was never very formal. It was, in fact, disorganized and haphazard; this feature hampered her scientific progress throughout her life.

Sophie never lacked for initiative and daring. After reading his *Essai sur le Théorie des Nombres*, she initiated a correspondence with Adrien-Marie Legendre about some of the problems posed therein. The subsequent exchange of ideas can be considered as no less than a collaboration. In fact Legendre included some of her results in a supplement to the second edition of his *Essai*.

After Carl Friedrich Gauss published his book *Disquisitiones Arithmeticae* in 1804, Sophie Germain became fascinated by the subject of number theory. At the age of 28, she began corresponding with the great man. Gauss was not only the most distinguished mathematician of the day—he was perhaps the greatest mathematician of all time. Commonly hailed as the "Prince of Mathematics", Gauss held a god-like status in scientific circles. His accomplishments were copious and, indeed, earthshaking. Just as one instance, Gauss was the person who had finally proved the Fundamental Theorem of Algebra—in fact he offered five different proofs in his doctoral dissertation. Recall that we discussed this central theorem of algebra in Chapter 9 of the present text.

Painfully aware of the social tenor of the times, Sophie Germain conducted her correspondence with Gauss under the pseudonym "J. LeBlanc". She sent Gauss some of her results in number theory. A vigorous and lively correspondence was the result. In fact, between 1804 and 1809, she wrote a dozen letters to Gauss. Gauss gave Germain's work high praise, and even repeated that praise to his colleagues. In 1807 Gauss learned, only because of the French occupation of his hometown of Braunschweig, that his talented correspondent was a woman. Recalling the nasty fate that befell Archimedes, and fearing for Gauss's life, Sophie Germain contacted a French commander who was a friend of her family and asked for protection for Professor Gauss. When Gauss learned of Sophie's intervention on his behalf, he was lavishly grateful.

On learning that his talented mathematical correspondent was a woman, Professor Gauss was both delighted and thrilled.[1] Gauss's letter to Germain, after he discovered her true identity, reveals something about the man:

> But how can I describe my astonishment and admiration on seeing my esteemed correspondent Monsieur LeBlanc metamorphosed into this celebrated person, yielding a copy so brilliant it is hard to believe? The taste for the abstract sciences in general and, above all, for the mysteries of numbers, is very rare: this is not surprising, since the charms of this sublime science in all their beauty reveal themselves only to those who have the courage to fathom them. But when a woman, because of her sex, our customs and prejudices, encounters infinitely more obstacles than men, in familiarizing herself with their knotty problems, yet overcomes these fetters and penetrates that which is most hidden, she doubtless has the most noble courage, extraordinary talent, and superior genius. Nothing could prove to me in a more flattering and less equivocal way that the attractions of that science, which have added so much joy to my life, are not chimerical, than the favour with which you have honoured it.

Gauss did provide Sophie Germain with notable guidance for her research. But it was around this time that he accepted a position as Professor of Astronomy at the University of Göttingen, and his interest in number theory had waned. As previously mentioned, she subsequently initiated contact with Adrien-Marie Legendre, and sent him a description of what would turn out to be some of her most seminal work in the subject of number theory. Her communications with Legendre became quite sophisticated and extensive.

The assertion that if p and $2p + 1$ are both prime, then any solution of $x^p + y^p = z^p$ must satisfy the conclusion p divides at least one of x, y, z, is known as Sophie Germain's

[1] This eventuality is really to Gauss's credit. In spite of his many fine attributes, Gauss was not noted for lending support to struggling young mathematicians. But he made special efforts on behalf of Sophie Germain.

theorem.[2] It was included in a letter from Germain to Legendre in the early 1820s, and he presented it in a paper to the French Academy of Sciences in 1823.

As noted earlier in this book, Pierre de Fermat (1601–1665) was perhaps the most talented mathematical amateur of all time. We say "amateur" not to downgrade his efforts in any way, but instead to acknowledge the fact that he was a judge in Toulouse—that was his full-time avocation. He practiced his mathematics strictly as a hobby. But his work was so profound that he was part of the most distinguished scientific circles, and he corresponded with the important mathematicians of his day. In the late 1630s, Fermat wrote a marginal note in his copy of Claude Bachet's Latin translation of Diophantus's *Arithmetica* that, it would turn out, intrigued mathematicians for the next 360 years. Such eminent scholars as Euler, Legendre, Gauss, Abel, Dirichlet, Kummer, Cauchy, and of course Sophie Germain herself worked on this problem.

What Fermat said in the margin of his text was

> It is impossible to separate a cube into two cubes, or a biquadrate into two biquadrates, or in general any power higher than the second into two powers of like degree; I have discovered a truly remarkable proof which this margin is too small to contain.

In modern language—language, in fact, which Fermat could not have known—we can express Fermat's marginal assertion as follows:

> The equation $x^n + y^n = z^n$ has no positive integer solutions x, y, z when the integer exponent $n > 2$.

We know from our studies in Chapter 1 of the present text that, when $n = 2$, the equation has infinitely many triples (x, y, z) of solutions. It was Sophie Germain who proved one of the very first general results about Fermat's problem.[3]

In any event, Sophie Germain studied Fermat's problem and made this contribution:

> If x, y, z are positive integers and $x^5 + y^5 = z^5$ then one of x, y, or z must be divisible by 5.

Sophie later generalized this result to all exponents less than 100. Again, this was one of the first truly general results about Fermat's problem. It was the best work in the area until the contributions of Kummer in 1840.

We will discuss some of the details of Fermat's last problem, and of Sophie Germain's particular contributions to it, in a later section. For now we will continue with the story of her life.

Sophie Germain's education was erratic and uneven. Yet she never feared to forge into unknown territory. In 1808, the German physicist Ernst F. F. Chladni visited Paris; there

---

[2] It must be noted that a prime number p such that $2p + 1$ is also prime is known to this day as a "Sophie Germain prime". This is an important idea in modern number theory.

[3] It must be clearly understood here that Fermat never published nor communicated the details of his idea. His credibility was so strong that scholars of subsequent generations believed that he must have, indeed, had a solution to the problem. Fermat did in fact use his method of infinite descent to prove the result when $n = 4$. Those details were made public. But his general solution has never been seen.

The problem was finally solved by Andrew Wiles—he announced his solution (joint with his student Richard Taylor) in 1993 and published it in 1995 (see [WIL]). Wiles's solution is so complex, and so sophisticated, that it really suggests that Fermat must have been mistaken. And in fact Wiles himself has said that he believes that Fermat made an error.

he conducted experiments on vibrating plates, exhibiting the so-called Chladni figures. Subsequently the Institut de France set a prize competition with the following challenge:

> Formulate a mathematical theory of elastic surfaces and indicate just how it agrees with empirical evidence.

The Academy set a two-year deadline for contest submissions. Most mathematicians were frightened away from Chladni's problem because the great Lagrange had declared that the available methods were inadequate to attack such a situation. Thus in 1811 Sophie Germain submitted the one and only entry to the contest. In spite of her great raw talent, Sophie lacked the sophistication that would have been the product of a true formal education. She had no training in physics and did not know the calculus of variations. Her naivete and inexperience showed in her written work, and the judges downgraded it. The fact that she submitted the paper anonymously may have worked against her also. She was not awarded the prize. Clearly Sophie had much to learn. But the judges extended the term for the contest. She still had a chance to make her mark.

Her friend Professor Lagrange was in fact one of the judges for the French Prize, and he could see the merit and originality in her work. He came to Sophie's aid, and he helped her to correct various errors and to bring the paper up to acceptable scientific quality. She again submitted her efforts to the French Prize Committee. She was able to demonstrate that Lagrange's equation (from the calculus of variations) *did* yield Chladni's patterns in several instances, but she was unable to give a satisfactory derivation of the particular Lagrange equation for this physical problem from the first principles of the physics of elasticity. This time she received an Honorable Mention.

In 1816 Sophie Germain made a third submission to the French Academy's Mathematical Physics Contest. This time she won with her paper entitled *Memoir on the Vibrations of Elastic Plates*. She was awarded one kilogram of gold.

This was one of the seminal scientific achievements of Sophie's scientific career. Sadly, she did not attend the awards ceremony as she feared for the scandal that might result. Still, she was heartened by the recognition. The judges were still firm in pointing out that the paper had shortcomings, and many of these shortcomings would not be corrected for decades to come. Some highly placed scientists were still exceedingly critical; Poisson (1781–1840), for example, altogether rejected her efforts.

As one biographer phrases it:

> Although it was Germain who first attempted to solve a difficult problem, when others of more training, ability and contact built upon her work, and elasticity became an important scientific topic, she was closed out. Women were simply not taken seriously.

Sophie continued her research in the subject area, and published several more memoirs. She had considerable impact on the field. In fact the research of Sophie Germain is applied today in the construction of skyscrapers, and it also has applications in acoustics and elasticity. Sophie sensed that the judges did not fully appreciate her work. She also felt, with good reason, that the scientific community did not accord her the respect that she was due. Siméon Poisson was one of the judges for the prize, and he was particularly diffident in offering any appreciation for her insights.

In fact Sophie Germain submitted in 1825 a paper to a commission of the Institut de France. The membership at that time included Poisson, Gaspard Clair François Marie Riche de Prony (1755–1839), and Pierre-Simon Laplace (1749–1827). The work was due some criticisms, but the editorial board did not report these to the author. Instead it just ignored the submission. The paper did not see the light of day until it was recovered from the papers of de Prony and published in 1880.

One of the important consequences of Sophie's prize is that it introduced her into the first rank of academic circles. With the help of Jean-Baptiste Fourier, she became the first woman who was not the wife of an academy member (and therefore a courtesy guest) to be allowed to attend the sessions of the Academy of Sciences. She received praise and attention from the Institut de France and was also invited to attend their meetings. This was in fact the highest honor that the Institut ever bestowed on a woman.

Sophie Germain died at the age of 55 from complications due to breast cancer. Just before her death, Gauss had convinced the University of Göttingen to grant her an honorary degree. Sadly, she died before she was able to receive the honor.

Sophie Germain worked on mathematics and physics until the time of her death. Not long before she got sick and died, she outlined a philosophical essay entitled *Considé rations générales sur l'é tat des sciences ed des lettres*. The work was published posthumously in the journal *Oeuvres philosophiques*. The paper was most highly praised by August Comte (1798–1857). Even after she was diagnosed with breast cancer in 1829, Sophie Germain continued her work on number theory and the curvature of surfaces. She completed papers in both fields before her passing.

It is heartening to note that Sophie Germain has received perhaps more recognition for her work in recent times than during her brief lifetime. The street *Rue Sophie Germain* in Paris is named for her, as is the École Sophie Germain. There is a statue of Sophie standing in the courtyard of that institution of learning. There is also a Sophie Germain Hotel standing on the street named after her. The house in which she died, at 13 rue de Savoie, has been designated a historical landmark. Perhaps most significant for a mathematician, certain prime numbers—the prime numbers p such that $2p + 1$ is also prime—are now called "Sophie Germain primes". As examples, 2, 5, 11, 23 are Sophie Germain prime numbers.

Sophie Germain died in June 1831, and her death certificate listed her not as mathematician or scientist, but *rentier* (property holder).

Further Reading

Debra Charpentier, Women mathematicians, *The Two-Year College Mathematics Journal* 8(1977), 73–79.

John Ernest, Mathematics and sex, *The American Mathematical Monthly* 83(1976), 595–614.

11.2 Sophie Germain's Work on Fermat's Problem

We begin with some foundational ideas.

It was realized early on that Fermat's Last Problem need only be studied for prime number exponents. In studying the equation

$$x^n + y^n = z^n \qquad (*)$$

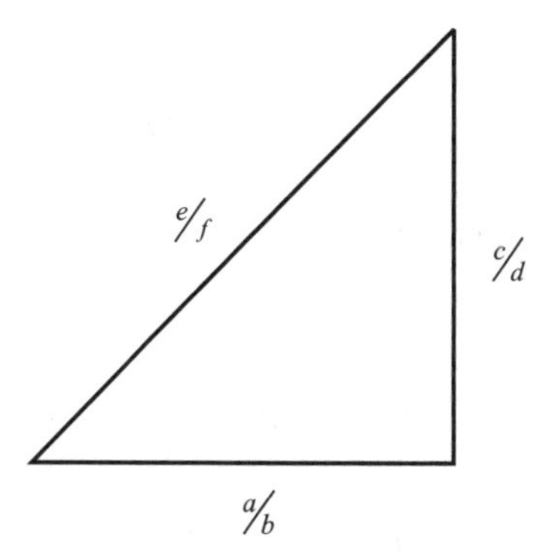

Figure 11.1. Not half of a perfect square.

from Fermat's point of view (the point of view of Diophantine equations, that is, equations for which we seek only integer roots), it is sufficient to consider only prime number exponents n. For if the equation $(*)$ has a set of integer roots x, y, z for some composite number exponent n, then say that $n = p \cdot r$, with p prime. We may write

$$(x^r)^p + (y^r)^p = (z^r)^p$$

thus revealing that the equation

$$\alpha^p + \beta^p = \gamma^p \tag{$**$}$$

has the integer roots x^r, y^r, z^r. Thus, if we can show that $(**)$ has no solutions, then it follows that $(*)$ has no solutions.

We have already noted that Fermat published a proof of his problem, using the method of infinite descent, for the case $n = 4$. In fact what Fermat showed more precisely is that if T is a right triangle, all of whose sides have rational length, then twice the area of the triangle cannot be a perfect square. Let us examine this assertion. Consider Figure 11.1. Let the lengths of the sides of the triangle be a/b, c/d, e/f—all rational numbers. Since the triangle is a right triangle, we know that

$$\left(\frac{a}{b}\right)^2 + \left(\frac{c}{d}\right)^2 = \left(\frac{e}{f}\right)^2. \tag{$\star$}$$

If it were the case that twice the area of the triangle is a perfect square, then we would have

$$2 \cdot \left[\frac{1}{2} \cdot \frac{a}{b} \cdot \frac{c}{d}\right] = \alpha^2$$

for some integer α. But then

$$\frac{a}{b} = \alpha^2 \cdot \frac{d}{c}.$$

Substituting this last identity into equation $(\star)$ gives

$$\alpha^4 \left(\frac{d}{c}\right)^2 + \left(\frac{c}{d}\right)^2 = \left(\frac{e}{f}\right)^2.$$

Multiplying through by $(c/d)^2$ and simplifying gives

$$\alpha^4 + \left(\frac{c}{d}\right)^4 = \left(\frac{ce}{df}\right)^2.$$

We may multiply through by $d^4 f^4$ to obtain

$$(\alpha df)^4 + (cf)^4 = (cdef)^2 .$$

But this just says that the equation

$$m^4 + n^4 = p^2$$

has a set of integer solutions—namely $m = \alpha df$, $n = cf$, $p = cdef$. And that in turn says that the equation

$$m^4 + n^4 = q^4 \tag{$\star\star$}$$

has a set of integer solutions. For we simply let $e = cdf$ and then the integer solutions to ($\star\star$) are $m = \alpha^2 d^2 f^2, n = c^2 f^2$, and $q = cdf$. Of course the argument can be reversed. Hence Fermat's statement about right triangles with rational sides is equivalent to Fermat's last problem for the exponent $n = 4$.

In 1770 Leonhard Euler (1707 C.E.–1783 C.E.) published a proof of Fermat's last theorem for $n = 3$, but it was considered to be incomplete (i.e., there was an error about divisibility properties of certain integers). Gauss also produced a proof for $n = 3$ that was published posthumously. What Sophie Germain wrote to Carl Friedrich Gauss on November 21, 1804 was:

> I add to this art some other considerations which relate to the famous equation of Fermat $x^n + y^n = z^n$ whose impossibility in integers has still only been proved for $n = 3$ and $n = 4$; I think I have been able to prove it for $n = p - 1$, p being a prime number of the form $8k + 7$. I shall take the liberty of submitting this attempt to your judgment, persuaded that you will not disdain to help with your advice an enthusiastic amateur in the science which you have cultivated with such brilliant success.

There is no extant record of either Sophie's proof or of Gauss's reply. It is possible that her proof was in error, and the result hence forgotten.

Many years later, in a letter penned in May of 1819, Sophie wrote to Gauss that

> Although I have labored for some time on the theory of vibrating surfaces (to which I have much to add if I had the satisfaction of making some experiments on cylindrical surfaces I have in mind), I have never ceased to think of the theory of numbers ... A long time before our Academy proposed as the subject of a prize the proof of the impossibility of Fermat's equation, this challenge ... has often tormented me.

Sophie Germain was corresponding with Legendre at about this same time. In one of her letters to Legendre from the early 1820s, she proved that if p is a "Sophie Germain prime" then any solution of Fermat's equation with exponent p must have the property that one of x, y, or z is divisible by p. As a concrete instance, if $p = 11$ then $2p + 1 = 23$. Since 23 is prime, we see that 11 is a Sophie Germain prime. Thus if x, y, z are integer solutions of the equation

$$x^{11} + y^{11} = z^{11}$$

then one of x, y, z must be divisible by 11.

In fact what she proved was a bit more, and a bit more sophisticated:

Theorem. *Let p be an odd prime. If there is another prime q with the properties that*

1. *The equation $x^p + y^p + z^p = 0 \bmod q$ implies that either x or y or z has residue 0 mod q;*
2. *The equation $n^p = p \bmod q$ is impossible for any integer n;*

then any integer solutions x, y, z of $x^p + y^p = z^p$ has the property that one of x, y, z is divisible by p.

Let us consider why this more general result implies the theorem that we attributed above to Sophie Germain. A classical result known as "Fermat's Little Theorem" says that if q is prime and $0 < a < q$, then $a^{q-1} = 1 \bmod q$. We will discuss why Fermat's Little Theorem is true in just a little while. If we take Fermat's result for granted, then we may reason as follows.

First of all, Fermat tells us that, with $q = 2p + 1$ with p and q both primes, and if $0 < a < q$, then

$$(a^p)^2 = a^{2p} = a^{q-1} = 1 \bmod q .$$

As a result,

$$(a^p - 1)(a^p + 1) = a^{2p} - 1 = 1 - 1 = 0 \bmod q .$$

We conclude that $a^p = \pm 1 \bmod q$. Thus, if x, y, and z are *not* congruent to 0 mod q then we may think of these three numbers as lying between 0 and q (mod q) and hence

$$x^p + y^p + z^p = \pm 1 \pm 1 \pm 1$$

which certainly cannot equal 0 mod q. This establishes the simpler version of Sophie Germain's theorem that we discussed above.

For each odd prime p less than 100, Sophie Germain gave a prime q to which the more general version of her theorem applies. For instance

Value of p	Value of q
3	7
5	11
7	29
11	23
13	53
17	137
19	191

Further Reading

Glenn James, On Fermat's last theorem, *The American Mathematical Monthly* 41(1934), 419–424.

John McCleary, How not to prove Fermat's last theorem, *The American Mathematical Monthly* 96(1989), 410–420.

Lee Whitt, Medical cozenage on Fermat's last theorem, *The College Mathematics Journal* 16(1985), 55–56.

Exercises

1. Add some meat to the discussion in the text: If a, b, c were solutions to the Fermat-type equation
$$x^{18} + y^{18} = z^{18},$$
then there would also exist solutions to the equations
$$x^9 + y^9 = z^9$$
and
$$x^6 + y^6 = z^6$$
and
$$x^3 + y^3 = z^3.$$
Discuss this problem in class.

2. A *Diophantine equation* is a polynomial equation for which we seek integer solutions. Such equations are named after Diophantus, who considered problems of this type more than 2000 years ago.
Certainly Pythagoras's equation
$$x^2 + y^2 = z^2$$
is an instance of a Diophantine equation. Another example is
$$2x + y^2 = z^2. \qquad (*)$$
How many solutions can you find to $(*)$? Are there infinitely many? Discuss this question in class.

3. Refer to problem 2 for terminology. One of the most famous Diophantine equations is *Pell's equation*. This is an equation of the form
$$x^2 - dy^2 = 1.$$
The idea is to fix an integer value for d and then seek solutions x, y. In the present exercise we explore what happens when d is a perfect square. Discuss this problem in class.
Say that $d = n^2$, for n some positive integer. Then Pell's equation becomes
$$x^2 - n^2y^2 = 1.$$
We may factor the left-hand side as
$$(x - ny)(x + ny) = 1.$$
Now the only way that the product of two integers can equal 1 is if they are both equal to $+1$ or they are both equal to -1. In either event, we have
$$x - ny = x + ny.$$
From this we conclude that $ny = 0$. But then $x^2 = 1$ so $x = \pm 1$. Thus the problem is solved and the solution is trivial.

4. We learned in Exercise 3 that there is no interest in studying Pell's equation $x^2 - dy^2 = 1$ when d is a perfect square. So let us examine some other cases.

Consider the Pell equation for $d = 5$. Verify that $x = 9$, $y = 4$ is a solution. It happens that there is also a solution when $x = 161$. Can you find it? There are also solutions when $x = 2889$ and when $x = 51841$ and when $x = 930248$. Use the computer to aid your search.

Now examine the Pell equation for $d = 7$. There are solutions when $x = 8$ and also when $x = 127$. See whether you can find them.

5. The *Waring problem*, which was ultimately solved by David Hilbert in 1909, asserts that every positive integer can be written as the sum of at most four perfect squares. As an example, $22 = 4^2 + 2^2 + 1^2 + 1^2$. We discussed this problem in Exercise 9 of Chapter 2. Find a positive integer that has two different Waring decompositions (i.e., decompositions as the sum of at most four perfect squares). Can you find a positive integer that has three Waring decompositions?

6. In 1964, J. M. Gandhi proposed that the Diophantine equation

$$x^5 + y^5 = 2z^5$$

will only have solutions when $x = y$ or $x = -y$. Explain why there will always be solutions under these conditions. What about the equation

$$x^3 + y^3 = 2z^3\,?$$

7. Consider the Diophantine equation

$$x^2 + xy + y^2 = z^2\,. \qquad (\star)$$

Verify that $x = -7$, $y = 15$, $z = 13$ is a solution. It is known that if an equation of type $(\star)$ has one solution then it has infinitely many. Can you find any other solutions of $(\star)$?

8. Andrew Wiles's solution to Fermat's last problem asserts that the Diophantine equation

$$x^n + y^n = z^n$$

has no integer solutions x, y, z when the exponent n is any integer greater than 2. In particular, the equation

$$x^3 + y^3 = z^3$$

has no integer solutions. But there are some near misses. For example, the equation

$$3086^3 + 21588^3 = 21609^3$$

is only off by 1. Do the calculation and explain what this means.

Now verify the equation

$$(9u^3 + 1)^3 + (9u^4)^3 = (9u^4 + 3u)^3 + 1$$

for u a positive integer. Explain why this will generate infinitely many "near misses" for Fermat's equation with exponent 3.

9. The television show *The Simpsons*, in the "Treehouse of Horror" episode from the sixth season, revealed the following counterexample to Fermat's Last Theorem:

Take your TI-83 calculator and compute

$$\left[1782^{12} + 1841^{12}\right]^{1/12} .$$

You will find the answer to be 1922. Thus

$$1782^{12} + 1841^{12} = 1922^{12} .$$

This apparently contradicts Andrew Wiles's result that Fermat's equation with exponent 12 has no solutions. [Note right away that the left-hand side is the sum of an even number and an odd number, while the right-hand side is plainly even. So something must be wrong here.]

The example illustrates why calculators must be used with caution—especially when you are dealing with large numbers. *Because the calculator has the capacity to handle only so many digits*—so it rounds off the answers. The conundrum described in this problem is the result of roundoff error.

Discuss this problem in class.

10. A variant of Pell's equation (see Exercises 3 and 4) is

$$x^2 - dy^2 = -4 . \qquad (\dagger)$$

Show that equation (†) has a solution when $d = 8$. What can you determine when $d = 20$ or $d = 40$? It turns out that the Pell-type equation

$$x^2 - dy^2 = -1$$

has no solution in these cases. Discuss the situation in class.

11. Project: We say that a prime number p is a *Sophie Germain prime* if both p is prime and $2p + 1$ is prime. List the first ten Sophie Germain primes.

It is not known whether there are infinitely many Sophie Germain primes or not. The largest known such prime has 51910 decimal digits. In 1825 Sophie Germain proved that Fermat's last theorem (that the equation $x^n + y^n = z^n$ has no integer solutions) is true when the exponent n is a Sophie Germain prime.

A Sophie Germain prime $p > 3$ is of the form $6k - 1$ or, equivalently, $p = 5 \mod 6$, as is its matching safe prime $2p+1$ (a safe prime is one of the form $2p+1$). Prove these statements. We note that the other form for a prime $p > 3$ is $6k + 1$ or, equivalently, $p = 1 \mod 6$, and that 3 divides $2p + 1$. Thus such a p cannot be a Sophie Germain prime. Prove these statements as well. See [HAW] for more on this topic.

12. Project: Sophie Germain proved that if $x^5 + y^5 = z^5$ then one of x, y, z must be divisible by 5. This is a special case of the result discussed in Exercise 11. Verify this assertion for x, y, z all smaller than 100. The book [HAW] considers these matters.

13. Project: Prove Sophie Germain's identity

$$x^4 + 4y^4 = (x^2 + 2y^2 + 2xy)(x^2 + 2y^2 - 2xy) .$$

Explain the relevance of this identity to Fermat's last theorem. See [DED] for this topic.

12

Cauchy and the Foundations of Analysis

12.1 Introduction

Augustin Louis Cauchy (1789–1857) was born in Paris during a tumultuous period of French history. His father feared for the family's safety because of the political events connected with the French revolution, so he moved the family to Arcueil. There life was hard. The family often did not have adequate food.

As a consequence, the family moved back to Paris. This was good for young Cauchy, as his father (in addition to educating the young man himself) had Laplace and Lagrange as regular guests in the Cauchy home. Lagrange especially took a special interest in young Cauchy's development. He advised the senior Cauchy that his son should learn languages before engaging in mathematics. Thus Augustin Cauchy enrolled in 1802 in the École Centrale du Panthéon. He spent two years on the study of classical languages. Beginning in 1804, young Cauchy began his study of mathematics. He took the entrance exam for the École Polytechnique in 1805; he managed to place second. Cauchy graduated in 1807 and then entered engineering training at the École des Ponts et Chaussées. Cauchy excelled at engineering, and graduated in 1810. He then took up his first job in Cherbourg to develop the port facilities for Napoleon's English invasion fleet. Cauchy nevertheless maintained his interest in mathematics; he kept with him a copy of Laplace's *Mécanique Céleste* and one of Langrange's *Théorie des Fonctions.*

Cauchy was very busy in these days. He was also a devout Catholic, and devoted time to his religion. Evidently his religious beliefs were causing him trouble in his relationships with others. He wrote to his mother

> So they are claiming that my devotion is causing me to become proud, arrogant and self-infatuated. ... I am now left alone about religion and nobody mentions it to me anymore ...

Cauchy's continued devotion to mathematics led to some original research, and in 1811 he proved that the angles of a convex polyhedron are determined by its faces. Legendre and Malus encouraged him; he submitted that paper, and in 1812 produced another on the same topic.

Cauchy felt that, if he was to further his mathematical career, then he must return to Paris. This he did in 1812. He was ill at the time, though it appears that this illness was psychosomatic and stemmed from depression. Cauchy did manage to continue his mathematical activities, and he produced more papers. He was expected to return to Cherbourg after he recovered from his illness. But this was not consistent with Cauchy's mathematical ambitions. He was able to arrange to instead stay in Paris and work on the Ourcq Canal project. He applied for an associate professorship at the École des Ponts et Chausées but failed.

Cauchy continued to do strong research in mathematics, but his several applications for academic posts failed. In 1814 Cauchy published a tract on definite integrals that later became the basis for his theory of complex functions.

In 1815 Cauchy finally got lucky. He failed to obtain the mechanics chair at the École Polytechnique, but instead landed an assistant professorship. The following year he won the Grand Prix of the French Academy of Sciences for a research paper on waves. He really established his reputation when he wrote a paper solving one of Fermat's problems on polygonal numbers. Cauchy managed to obtain membership in the National Academy of Sciences when Carnot and Monge fell from political favor and were dismissed.

In 1817, Biot left Paris for a scientific expedition; thus Cauchy was able to fill his position at the Collége de France. At this time Cauchy engaged in detailed and rigorous studies of analysis—both of a real variable and a complex variable. He published a number of important tracts, and gave many lectures on different parts of the subject.

Cauchy did not have good relations with other mathematicians and scientists. His strict Catholic views put him on the side of the Jesuits against the Académie des Sciences. He would bring religion into his scientific work; for instance, he did so in a report on a paper on the theory of light in 1824. He attacked the author for his view that Newton had not believed that people had souls. He was described by a journalist who said:

> ...it is certainly a curious thing to see an academician who seemed to fulfil the respectable functions of a missionary preaching to the heathens. An example of how Cauchy treated colleagues is given by Poncelet whose work on projective geometry had, in 1820, been criticised by Cauchy:
>
> ...I managed to approach my too rigid judge at his residence ...just as he was leaving ...During this very short and very rapid walk, I quickly perceived that I had in no way earned his regards or his respect as a scientist ...without allowing me to say anything else, he abruptly walked off, referring me to the forthcoming publication of his Leçons à l'École Polytechnique where, according to him, 'the question would be very properly explored.'

His relationship with Galois and Abel during this period was unfortunate. Abel, who visited the Institute in 1826, wrote of him:

> Cauchy is mad and there is nothing that can be done about him, although, right now, he is the only one who knows how mathematics should be done.

By 1830 the political events in Paris and his many years of hard work had taken their toll. Augustin Cauchy decided to take a break. He left Paris in September 1830, after the revolution of July, and spent a short time in Switzerland. There he was an important assistant in setting up the Académie Helvétique, but this project collapsed as it became caught up in political events.

Political events in France meant that Cauchy was now required to swear an oath of allegiance to the new regime and when he failed to return to Paris to do so he lost all his positions there. In 1831 Cauchy went to Turin and after some time there he accepted an offer from the King of Piedmont as a chair of theoretical physics. He taught in Turin from 1832. The later-to-be-famous Italian mathematician and engineer Luigi Federico Menabrea (1809 C.E.–1896 C.E.) attended these courses in Turin and described them as follows:

> were very confused, skipping suddenly from one idea to another, from one formula to the next, with no attempt to give a connection between them. His presentations were obscure clouds, illuminated from time to time by flashes of pure genius. ... of the thirty who enrolled with me, I was the only one to see it through.

In 1833 Cauchy went from Turin to Prague in order to follow Charles X and to tutor his grandson. However he was not very successful in teaching the prince:

> ... exams were given each Saturday. ... When questioned by Cauchy on a problem in descriptive geometry, the prince was confused and hesitant. ... There was also material on physics and chemistry. As with mathematics, the prince showed very little interest in these subjects. Cauchy became annoyed and screamed and yelled. The queen sometimes said to him, soothingly, smilingly, "too loud, not so loud."

Cauchy had meetings with Bolzano in 1834, and this evidently had some influence over his definition of continuity. He returned to Paris in 1838. Cauchy continued to refuse to take the oath of allegiance, and this worked to his detriment. He could not teach, could not attend meetings, and could not draw a salary (even though he had a position). Cauchy was a candidate for a chair at the Collége de France, but failed because of his political and religious beliefs. Although Cauchy's mathematical output was diminished during this period, he did important work on differential equations, mathematical physics, and astronomy.

King Louis Philippe was overthrown in 1848 and Cauchy regained his university position. But he continued to be a thorn in the side of his colleagues. Liouville beat him out for an important chair, and Cauchy was quite bitter. Their relationship suffered.

Unfortunately the final period of Cauchy's life was marred by a dispute with Duhamel over the priority to a discovery concerning shock waves. Poncelet sided with Duhamel, and Cauchy was proved to be wrong. But he would never admit his error, and he went to his grave bitter over the matter.

Cauchy's final hour is described by his daughter as follows:

> Having remained fully alert, in complete control of his mental powers, until 3:30 A.M., my father suddenly uttered the blessed names of Jesus, Mary and Joseph. For the first time, he seemed to be aware of the gravity of his condition. At about four o'clock, his soul went to God. He met his death with such calm that made us ashamed of our unhappiness.

Augustin Cauchy's life was somewhat chaotic. But his effect on mathematics was profound and undeniable. He did seminal work on complex variables, the foundations of calculus and analysis, and many other areas of mathematics and physics.

Cauchy is often credited with making calculus rigorous. His difficult and profound work laid to rest 150 years of doubts about the foundations of Newton and Leibniz's seminal new tool. But nobody is perfect. Galois's personal torment was due in no small part to the fact that Cauchy failed to promote the young man's work—as he had *promised* to do. Some of Abel's failures and frustrations can similarly be laid at the feet of Cauchy.

In the present chapter we shall explore some of Cauchy's ideas about the structure of the real number system.

Further Reading

Walter Felscher, Bolzano, Cauchy, epsilon, delta, *The American Mathematical Monthly* 107(2000), 844–862.

Israel Kleiner, Evolution of the function concept: A brief survey, *The College Mathematics Journal* 20(1989), 282–300.

12.2 Why Do We Need the Real Numbers?

In everyday life—buying groceries, or machining a dowel, or keeping the books for a business, or building a bookshelf—the rational number system is more than adequate for our needs. You would never walk into the hardware store and say, "I need a board that is $\sqrt{3}$ feet long" or "I need a bolt of length $2/\pi$ centimeters." In the first place, nobody would know what you were talking about. In the second place, there is nothing in our sentient world that requires that kind of precision. So why are mathematicians not content with the rational numbers?

There are at least two reasons, and we have seen one of them already. One reason is that there are very elementary constructions in geometry that give rise to numbers that are *not* rational. For example (see Figure 12.1), the diagonal of a square of side 1 has length that is not rational. In fact, we showed in Chapter 1 that its length is $\sqrt{2}$ and that $\sqrt{2}$ is an irrational number.

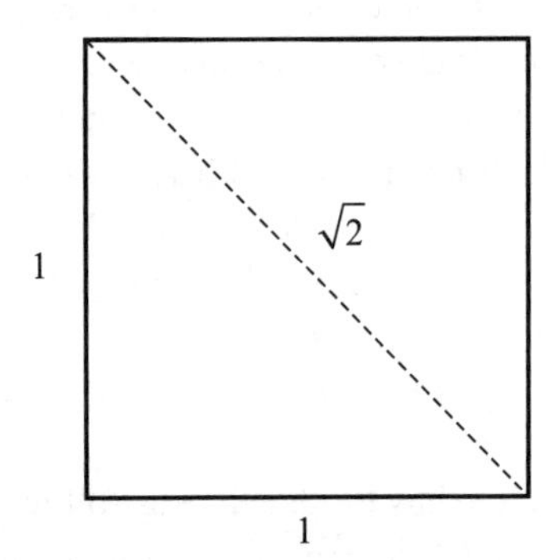

Figure 12.1. Irrational numbers arising from geometry.

The more profound and subtle reason that we find the rational numbers inadequate has to do with the notion of *completeness*. Consider the sequence of rational numbers

$$1, 1.4, 1.41, 1.414, 1.4142, 1.41421, 1.414213, 1.4142135, 1.41421356, \ldots.$$

In fact these numbers represent the decimal expansion of $\sqrt{2}$. They *converge* (i.e., become ever closer to) $\sqrt{2}$. But we know that $\sqrt{2}$ is *not* rational. And that is the rub: it is possible for a sequence of rational numbers to become ever closer together, to *seem* to converge to some number—but the number is not there! It is not part of the rational number system.

Of course the same problem occurs with the sequence of rational numbers (i.e., the decimal expansion) that represents $\sqrt{7}$ or π or any of the other irrational numbers.[1] In calculus, and more generally in mathematical analysis, we frequently find ourselves wanting

[1] We shall learn in Chapter 16 that there are, in a precise sense, more irrational numbers than rational. So the irrational numbers are something that we *must* reckon with.

to pass to limits. We need to know that the number which will play the role of limit will actually be there waiting. We cannot have *gaps* in the number system. This is why we need the real numbers.

In fact, historically, mathematicians used the real numbers before they actually knew what they were or how to construct them. Today we have a much firmer grasp of why the real numbers *exist*. Part of the purpose of this chapter is to share these ideas with you.

Further Reading

E. A. Maier and David Maier, A construction of the real numbers, *The Two-Year College Mathematics Journal* 4(1973), 31–35.

Richard Steiner, Teaching about the real numbers, *The American Mathematical Monthly* 91(1984), 202–203.

12.3 How to Construct the Real Numbers

In Chapter 9 we constructed the complex numbers. This was a little tricky, but it could be done in a page or so. The construction of the reals is quite a bit more difficult. There are actually a number of different methods for constructing the reals. The method of Dedekind cuts (see [KRA1]) is rather technical, but yields the necessary properties rather quickly. One can also invoke the surreal numbers (described in Exercise 15 of Chapter 17) to get a construction of the reals. We shall instead describe below a technique involving equivalence classes of Cauchy sequences.

We cannot give a completely rigorous treatment here, but shall content ourselves with a heuristic description of the process. And it is, truly, a matter of necessity being the mother of invention. We shall use the *fault* of the rational numbers that we have just described as the foothold on which to build our new number system.

We begin by considering the collection of all sequences of rational numbers that "appear to" tend to a limit. Thus the sequence

$$1, \frac{1}{2}, \frac{1}{3}, \ldots$$

is one that we shall want to treat while the sequence

$$1, -1, 1, -1, \ldots$$

is not.

A sequence will be denoted, in short form, by $\{a_j\} = \{a_j\}_{j=1}^{\infty}$. This stands for the list

$$a_1, a_2, a_3, \ldots .$$

A sequence is said to be *Cauchy*, or to satisfy the *Cauchy condition*, if its elements seem to get closer and closer together. So, for instance, the sequence

$$1, \frac{1}{2}, \frac{1}{3}, \ldots$$

is certainly Cauchy. Also the sequence

$$3, 3.1, 3.14, 3.141, 3.1415, 3.14159, \ldots$$

is Cauchy. Notice that the notion of being Cauchy has nothing to do with whether the sequence actually has a limit. All we care about is whether the sequence is getting close together—and therefore seems to *demand* a limit. Whether or not the limit is actually there is a separate issue.

We will say that two Cauchy sequences are *equivalent* if they seem to tend to the same limit. A convenient way to say this precisely is that $\{a_j\}$ and $\{b_j\}$ are equivalent if $a_j - b_j \to 0$. As an example, the sequences

$$1, \frac{1}{2}, \frac{1}{3}, \frac{1}{4}, \ldots$$

and

$$1, \frac{1}{2}, \frac{1}{4}, \frac{1}{8}, \frac{1}{16}, \ldots$$

are equivalent. We can see this precisely because the jth term of the first sequence is $a_j = 1/j$ and the jth term of the second sequence is $b_j = 1/2^{j-1}$ and

$$a_j - b_j = \frac{1}{j} - \frac{1}{2^{j-1}} \longrightarrow 0 .$$

Now we form *equivalence classes* of sequences. An equivalence class of sequences is a collection of sequences, all of which are equivalent to one another. It is most convenient to describe an equivalence class in this way:

Let $\mathcal{A}$ be the collection of sequences which are equivalent to

$$1, 0, 0, 0, \ldots .$$

What other sequences are in $\mathcal{A}$? Well, the sequence

$$1, 2, 3, 0, 0, 0, \ldots$$

is certainly in $\mathcal{A}$ (check this for yourself). The sequence

$$1, \frac{1}{2}, \frac{1}{3}, \frac{1}{4}, \ldots$$

is in $\mathcal{A}$. In fact any sequence that tends to 0 is in $\mathcal{A}$.

For the record, let us enunciate precisely what we mean for a sequence to be Cauchy. Let $\{a_j\}$ be a sequence. This sequence is *Cauchy* if, for each $\epsilon > 0$, there is an $N > 0$ such that when $j, k > n$ then

$$|a_j - a_k| < \epsilon .$$

This says quite directly that, for any degree of closeness ϵ, we can achieve that closeness in our sequence by simply going far enough out in the sequence. By the same token, two sequences $\{a_j\}$ and $\{b_j\}$ are equivalent provided that, for any $\epsilon > 0$, there is an $N > 0$ such that if $j > N$ then

$$|a_j - b_j| < \epsilon .$$

Again, this says that we can achieve any degree ϵ of closeness between the two sequences just by going far enough out in the sequences.

Another equivalence class is the set $\mathcal{B}$ of all sequences that are equivalent to

$$1, 1, 1, 1, 1, \ldots.$$

Which other sequences are in $\mathcal{B}$? Well, the sequence

$$1 + 1, 1 + \frac{1}{2}, 1 + \frac{1}{3}, 1 + \frac{1}{4}, \ldots$$

lies in $\mathcal{B}$. In fact any sequence that tends to 1 lies in $\mathcal{B}$.

Let us give the collection of equivalence classes that we have been describing a name. Let us call the collection $\mathcal{R}$. We claim that $\mathcal{R}$ forms a number system. And the number system $\mathcal{R}$ has operations of addition, subtraction, multiplication, and division. That number system contains the rational numbers. And it contains some interesting new numbers as well. Finally, the number system $\mathcal{R}$ (unlike the rational number system) *has no holes in it* (as we shall learn below). In fact the new number system $\mathcal{R}$ has the important feature that *any sequence in it that seems to tend to a limit actually has a limit in that number system.* We say that $\mathcal{R}$ *is complete.*

Let us now verify some of these assertions. Let $\mathcal{A}$ denote the equivalence class containing the sequence $\{a_j\}$ and let $\mathcal{B}$ denote the equivalence class containing $\{b_j\}$. Then we add these two equivalence classes according to the rule that $\mathcal{A} + \mathcal{B}$ is the equivalence class of all sequences equivalent to $\{a_j + b_j\}$. Likewise:

- The difference of these two equivalence classes, $\mathcal{A} - \mathcal{B}$, is the equivalence class of all sequences equivalent to $\{a_j - b_j\}$.
- The product of these two equivalence classes, $\mathcal{A} \cdot \mathcal{B}$, is the equivalence class of all sequences equivalent to $\{a_j \cdot b_j\}$.
- If the elements b_j tend to a nonzero limit, then $\mathcal{A}/\mathcal{B}$ is the equivalence class of all sequences equivalent to $\{a_j/b_j\}$.

In mathematics we say that the arithmetic operations are defined on sequences *termwise.* So $\mathcal{R}$ forms a number system.

We have asserted that $\mathcal{R}$ contains the rational numbers. In fact we will exhibit a mapping

$$\Phi : \mathbb{Q} \longmapsto \mathcal{R}$$

that takes each rational number to a unique representative in our new number system. Let $q \in \mathbb{Q}$. Define $\Phi(q)$ to be the collection of all sequences equivalent to $\{q, q, q, \ldots\}$. So q is represented in our new number system by the sequences that are equivalent to a constant sequence modeled on q. The mapping Φ gives us a means to think of the rational numbers as a "sub-number-system" of the real numbers.

We have also asserted that $\mathcal{R}$ will contain new numbers—besides the rationals. An example of such a "new number" is the collection of all sequences equivalent to

$$\{1, 1.4, 1.41, 1.414, 1.4142, 1.41421, 1.414213, 1.4142135, 1.41421356, \ldots\}.$$

This equivalence class does *not* correspond to any rational number—in the manner we have just indicated. In fact, as we know from earlier discussions, it corresponds to $\sqrt{2}$. Another example of a "new number" is the equivalence class corresponding to

$$\{3, 3.1, 3.14, 3.141, 3.1415, 3.14159, 3.141592, 3.1415926, 3.14159265, \ldots\}.$$

This equivalence class corresponds to π. Of course there are many more examples.

Let us conclude this discussion by saying a few words about completeness. This is a subtle matter, and we cannot tell the whole story.

Let $\mathcal{A}_1, \mathcal{A}_2, \mathcal{A}_3, \ldots$ be a sequence of elements of the set $\mathcal{R}$. So each $\mathcal{A}_j$ is an equivalence class of sequences. And assume that the $\mathcal{A}_j$ are getting closer and closer together—in the sense that when j and k are large, then the terms of $\mathcal{A}_j$ and $\mathcal{A}_k$ are getting very close together. Then we would like to say that there is a limiting sequence. Where would we find such a sequence? We can produce it by taking the first term of a sequence in $\mathcal{A}_1$, the second term of a sequence in $\mathcal{A}_2$, the third term of a sequence in $\mathcal{A}_3$, and so forth. Call this manufactured sequence $\mathcal{X}$. Then it will be the case that

$$\mathcal{A}_j \longrightarrow \mathcal{X}.$$

So we have achieved our goal: Our new number system $\mathcal{R}$ has no holes in it. Sequences that are getting closer and closer together always converge; in other words, the number system is complete. We honor our achievement by calling this new number system the *real numbers* and denoting it by the special letter $\mathbb{R}$.

It is well to review the subtle construction that we have just performed:

- We first observed that the rational number system $\mathbb{Q}$ is inadequate because it is not complete. This means that it has holes: there are sequences of rational numbers that get ever closer together but that tend to no limit *within the rational numbers*.

- We constructed a new number system by considering equivalence classes of sequences of rational numbers that seem to get ever closer together.

- We confirmed that this new collection of objects has operations of addition, subtraction, multiplication, and divison. It is a number system.

- Our new collection of objects also has the special feature that any sequence of these objects that seems to get ever closer together will actually have a limit (*unlike the rational numbers*) in that number system. Thus the new number system repairs the flaw that the rational number system $\mathbb{Q}$ suffered. This development is so important that we give the new number system a name. We call these the *real numbers* $\mathbb{R}$.

Further Reading

David Hecker, Constructing a Poincaré line with straightedge and compass, *The College Mathematics Journal* 34(2003), 362–366.

Shlomo Vinner and David Tall, Existence statements and constructions in mathematics and some consequences to mathematics teaching, *The American Mathematical Monthly* 89(1982), 752–756.

12.4 Properties of the Real Number System

Most any construction of the real number system is rather opaque. The main thing for you to understand is that we now have a number system that

(a) Contains the rational numbers $\mathbb{Q}$;

(b) Is an ordered field;

(c) Is complete in the sense that every Cauchy sequence has a limit *in the number system.*

In fact the real number system is the smallest ordered field containing the rationals that satisfies these properties,[2] and it is unique.

In practice, when we use the real numbers, we do not think about how they were constructed. We just use them. Classical mathematicians like Cauchy and Weierstrass and Gauss also used the real numbers with such blissful abandon; but *they did not know how the real numbers were constructed.* They were not sure of the axioms of the real numbers. In other words, they were not certain of the fundamental properties of this number system. As a result, they were frequently hamstrung by paradoxes and contradictions. Our rigorous construction of the reals puts our study on a firm foundation and avoids logical traps and pitfalls.

In the present section we shall explore some of the deeper and more important properties of the real number systems. Some of these will have foundations in "intuitively obvious" statements that you have used (perhaps rather vaguely) in the past. Others will be new to you. We begin with a property of sequences that will be key to all of our other results.

12.4.1 Bounded Sequences

One of the useful operations in mathematical analysis is to extract a convergent subsequence from a given sequence. It is here that the special properties of the real numbers (especially completeness) play a role. We will use the result of this subsection in all of our further discussions.

In what follows we shall use the concept of *subsequence*. If $\{x_j\}$ is a given sequence of numbers, then a subsequence is denoted by $\{x_{j_k}\}$. This is a list of numbers taken from the original sequence in order (but omitting some terms). A concrete example makes the idea clear:

$$\left\{\frac{1}{j}\right\} = \frac{1}{1}, \frac{1}{2}, \frac{1}{3}, \ldots$$

is a sequence of real numbers. An example of a subsequence—one of many possible—is

$$\frac{1}{2}, \frac{1}{5}, \frac{1}{13}, \frac{1}{47}, \ldots.$$

Notice that the subsequence contains certain elements of the original sequence—*in order*—but not necessarily all of the original elements.

Theorem 12.1 (Bolzano-Weierstrass). *Let $[a, b]$ be a closed, bounded interval. Let $\{x_j\}$ be a sequence contained in $[a, b]$. Then there is a subsequence $x_{j_1}, x_{j_2}, \ldots$ and a limit point*

[2]That is to say, any other ordered field with all these properties, and that contains the rationals, will also contain the reals.

$\ell \in [a, b]$ *such that*

$$\lim_{k\to\infty} x_{j_k} = \ell .$$

The reasoning behind this result is very natural. Let us assume for simplicity that the interval is actually $[0, 1]$. So $\{x_j\}$ is a sequence that lies in $[0, 1]$. Now divide the interval into two pieces: $[0, 1] = [0, 1/2] \cup [1/2, 1]$. Clearly one of those pieces must contain infinitely many elements of the sequence. Say it is the first piece $[0, 1/2]$. Now divide this piece into two further pieces: $[0, 1/2] = [0, 1/4] \cup [1/4, 1/2]$. Since there are infinitely many elements of the sequence in $[0, 1/2]$, therefore one of the two pieces $[0, 1/4]$ or $[1/4, 1/2]$ must contain infinitely many elements of the sequence. Say that it is $[1/4, 1/2]$. For the third step, subdivide the interval again. And continue.

We end up with a nested, decreasing collection of intervals whose intersection is the limit point ℓ that we seek. See Figure 12.2.

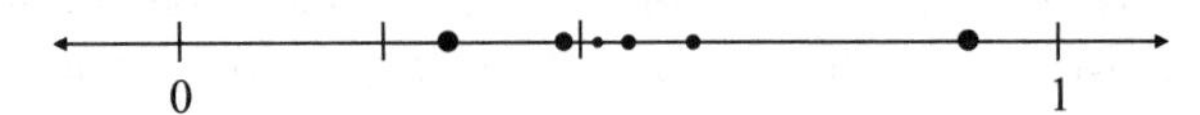

Figure 12.2. Intersection of a decreasing collection of intervals.

In the next subsection we shall immediately see a dramatic application of Theorem 12.1.

12.4.2 Maxima and Minima

Let f be a function with domain the closed interval $[a, b]$. Is there a point M such that f takes its maximal value at M (i.e., $f(M)$ is as large as possible)? Is there a point m such that f takes its minimal value at m (i.e., $f(m)$ is as small as possible)?

In general, the answer to these questions is "no". Have a look at Figure 12.3. It shows a function with domain $[0, 1]$ that has no maximum and no minimum. Often in mathematics, if we look hard at what is going wrong, we can figure out how to make it right. We realize that the function exhibited in Figure 12.3 is *not* the sort of function that we want to be thinking about. We wish to consider functions whose graphs do not have jumps or gaps. In other words, we want functions whose graphs can be drawn in an uninterrupted sweep, without lifting the pencil from the paper. These are the *continuous functions*. The theorem is:

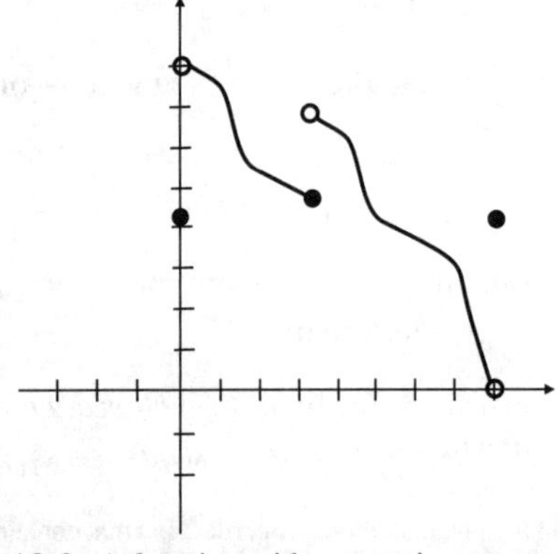

Figure 12.3. A function with no maximum or minimum.

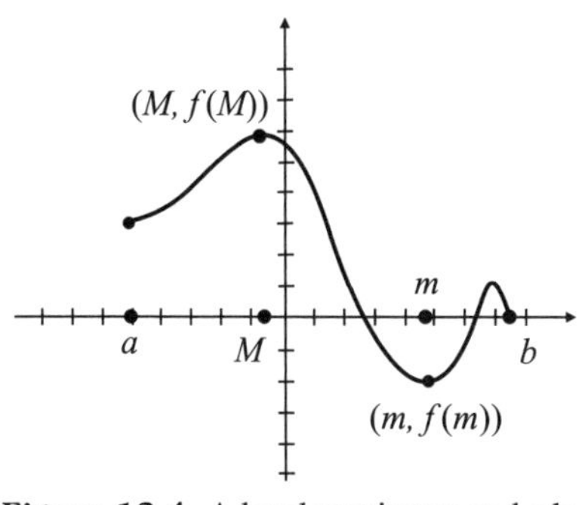

Figure 12.4. A local maximum and a local minimum for f.

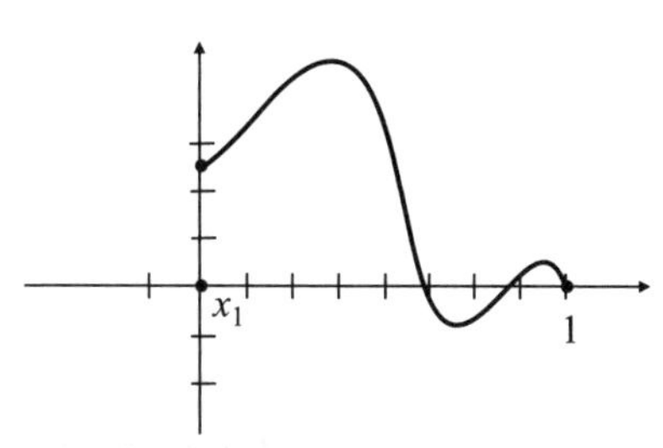

Figure 12.5. A step in the algorithm for finding M.

Theorem 12.2. *Let f be a continuous function with domain the closed, bounded interval $[a, b]$. Then there is a point $m \in [a, b]$ such that*

$$f(m) \leq f(x)$$

for every $x \in [a, b]$. Also there is a point $M \in [a, b]$ such that

$$f(M) \geq f(x)$$

for every $x \in [a, b]$. See Figure 12.4.

We should like to discuss how the special properties of the real number system make this result true.

Now let us give an algorithm for finding M. First, it simplifies our notation and helps us to see things more clearly if we assume that the interval is $[0, 1]$. Consider the net[3] of two points $\{0, 1\}$. Choose the one of these two points at which f takes the greater value. Call it x_1. Refer to Figure 12.5.

Now look at the net $\{0, 1/2, 1\}$. Choose the one of these three points at which f takes the greater value. Call it x_2. Refer to Figure 12.6.

Now look at the net $\{0, 1/4, 1/2, 3/4, 1\}$. Choose the one of these five points at which f takes the greater value. Call it x_3. Refer to Figure 12.7.

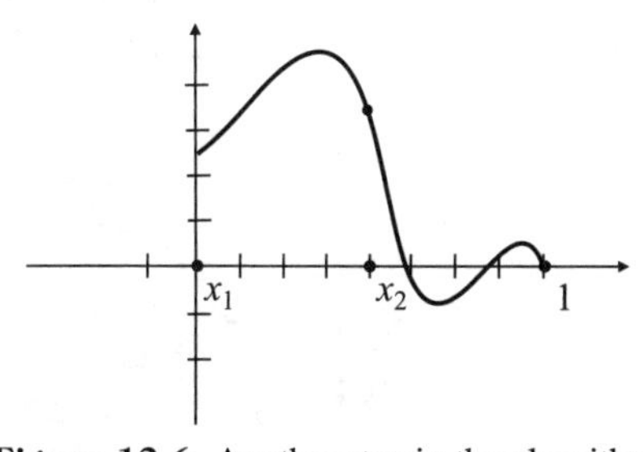

Figure 12.6. Another step in the algorithm.

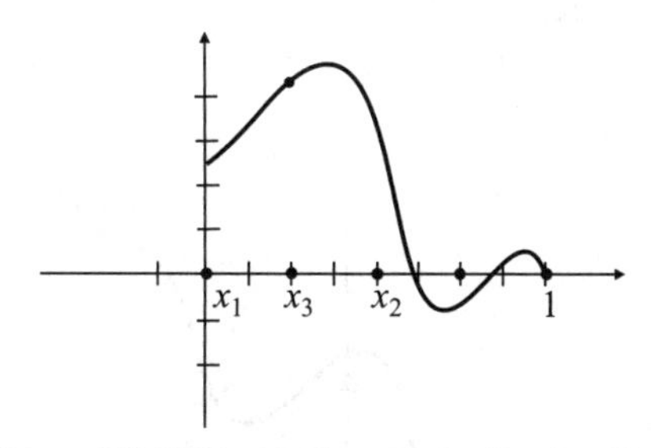

Figure 12.7. Yet another step in the algorithm.

Continue in this manner. The result is a sequence $\{x_j\}$. We may now apply Theorem 12.1 to obtain a subsequence $\{x_{j_k}\}$ that converges to a limit point M in $[a, b]$. That is the point where f takes its maximum value.

Of course a completely analogous discussion produces the point m where f takes its minimum value.

[3]Here, by a "net," we simply mean a finite list of points.

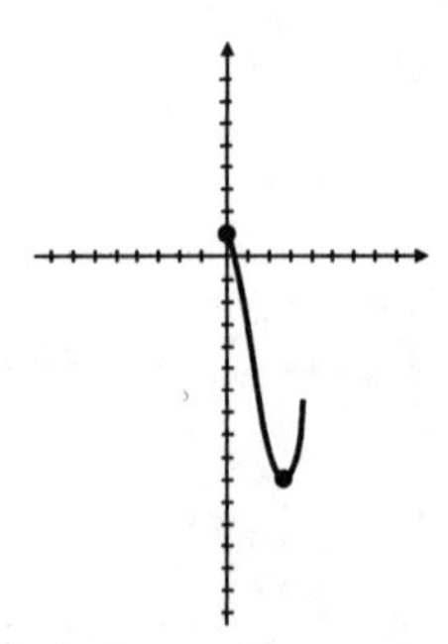

Figure 12.8. Extrema of a cubic function.

EXAMPLE 12.1. The continuous function $f(x) = x^3 - 3x^2 - 8x + 2$ on the closed, bounded interval $[0, 4]$ takes its minimum value on the interval at the point $m = 1 + \sqrt{33}/3$ and its maximum value on the interval at the point $M = 0$. Neither of these statement is at all obvious (they require ideas from calculus). But our theorem guarantees that m and M will exist. And these are their values. See Figure 12.8.

For You to Try In fact we learned some of the basic ideas of calculus in Section 7.3. One of the more useful techniques was Fermat's test. Use that test now to confirm the values for m and M in Example 12.1.

For You to Try Consider the continuous function $g(x) = x \sin[1/x]$ on the interval $[0, \pi]$ (where it is understood that $g(0) = 0$). Discuss in class where g will take its minimum value and where g will take its maximum value.

12.4.3 The Intermediate Value Property

Let f be a function with domain the interval $[a, b]$. Say that $f(a) = \alpha$ and $f(b) = \beta$. See Figure 12.9. Let γ be a number that lies between α and β. Is there a number c between a and b such that $f(c) = \gamma$?

In general, the answer to this question is "no". See Figure 12.10. But, as before, if we look hard at what is going wrong then we can sometimes see how to make it right. The

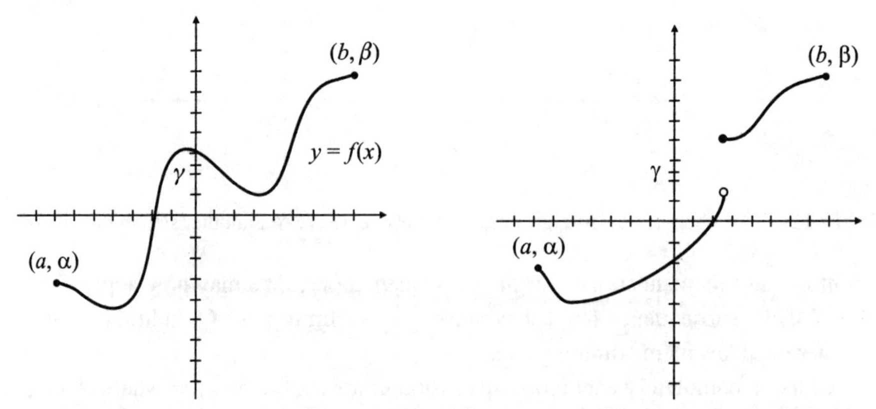

Figure 12.9. The Intermediate Value Property. **Figure 12.10.** Failure of the Intermediate Value Property.

function in Figure 12.10 is clearly not the sort of function that we want to be considering. The graph has a break, or jump, in it. That is to say, the function is *discontinuous*. What we want is a function whose graph is unbroken, *one whose graph can be drawn without lifting our pencil from the paper*. The correct statement of the Intermediate Value Property is as follows:

Theorem 12.3. *Let f be a continuous function with domain the interval $[a, b]$. Let $f(a) = \alpha$ and $f(b) = \beta$. If γ is a number between α and β then there is a number c between a and b such that $f(c) = \gamma$.*

We should like to give an indication of why this result is true; particularly, we want to see why the special properties of the real number system—especially completeness—play a role. For convenience and simplicity, let us suppose that $\alpha < \gamma < \beta$. We shall use the same mechanism that we used to find maxima and minima.

First, it simplifies our notation and helps us to see things more clearly if we assume that the interval is $[0, 1]$. Consider the net of two points $\{0, 1\}$. Moving from left to right, choose the last one of these two points at which f takes a value less than γ. Call it x_1^1. The next point in the net will be a point where f takes a value greater than γ. Call it x_2^1. Refer to Figure 12.11. [In fact for this simple, two-point net, it will certainly be true that $x_1^1 = 0$ and $x_2^1 = 1$. There are no other choices; and certainly $f(0) = \alpha < \gamma$ and $f(1) = \beta > \gamma$.]

Now look at the net $\{0, 1/2, 1\}$. Moving from left to right, choose the last one of these three points at which f takes a value less than γ. Call it x_1^2. The next value in the net will be a point where f takes a value greater than γ. Call it x_2^2. Refer to Figure 12.12.

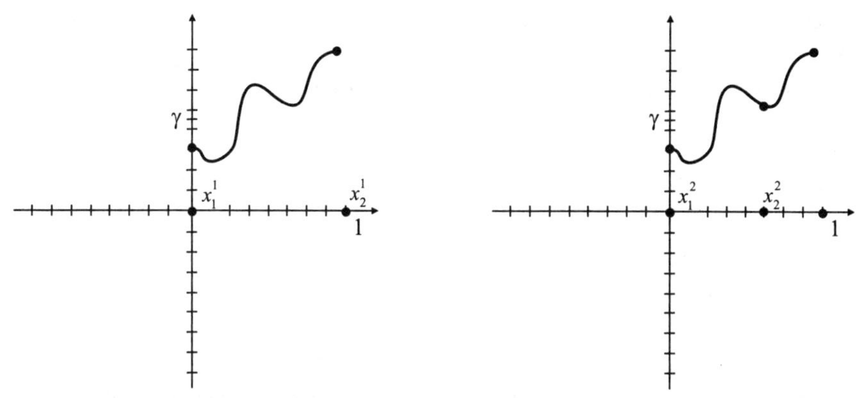

Figure 12.11. A two-point net. **Figure 12.12.** Another point in the net.

Now look at the net $\{0, 1/4, 1/2, 3/4, 1\}$. Moving from left to right, choose the last one of these five points at which f takes a value less than γ. Call it x_1^3. The next value in the net will be a point x_2^3 where f takes a value greater than γ. Refer to Figure 12.13.

Continue in this manner. The result is a pair of sequences $\{x_1^j\}$ and $\{x_2^j\}$ in $[0, 1]$. By Theorem 10.1, we may find a subsequence of $\{x_1^j\}$ that converges to a point c. Correspondingly, there is a subsequence of $\{x_2^j\}$ that converges to c. By the way that we chose these points, these sequences are clamping down on a point (namely c) at which f takes the intermediate value γ. [This is where we are using the fact that f is continuous.]

The intermediate value theorem is particularly satisfying and clear when it is used in practice. So let us look at some examples.

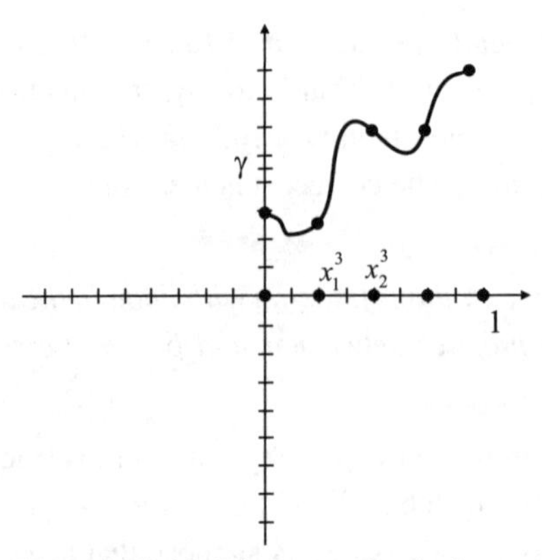

Figure 12.13. The next value in the net.

EXAMPLE 12.2. Explain why every positive real number has a square root.

Solution. Let r be a fixed, positive real number. Consider the function $f(x) = x^2 - r$. Then $f(0) = -r < 0$. And $f(r+1) = [r^2 + 2r + 1] - r = r^2 + r + 1 > 0$. By the intermediate value property, there is some number c between 0 and $r + 1$ such that $f(c) = 0$. But this says that $c^2 - r = 0$ or $c^2 = r$. We have produced the required square root for r.

For You to Try Let k be a positive integer. Show that every positive real number has a kth root.

It must be stressed that the square root that we produced in Example 12.2, and the kth root that you produced in the last **For You to Try** are determined by way of existence proofs. Our arguments are *nonconstructive*, and give no clue of the actual values of the roots.

EXAMPLE 12.3. Show that the polynomial $p(x) = x^5 - 3x^2 + 1$ has a root between 0 and 1.

Solution. We notice that $p(0) = 1 > 0$ and $p(1) = -1 < 0$. By the intermediate value property, there is a number c between 0 and 1 such that $p(c) = 0$. That is the required result.

For You to Try Show that the function $f(x) = x^{11} - 9x^7 + 5x^3 + 2$ has at least one real root.

Remark In our discussion of the Brouwer fixed point theorem in Section 18.4 we shall use the Intermediate Value Property to prove that if $f : [0, 1] \to [0, 1]$ is continuous, then there is a fixed point $p \in [0, 1]$ such that $f(p) = p$. This is a dramatic geometric application of the idea.

Further Reading

R H Bing, Properties of the real numbers, *The American Mathematical Monthly* 67(1960), 35–37.

Peter Renz, The path to hell, *The College Mathematics Journal* 16(1985), 9–11.

Exercises

1. The product of two rational numbers is always rational. But the product of two irrational numbers need not be irrational. Explain these two statements, and give examples where appropriate.

2. The sum of two rational numbers is always rational. But the sum of two irrational numbers need not be irrational. Explain these two statements, and give examples where appropriate.

3. The quotient of two rational numbers is always rational. But the quotient of two irrational numbers need not be irrational. Explain these two statements, and give examples where appropriate.

4. Prove that the sum of a rational number and an irrational number is irrational.

5. Prove that the product of a nonzero rational number and an irrational number is irrational.

6. A version of Gauss's lemma states that if a positive, whole number has a rational square root, then in fact it has a whole number (integer) square root. Explain this statement.

7. The square root of a rational number can be irrational; give an example. Can the square root of an irrational number be rational?

8. Show that there is a number $x > 0$ such that $x^2 = \sin x$.

9. Suppose that f is a continuous function with domain $[0, 1]$ and range $[0, 1]$. Show that there is a point $c \in [0, 1]$ such that $f(c) = c$. [*Hint:* Consider the function $g(x) = f(x) - x$ and apply the intermediate value theorem in an appropriate manner.] We will discuss this fixed point result of Brouwer and its generalizations in Chapter 18.

10. A certain business has assets that begin at time $t = 0$ by equaling 0. After a certain amount of time, at time $t = K$, the business goes bust and the total assets are again 0. By applying the maximum/minimum theorem to the asset function $A(t)$ on the interval $[0, K]$, we may conclude that there is a time when the assets are a maximum. Explain this reasoning. What are the guaranteed points where the assets A are a minimum?

11. Prove that between every two irrational numbers on the line there is a rational number. [*Hint:* Observe that $\sqrt{2} < 2 < \sqrt{5}$.]

12. Prove that between every two rational numbers on the line there is an irrational number. [*Hint:* Observe that $1 < \sqrt{2} < 2$.]

13. **Project:** Prove that every positive real number has a natural logarithm. See [KRA2] for the basis of these ideas.

14. **Project:** Prove that every real number is the limit of a sequence of rational numbers having denominator a power of 2. Again see [KRA1].

15. **Project:** Prove that the numbers of the form $a + b\sqrt{7}$, with a and b rational, are dense in the real numbers (that is to say, each real number is the limit of a sequence of such numbers). Refer to [KRA6].

16. **Project:** Prove that every rational number is the limit of a sequence of numbers of the form $a + b\pi$, where a and b are rational. The book [KRA6] is a good source for these ideas.

13

The Prime Numbers

13.1 The Sieve of Eratosthenes

The prime numbers are the units of arithmetic. A *prime number* is defined to be a positive whole number (i.e., an integer) that has no divisors except 1 and itself. By convention, we do not consider 1 to be a prime. Thus

- 2 is prime because the only divisors of 2 are 1 and 2.
- 3 is prime because the only divisors of 3 are 1 and 3.
- 4 is *not* prime because 2 divides 4.
- 5 is prime because the only divisors of 5 are 1 and 5.
- 6 is *not* prime because 2 and 3 divide 6.
- 7 is prime because the only divisors of 7 are 1 and 7.
- 8 is *not* prime because 2 divides 8.
- 9 is *not* prime because 3 divides 9.

and so forth. The *Fundamental Theorem of Arithmetic* states that every positive integer can be factored into primes in one and only one way. For example,

$$98 = 2 \cdot 7^2$$

and

$$12745656 = 2^3 \cdot 3^2 \cdot 7 \cdot 11^3 \cdot 19 .$$

Except for rearranging the order of the factors, there is no other way to factor either of these two numbers.

The ancient Greeks had a particular fascination with the primes. One such was Eratosthenes (276 B.C.E.–194 B.C.E.). Eratosthenes was born in Cyrene, Libya, North Africa. His teachers included Lysanias of Cyrene and Ariston of Chios. The latter made Eratosthenes part of the stoic school of philosophy. Around 240 B.C.E., Eratosthenes became the third librarian of the great library of Alexandria (founded about 300 B.C.E., this library was later destroyed and all the books lost). One of Eratosthenes's most important works was the *Platonicus*, a tract that dealt with the mathematics underlying Plato's *Republic*.

Eratosthenes devised a *sieve* for creating a list of the primes. In fact sieve methods are still used today to attack such celebrated problems as the Goldbach conjecture.[1] Here is how Eratosthenes's method works.

We begin with an array of the positive integers:

1	2	3	4	5	6	7	8	9	10	11	12	13	14	15	16
17	18	19	20	21	22	23	24	25	26	27	28	29	30	31	32
33	34	35	36	37	38	39	40	41	42	43	44	45	46	47	48
49	50	51	52	53	54	55	56	57	58	59	60	61	62	63	64
65	66	67	68	69	70	71	72	73	74	75	76	77	78	79	80
81	82	83	84	85	86	87	88	89	90	91	92	93	94	95	96

...

First we cross out 1. Then we cross out all the multiples of 2 (but not 2 itself):

~~1~~	2	3	~~4~~	5	~~6~~	7	~~8~~	9	~~10~~	11	~~12~~	13	~~14~~	15	~~16~~
17	~~18~~	19	~~20~~	21	~~22~~	23	~~24~~	25	~~26~~	27	~~28~~	29	~~30~~	31	~~32~~
33	~~34~~	35	~~36~~	37	~~38~~	39	~~40~~	41	~~42~~	43	~~44~~	45	~~46~~	47	~~48~~
49	~~50~~	51	~~52~~	53	~~54~~	55	~~56~~	57	~~58~~	59	~~60~~	61	~~62~~	63	~~64~~
65	~~66~~	67	~~68~~	69	~~70~~	71	~~72~~	73	~~74~~	75	~~76~~	77	~~78~~	79	~~80~~
81	~~82~~	83	~~84~~	85	~~86~~	87	~~88~~	89	~~90~~	91	~~92~~	93	~~94~~	95	~~96~~

...

Now we proceed by crossing out all the multiples of 3 (but not 3 itself):

~~1~~	2	3	~~4~~	5	~~6~~	7	~~8~~	~~9~~	~~10~~	11	~~12~~	13	~~14~~	~~15~~	~~16~~
17	~~18~~	19	~~20~~	~~21~~	~~22~~	23	~~24~~	25	~~26~~	~~27~~	~~28~~	29	~~30~~	31	~~32~~
~~33~~	~~34~~	35	~~36~~	37	~~38~~	~~39~~	~~40~~	41	~~42~~	43	~~44~~	~~45~~	~~46~~	47	~~48~~
49	~~50~~	~~51~~	~~52~~	53	~~54~~	55	~~56~~	~~57~~	~~58~~	59	~~60~~	61	~~62~~	~~63~~	~~64~~
65	~~66~~	67	~~68~~	~~69~~	~~70~~	71	~~72~~	73	~~74~~	~~75~~	~~76~~	77	~~78~~	79	~~80~~
~~81~~	~~82~~	83	~~84~~	85	~~86~~	~~87~~	~~88~~	89	~~90~~	91	~~92~~	~~93~~	~~94~~	95	~~96~~

...

You can see that the numbers we are crossing out *cannot* be prime since, in the first instance, they are divisible by 2, and in the second instance, they are divisible by 3. Now we will cross out all the numbers that are divisible by 5 (why did we skip 4?) but not 5 itself. The result is:

~~1~~	2	3	~~4~~	5	~~6~~	7	~~8~~	~~9~~	~~10~~	11	~~12~~	13	~~14~~	~~15~~	~~16~~
17	~~18~~	19	~~20~~	~~21~~	~~22~~	23	~~24~~	~~25~~	~~26~~	~~27~~	~~28~~	29	~~30~~	31	~~32~~
~~33~~	~~34~~	~~35~~	~~36~~	37	~~38~~	~~39~~	~~40~~	41	~~42~~	43	~~44~~	~~45~~	~~46~~	47	~~48~~
49	~~50~~	~~51~~	~~52~~	53	~~54~~	~~55~~	~~56~~	~~57~~	~~58~~	59	~~60~~	61	~~62~~	~~63~~	~~64~~
~~65~~	~~66~~	67	~~68~~	~~69~~	~~70~~	71	~~72~~	73	~~74~~	~~75~~	~~76~~	77	~~78~~	79	~~80~~
~~81~~	~~82~~	83	~~84~~	~~85~~	~~86~~	~~87~~	~~88~~	89	~~90~~	91	~~92~~	~~93~~	~~94~~	~~95~~	~~96~~

...

[1] The Goldbach conjecture is the problem of showing that any even number greater than 4 can be written as the sum of two odd primes. It is one of the great unsolved problems of modern number theory.

Let us perform this procedure just one more time, by crossing out all multiples of 7 (why can we safely skip 6?), but not 7 itself:

~~1~~	2	3	~~4~~	5	~~6~~	7	~~8~~	~~9~~	~~10~~	11	~~12~~	13	~~14~~	~~15~~	~~16~~
17	~~18~~	19	~~20~~	~~21~~	~~22~~	23	~~24~~	~~25~~	~~26~~	~~27~~	~~28~~	29	~~30~~	31	~~32~~
~~33~~	~~34~~	~~35~~	~~36~~	37	~~38~~	~~39~~	~~40~~	41	~~42~~	43	~~44~~	~~45~~	~~46~~	47	~~48~~
~~49~~	~~50~~	~~51~~	~~52~~	53	~~54~~	~~55~~	~~56~~	~~57~~	~~58~~	59	~~60~~	61	~~62~~	~~63~~	~~64~~
~~65~~	~~66~~	67	~~68~~	~~69~~	~~70~~	71	~~72~~	73	~~74~~	~~75~~	~~76~~	~~77~~	~~78~~	79	~~80~~
~~81~~	~~82~~	83	~~84~~	~~85~~	~~86~~	~~87~~	~~88~~	89	~~90~~	~~91~~	~~92~~	~~93~~	~~94~~	~~95~~	~~96~~

...

And now here is the punchline: The numbers that remain (i.e., that are *not* crossed out) are those that are *not* multiples of 2, nor multiples of 3, nor multiples of 5, nor multiples of 7. In fact those that remain are not multiples of anything. They are the primes:

$$2\,,\,3\,,\,5\,,\,7\,,\,11\,,\,13\,,\,17\,,\,19\,,\,23\,,\,29\,,\,31\,,\,37\,,\,41\,,\,43\,,$$

$$47\,,\,53\,,\,59\,,\,61\,,\,67\,,\,71\,,\,73\,,\,79\,,\,83\,,\,89\ldots$$

And on it goes. No prime was missed. The sieve of Eratosthenes will find them all.

But a number of interesting questions arise. We notice that our list contains a number of *prime pairs*: $\{3,5\}$, $\{5,7\}$, $\{11,13\}$, $\{17,19\}$, $\{29,31\}$, $\{41,43\}$, $\{71,73\}$. These are primes in sequence that differ by just 2. How many such pairs are there? Could there be infinitely many prime pairs? To date, nobody knows the answer to this question. Another old problem is whether the list of primes contains arbitrarily long arithmetic sequences, that is, sequences that are evenly spaced. For example, $3, 5, 7$ is a list of primes that is evenly spaced (by units of 2). Also $41, 47, 53, 59$ is evenly spaced (by units of 6). It was only just proved in 2004 by Green and Tao [TAG] that the primes *do contain* arbitrarily long arithmetic sequences.

An even more fundamental question is this: How many prime numbers are there altogether? Perhaps 100? Or 1000? Or 1,000,000? In fact it was Euclid (330–275 B.C.E.) who determined that there are infinitely many prime numbers. We shall discuss his argument in the next section.

Further Reading

Karl Greger, Random sieving and the prime number theorem, *The Two-Year College Mathematics Journal* 5(1974), 41–46.

Fred Gross, On the sieve of Eratosthenes, *The American Mathematical Monthly* 70(1963), 1079–1081.

13.2 The Infinitude of the Primes

Euclid's reasoning constitutes one of the earliest number-theoretic proofs in all of mathematics. Furthermore, it was a proof by contradiction—a method that did not take a firm hold in mathematics until the late nineteenth century (thanks to David Hilbert).

It is worth contemplating the significance of what Euclid did. He wanted to prove something rather abstract: that the collection of all primes is not finite. In the language of Georg

Cantor (see Chapter 16), which was not developed until 2000 years later, one might achieve this goal by setting up a one-to-one correspondence with another set that is known to be infinite. But this technique did not exist in Euclid's time. What he did was quite daring: he played a game of chess[2] with the universe. Euclid hypothesized that the collection of all prime integers is finite, and he showed that that led to a contradiction. The conclusion that one can derive from this argument is this: If one adjoins to all the known true statements in mathematics the statement

There are only finitely many prime integers.

then a contradiction results. If we take it for granted that the known body of mathematics is consistent and correct, then the only possible conclusion is that the contradiction lies in the sentence that we adjoined. That is a rather simple sentence. The only thing that could be wrong with it is that it is false. Thus there must be infinitely many primes. We now provide the details of Euclid's proof.

What Euclid said is this: Suppose to the contrary that there are only finitely many primes. Call them $p_1, p_2, \ldots, p_K$. This is alleged to be a *complete list of all the primes.* But now consider the number

$$p^* = p_1 \cdot p_2 \cdot \cdots \cdot p_{K-1} \cdot p_K + 1\,.$$

Then p^* is greater than all the primes we have listed. So p^* cannot be a prime. Thus it must be what we call a *composite number*. That means that it factors into primes. So it is divisible by some of our primes. Well, if we divide p^* by p_1 then p_1 goes evenly into $p_1 \cdot p_2 \cdot \cdots \cdot p_{K-1} \cdot p_K$, but there is a remainder because of the $+1$. So, when we divide p^* by p_1 there is a remainder 1. Thus p_1 is not a prime factor of p^*. Likewise, when we divide p^* by p_2 there is a remainder 1. So p_2 is not a prime factor of p^*. Proceeding iteratively, we see in fact that none of the primes on our list is a prime factor of p^*. And those are all the primes that there are! Thus p^* is *not* composite. It must be prime. But that is a contradiction, because we claimed at the outset to have listed all the primes as $p_1, \ldots, p_K$. The only possible conclusion is that there are infinitely many primes.

The reader will want to review this proof a few times in order to become comfortable with it. At first it all seems like a trick. But it is definitely not. It is strict and rigorous mathematical reasoning. And it shows that there is an unlimited supply of prime numbers.

Further Reading

Michael W. Ecker, The alluring lore of cyclic numbers, *The Two-Year College Mathematics Journal* 14(1983), 105–109.

Michael Rubinstein, A formula and a proof of the infinitude of the primes, *The American Mathematical Monthly* 100(1993), 388–392.

13.3 More Prime Thoughts

The prime numbers have inspired mountains of research by mathematicians. How are they distributed? What are their relationships? Today, the theory of prime numbers plays a decisive role in the making and breaking of secret codes (see Chapter 22).

[2]In chess one sometimes threatens to sacrifice his queen in order to gain an advantage and win the game. Euclid took this a step further and threatened to sacrifice *everything*.

We give in this section an example of one of the first profound facts that was ever discovered about the primes. Its author was Fermat. His result says this. Let p be any prime. Let n be any integer greater than 1. Then p evenly divides $n^p - n$. In fact Fermat stated this result but did not prove it. A formal proof was given years later by Gottfried Wilhelm von Leibniz (1646–1716) around 1683. We shall give a very elementary verification of the fact here, using little more than high school algebra. The one big idea that we shall invoke is the binomial expansion—see Section 10.2.

Now let us turn our attention to Fermat's little result. First let us test it out for $n = 2$. Fix any prime p. The claim is that p divides $2^p - 2$. It is convenient to write

$$2^p - 2 = (1+1)^p - 2$$

and then to apply the binomial theorem. Thus

$$\begin{aligned} 2^p - 2 = \Bigg[& 1^p + p \cdot 1^{p-1} \cdot 1 + \frac{p\cdot(p-1)}{2\cdot 1} \cdot 1^{p-2} \cdot 1^2 \\ & + \frac{p\cdot(p-1)\cdot(p-2)}{3\cdot 2\cdot 1} \cdot 1^{p-3} \cdot 1^3 \\ & + \cdots + \frac{p\cdot(p-1)\cdot(p-2)\cdot\cdots\cdot 4\cdot 3}{(p-2)\cdot(p-1)\cdot\cdots\cdot 3\cdot 2\cdot 1} \cdot 1^2 \cdot 1^{p-2} \\ & + \frac{p\cdot(p-1)\cdot(p-2)\cdot\cdots\cdot 4\cdot 3\cdot 2}{(p-1)\cdot(p-2)\cdot\cdots\cdot 3\cdot 2\cdot 1} \cdot 1^1 \cdot 1^{p-1} \\ & + \frac{p\cdot(p-1)\cdot(p-2)\cdot\cdots\cdot 3\cdot 2\cdot 1}{p\cdot(p-1)\cdot(p-2)\cdot\cdots\cdot 3\cdot 2\cdot 1} \cdot 1 \cdot 1^p \Bigg] - 2. \end{aligned}$$

Now we perform some simple arithmetic to rewrite this as

$$\begin{aligned} 2^p - 2 = \Bigg[& 1 + p \cdot 1 + \frac{p\cdot(p-1)}{2\cdot 1} \cdot 1 \\ & + \frac{p\cdot(p-1)\cdot(p-2)}{3\cdot 2\cdot 1} \cdot 1 \\ & + \cdots + \frac{p\cdot(p-1)}{2\cdot 1} \cdot 1 + \frac{p}{1} \cdot 1 + 1 \Bigg] - 2. \end{aligned}$$

Of course the two additive 1s and the 2 cancel. We are left with

$$2^p - 2 = p + \frac{p\cdot(p-1)}{2\cdot 1} + \frac{p\cdot(p-1)\cdot(p-2)}{3\cdot 2\cdot 1} + \cdots + \frac{p\cdot(p-1)}{2\cdot 1} + \frac{p}{1}.$$

Lo and behold, each summand on the right is a binomial coefficient and hence (as previously noted) a whole number. And each is divisible by p. In fact, the factor of p appears explicitly in each term. Thus the right-hand side is divisible by p, hence so is the left.

Will this trick work again for $n = 3$? Let us try. We calculate

$$\begin{aligned}
3^p - 3 &= (2+1)^p - 3 \\
&= \Bigg[2^p + p \cdot 2^{p-1} \cdot 1 + \frac{p \cdot (p-1)}{2 \cdot 1} \cdot 2^{p-2} \cdot 1^2 \\
&\quad + \frac{p \cdot (p-1) \cdot (p-2)}{3 \cdot 2 \cdot 1} \cdot 2^{p-3} 1^3 \\
&\quad + \cdots + \frac{p \cdot (p-1) \cdot (p-2) \cdot \cdots \cdot 4 \cdot 3}{(p-2) \cdot (p-1) \cdot \cdots \cdot 3 \cdot 2 \cdot 1} \cdot 2^2 \cdot 1^{p-2} \\
&\quad + \frac{p \cdot (p-1) \cdot (p-2) \cdot \cdots \cdot 4 \cdot 3 \cdot 2}{(p-1) \cdot (p-2) \cdot \cdots \cdot 3 \cdot 2 \cdot 1} \cdot 2^1 \cdot 1^{p-1} \\
&\quad + \frac{p \cdot (p-1) \cdot (p-2) \cdot \cdots \cdot 3 \cdot 2 \cdot 1}{p \cdot (p-1) \cdot (p-2) \cdot \cdots \cdot 3 \cdot 2 \cdot 1} \cdot 1^p \Bigg] - 3 .
\end{aligned}$$

Now we perform some simple arithmetic to rewrite this as

$$\begin{aligned}
3^p - 3 = \Bigg[& 2^p + p \cdot 2^{p-1} + \frac{p \cdot (p-1)}{2 \cdot 1} \cdot 2^{p-2} \\
& + \frac{p \cdot (p-1) \cdot (p-2)}{3 \cdot 2 \cdot 1} \cdot 2^{p-3} \\
& + \cdots + \frac{p \cdot (p-1)}{2 \cdot 1} \cdot 2^2 + \frac{p}{1} \cdot 2 + 1 \Bigg] - 3 .
\end{aligned}$$

We may rearrange this in order to take advantage of our result for $n = 2$. The result is

$$\begin{aligned}
3^p - 3 = \Big[2^p - 2 \Big] + \Bigg[& p \cdot 2^{p-1} + \frac{p \cdot (p-1)}{2 \cdot 1} \cdot 2^{p-2} \\
& + \frac{p \cdot (p-1) \cdot (p-2)}{3 \cdot 2 \cdot 1} \cdot 2^{p-3} \\
& + \cdots + \frac{p \cdot (p-1)}{2 \cdot 1} \cdot 2^2 + \frac{p}{1} \cdot 2 \Bigg] .
\end{aligned}$$

Now of course $2^p - 2$ is divisible by p because we have already proved that result. And all the other terms on the right have plainly exhibited a factor of p. Thus the right-hand side is divisible by p. We conclude that $3^p - 3$ is divisible by p.

Let us try this one more time with $n = 4$. We calculate that

$$\begin{aligned}4^p - 4 &= (3+1)^p - 4 \\ &= \Bigg[3^p + p\cdot 3^{p-1}\cdot 1 + \frac{p\cdot(p-1)}{2\cdot 1}\cdot 3^{p-2}\cdot 1^2 \\ &\quad + \frac{p\cdot(p-1)\cdot(p-2)}{3\cdot 2\cdot 1}\cdot 3^{p-3}1^3 \\ &\quad + \cdots + \frac{p\cdot(p-1)\cdot(p-2)\cdot\cdots\cdot 4\cdot 3}{(p-2)\cdot(p-1)\cdot\cdots\cdot 3\cdot 2\cdot 1}\cdot 3^2\cdot 1^{p-2} \\ &\quad + \frac{p\cdot(p-1)\cdot(p-2)\cdot\cdots\cdot 4\cdot 3\cdot 2}{(p-1)\cdot(p-2)\cdot\cdots\cdot 3\cdot 2\cdot 1}\cdot 3^1\cdot 1^{p-1} \\ &\quad + \frac{p\cdot(p-1)\cdot(p-2)\cdot\cdots\cdot 3\cdot 2\cdot 1}{p\cdot(p-1)\cdot(p-2)\cdot\cdots\cdot 3\cdot 2\cdot 1}\cdot 1^p\Bigg] - 4\,.\end{aligned}$$

Now we perform some simple arithmetic to rewrite this as

$$\begin{aligned}4^p - 4 = \Bigg[3^p &+ p\cdot 3^{p-1} + \frac{p\cdot(p-1)}{2\cdot 1}\cdot 3^{p-2} \\ &+ \frac{p\cdot(p-1)\cdot(p-2)}{3\cdot 2\cdot 1}\cdot 3^{p-3} \\ &+ \cdots + \frac{p\cdot(p-1)}{2\cdot 1}\cdot 3^2 + \frac{p}{1}\cdot 3 + 1\Bigg] - 4\,.\end{aligned}$$

We may rearrange this in order to take advantage of our result for $n = 3$. The result is

$$\begin{aligned}4^p - 4 = \Big[3^p - 3\Big] + \Bigg[p\cdot 3^{p-1} &+ \frac{p\cdot(p-1)}{2\cdot 1}\cdot 3^{p-2} \\ &+ \frac{p\cdot(p-1)\cdot(p-2)}{3\cdot 2\cdot 1}\cdot 3^{p-3} \\ &+ \cdots + \frac{p\cdot(p-1)}{2\cdot 1}\cdot 3^2 + \frac{p}{1}\cdot 3\Bigg].\end{aligned}$$

Now of course $3^p - 3$ is divisible by p because we have already proved that result. And all the other terms on the right have plainly exhibited a factor of p. Thus the right-hand side is divisible by p. We conclude that $4^p - 4$ is divisible by p.

The pattern is becoming painfully clear, is it not? If we want to prove Fermat's result for $n+1$, that is, that $(n+1)^p - (n+1)$ is divisible by p, we apply the binomial expansion to enable us to use the result for n that we have already established. This methodology is a time-honored technique of mathematics that is known as *mathematical induction*. We now turn to an informal description of the technique.

Suppose that you want to prove a statement or proposition $P(n)$ for each positive integer n. The statement might be

$$n^2 + 3 > 2n\,.$$

Of course there are algebraic techniques for proving the inequality directly. But suppose instead that we check the inequality for $n = 1$—after all it is clear that $1^2 + 3 > 2 \cdot 1$. Then imagine that we prove the abstract syllogism

If $P(n)$ is true then $P(n + 1)$ is true.

Now think about putting these two results together. We know $P(1)$. And, as an instance of the displayed syllogism, we know that $P(1) \Rightarrow P(2)$. As a result, we conclude $P(2)$. Now another instance of the displayed syllogism is $P(2) \Rightarrow P(3)$. Combined with $P(2)$ (which we now know), we may conclude $P(3)$. And so forth. In the end we obtain $P(n)$ for all n.

Formulated a bit more precisely, mathematical induction goes like this:

We wish to prove a proposition $P(n)$ for each positive integer n. We do so by first proving $P(1)$ and then proving the syllogism

$$P(n) \Rightarrow P(n + 1) .$$

Then we may conclude that $P(n)$ is valid for all n.

We shall treat mathematical induction, and other methods of proof as well, in Chapter 21. Pause a minute now to review the reasoning of this section and to realize that we have the perfect setup—in the context of Fermat's theorem—for applying mathematical induction. Our statement $P(n)$ is that $n^p - n$ is divisible by p. The assertion for $n = 1$ is trivial since $1^p - 1 = 0$. Now all we need to do is formalize the reduction procedure that we have already performed for $n = 1$, $n = 2$, $n = 3$, and $n = 4$. We will *assume* that the result is known for n. And we will use that hypothesis to prove the result for $n + 1$. Now let us carry out these steps.

We calculate that

$$\begin{aligned}(n+1)^p - (n+1) = \Bigg[n^p &+ p \cdot n^{p-1} \cdot 1 + \frac{p \cdot (p-1)}{2 \cdot 1} \cdot n^{p-2} \cdot 1^2 \\ &+ \frac{p \cdot (p-1) \cdot (p-2)}{3 \cdot 2 \cdot 1} \cdot n^{p-3} 1^3 \\ &+ \cdots + \frac{p \cdot (p-1) \cdot (p-2) \cdot \cdots \cdot 4 \cdot 3}{(p-2) \cdot (p-1) \cdot \cdots \cdot 3 \cdot 2 \cdot 1} \cdot n^2 \cdot 1^{p-2} \\ &+ \frac{p \cdot (p-1) \cdot (p-2) \cdot \cdots \cdot 4 \cdot 3 \cdot 2}{(p-1) \cdot (p-2) \cdot \cdots \cdot 3 \cdot 2 \cdot 1} \cdot n^1 \cdot 1^{p-1} \\ &+ \frac{p \cdot (p-1) \cdot (p-2) \cdot \cdots \cdot 3 \cdot 2 \cdot 1}{p \cdot (p-1) \cdot (p-2) \cdot \cdots \cdot 3 \cdot 2 \cdot 1} \cdot 1^p \Bigg] - (n+1) .\end{aligned}$$

Now we perform some simple arithmetic to rewrite this as

$$\begin{aligned}(n+1)^p - (n+1) = \Bigg[n^p &+ p \cdot n^{p-1} + \frac{p \cdot (p-1)}{2 \cdot 1} \cdot n^{p-2} \\ &+ \frac{p \cdot (p-1) \cdot (p-2)}{3 \cdot 2 \cdot 1} \cdot n^{p-3} \\ &+ \cdots + \frac{p \cdot (p-1)}{2 \cdot 1} \cdot n^2 + \frac{p}{1} \cdot n + 1 \Bigg] - (n+1) .\end{aligned}$$

We may rearrange this in order to take advantage of our result for n. The result is

$$(n+1)^p - (n+1) = \Big[n^p - n\Big] + \Big[p \cdot n^{p-1} + \frac{p \cdot (p-1)}{2 \cdot 1} \cdot n^{p-2}$$

$$+ \frac{p \cdot (p-1) \cdot (p-2)}{3 \cdot 2 \cdot 1} \cdot n^{p-3}$$

$$+ \cdots + \frac{p \cdot (p-1)}{2 \cdot 1} \cdot n^2 + \frac{p}{1} \cdot n\Big].$$

Now of course $n^p - n$ is divisible by p because we have assumed that result to be true. And all the other terms on the right have plainly exhibited a factor of p. Thus the right-hand side is divisible by p. We conclude that $(n+1)^p - (n+1)$ is divisible by p. The inductive step is complete, and we have proved Fermat's theorem for all n.

For You to Try Use mathematical induction to prove that, for each positive integer n, the number $n^2 + 7n + 12$ is even.

For You to Try Use mathematical induction to prove that, for any positive integer n,

$$1 + 2 + \cdots + (n-1) + n = \frac{n(n+1)}{2}.$$

Further Reading

S. W. Golomb, Combinatorial proof of Fermat's "little" theorem, *The American Mathematical Monthly* 63(1956), 718.

Thomas J. Osler, Fermat's little theorem from the multinomial theorem, *The College Mathematics Journal* 33(2002), 239.

13.4 The Concept of Relatively Prime

We have encountered at several junctures in this book the concept of "relatively prime". Two integers are relatively prime if they have no common prime factors. For instance, 8 and 27 are relatively prime because $8 = 2 \cdot 2 \cdot 2$ and $27 = 3 \cdot 3 \cdot 3$ and they share no prime factors. Also 315 and 418 are relatively prime because $315 = 3^2 \cdot 5 \cdot 7$ and $418 = 2 \cdot 11 \cdot 19$ and there are plainly no common prime factors.

A fundamental fact about two relatively prime integers a and b is that there exist integers x and y such that

$$xa + yb = 1.$$

This assertion is so important, and comes up so frequently, that we shall prove it here. We begin with a basic idea of Fermat.

If n is an integer, let $\mathcal{P}(n)$ be the set of integers less than n that are relatively prime to it. Let $\varphi(n)$ be the number of elements in $\mathcal{P}(n)$.

Theorem 13.1. *If n is a positive integer and k is relatively prime to n, then*

$$k^{\varphi(n)} = 1 \mod n.$$

Proof. The proof of this result is easy. For the collection $\mathcal{P}(n)$ of numbers relatively prime to n forms a group under multiplication. (Here a *group* is a set equipped with a binary multiplicative operation which is associative, has an identity element, and so that every group element has a multiplicative inverse.) That is, if a is relatively prime to n and b is relatively prime to n then logic dictates that $a \cdot b$ is also relatively prime to n. Now it is a fundamental fact—we cannot prove it here, but see [HER]—that if a group has m elements and g is an element of the group, then g^m is the group identity. Thus any element of the group $\mathcal{P}(n)$, raised to the power $\varphi(n)$ (the number of elements in $\mathcal{P}(n)$) will equal 1 modulo n. □

For later use, it is worth noting that if p, q are prime numbers and $n = p \cdot q$, then $\varphi(n) = (p-1)\cdot(q-1)$. The reason is that the only numbers less than or equal to n that are not relatively prime to n are $p, 2p, 3p, \ldots q \cdot p$ and $q, 2q, 3q, \cdots (p-1)q$. There are q numbers in the first list and $p-1$ numbers in the second list. The set $\mathcal{P}(n)$ of numbers relatively prime to n is the complement of these two lists, and it therefore has

$$pq - q - (p-1) = pq - q - p + 1 = (p-1)\cdot(q-1) \equiv \varphi(n)$$

elements.

Relatively Prime Integers

As we have noted, if a, b are relatively prime, then we can find other integers x and y such that

$$xa + yb = 1\,. \qquad (*)$$

For example, we note that $a = 72 = 2^3 \cdot 3^2$ and $b = 175 = 5^2 \cdot 7$ are relatively prime. The corresponding integers x, y are $x = -17$ and $y = 7$. Thus

$$(-17)\cdot 72 + 7 \cdot 175 = 1\,.$$

One can prove the result $(*)$ using Fermat's theorem above. For, since b is relatively prime to a, we see that

$$b^{\varphi(a)} = 1 \mod a\,.$$

But this just says that

$$b^{\varphi(a)} - 1 = k \cdot a$$

for some integer k. Unraveling this equation gives $(*)$.

In practice, one finds x and y by using the Euclidean algorithm (otherwise known as long division).

In the example of 72, 175, one calculates:

$$\begin{aligned} 175 &= 2\cdot 72 + 31\,,\\ 72 &= 2\cdot 31 + 10\,,\\ 31 &= 3\cdot 10 + 1\,. \end{aligned}$$

You know you are finished when the remainder is 1.

For now we have

$$\begin{aligned}1 &= 31 - 3 \cdot 10\\ &= 31 - 3 \cdot (72 - 2 \cdot 31)\\ &= 7 \cdot 31 - 3 \cdot 72\\ &= 7 \cdot (175 - 2 \cdot 72) - 3 \cdot 72\\ &= 7 \cdot 175 - 17 \cdot 72\,.\end{aligned}$$

That is the decomposition we seek.

Further Reading

Paulo Ribenboim, Are there functions that generate prime numbers?, *The College Mathematics Journal* 28(1997), 352–359.

M. V. Subbarao, On relatively prime sequences, *The American Mathematical Monthly* 73(1966), 1099–1102.

Exercises

1. Prove that if p is a prime number, then $p^2 + p$ is *not* a prime number.

2. Prove that it is impossible to have a sequence of integers $k, k + 1, k + 2$ all of which are primes.

3. If k is a prime number greater than 2, then $k^2 + 1$ is not a prime number. Prove the assertion. Discuss this question in class.

4. It is a recent spectacular result of Green and Tao that the prime numbers contain arbitrarily long arithmetic sequences. Thus, for any $N > 0$, there are lists of integers $n_1, n_2, n_3, \ldots, n_N$ such that the n_j are evenly spaced apart and all the n_j are primes. As a simple example, $5, 11, 17$ are all prime numbers that are spaced 6 apart. Find an example of four prime numbers (other than the one given in the text) that are evenly spaced. Find an example of five. Find an example of six. Use the computer to assist your search if you like. Discuss this question in class.

5. How many examples can you give of numbers of the form $k(k + 1) + 1$ that are prime? Are there infinitely many of these? Why?

6. The celebrated *Prime Number Theorem*—due to Hadamard and de la Vallée Poussin—states that the number of primes less than or equal to N is approximately $N/\ln N$. Thus the number of primes less than or equal to 1,000,000 is about $1,000,000/\ln 1,000,000 \approx 72.382$. This is an astonishing statement. How many primes are there less than or equal to 1000? How does this information fit with the prediction provided by the Prime Number Theorem? How many primes are there less than or equal to 10,000? How does this information fit with the prediction provided by the Prime Number Theorem? Use a computer to aid your searches if you wish. Discuss this problem in class.

7. A conjecture that has been researched is that any integer with all digits 1 is a prime. Well, 1 is not a prime by definition. But 11 is prime. Unfortunately 111 is not prime. What about 1111 or 11111? Investigate this question. Use the computer if you find it convenient. What is the largest prime with all digits 1 that you can find? Discuss this question in class.

8. Use mathematical induction to prove that the sum of the first N perfect squares: $1^2 + 2^2 + \cdots + N^2$ is equal to

$$\frac{2N^3 + 3N^2 + N}{6}.$$

9. Refer to Exercise 8. Find a formula for the sum of the first N perfect cubes. Verify using induction that it is correct. Discuss this problem in class.

10. Prove that if n is a positive integer, then n^2+3n+2 is even. Use mathematical induction.

11. A stronger form of mathematical induction is this:

> Let $P(j)$ be a statement for each positive integer j. If
>
> **(i)** $P(1)$ is true;
> **(ii)** Whenever $P(1), \ldots P(k-1)$ is true then $P(k)$ is true;
>
> then $P(j)$ is true for all j.

Use this "strong form of mathematical induction" to show that every positive integer has a decomposition into a product of prime factors.

12. Is the difference of two primes ever a prime? Discuss this question in class.

13. Use the Euclidean algorithm to find the greatest common divisor of 84 and 144.

14. **Project:** Use the statement of the prime number theorem to estimate how many primes there are between 1000 and 1000000. Refer to the book [HAW].

15. **Project:** Consider the quotient

$$\frac{\pi(N)}{N},$$

where, as usual, $\pi(N)$ denotes the number of primes less than or equal to N. Explain why the limit of this expression, as $N \to \infty$, is 0. We say that the primes are *not dense* in the positive integers. Explain what this means. Again see [HAW].

16. **Project:** A *prime pair* is a pair of numbers $p, p+2$ such that both these numbers are prime. List the first ten prime pairs. It is not known whether there are infinitely many prime pairs. The largest known prime pairs have 58711 decimal digits.

It is known that the sum

$$\sum_{p \text{ prime}} \frac{1}{p}$$

diverges. Yet Vigo Brun has shown that

$$\sum_{p \text{ is a twin prime}} \frac{1}{p}$$

converges. Calculate both these sums for primes smaller than 1000. In fact there are 35 twin primes less than 1000. Find them all. Only the number 5 occurs in two pairs. Refer to [WIKIPR] for more information about twin primes.

17. **Project:** Refer to Exercise 11 above. Explain why ordinary induction would not be useful for proving the statement at the end of that exercise. Explain why strong induction is logically equivalent with ordinary induction.

14

Dirichlet and How to Count

14.1 The Life of Dirichlet

Peter Gustav Lejeune Dirichlet (1805–1859) was one of the great number theorists of the nineteenth century. His father's first name was Lejeune, coming from "Le jeune de Richelet". This means "young man from Richelet." The Dirichlet family came from the neighborhood of Liége in Belgium. The father was postmaster of Düren, the town where young Peter was born. At a young age Dirichlet developed a passion for mathematics; he spent his pocket money on mathematics books. He entered the Gymnasium[1] in Bonn at the age of 12. There he was a model pupil. He exhibited an interest in history as well as mathematics. The remarkable life of Dirichlet is chronicled in [KUM] and [ORE2].

After two years at the Gymnasium, Dirichlet's parents decided that they would rather have him at the Jesuit College in Cologne. There he fell under the tutelage of the distinguished scientist Ohm. By age 16, Dirichlet had completed his school work and was ready for the university. German Universities were not very good, nor did they have very high standards, at the time. Hence Peter Dirichlet decided to study in Paris. It is worth noting that several years later the German Universities would set the worldwide standard for excellence; Dirichlet himself would play a significant role in establishing this pre-eminence.

Dirichlet always carried with him a copy of Gauss's *Disquisitiones arithmeticae*,[2] a work that he revered and kept at his side much as some people might keep the Bible. Thus he came equipped for his studies in Paris. Dirichlet contracted smallpox soon after his arrival in Paris. But he would not let this deter him from attending lectures at the Collége de France and the Faculté des Sciences. He enjoyed the teaching of some of the leading scientists of the time, including Biot, Fourier, Francoeur, Hachette, Laplace, Lacroix, Legendre, and Poisson.

Beginning in the summer of 1823, Dirichlet lived in the house of the retired General Maximilien Sébastian Foy. Dirichlet taught German to General Foy's wife and children. Dirichlet was treated very well by the Foy family, and he had time to study his mathematics. It was at this time that he published his first paper, and it brought him instant fame. For it dealt with Fermat's last theorem. The problem, as we know, is to show that the Diophantine equation

$$x^n + y^n = z^n$$

[1] In Germany, a *Gymnasium* is a preparatory or "prep" school.

[2] This book, on the subject of number theory, was Gauss's masterpiece. It is still read today for its deep insights.

has no integer solutions x, y, z when n is a positive integer greater than 2. The cases $n = 3, 4$ had already been handled by Euler and by Fermat himself. Dirichlet decided to attack the case $n = 5$ (we saw in Chapter 11 that Sophie Germain made notable contributions to this case of Fermat's theorem). The case $n = 5$ divides into two subcases, and Dirichlet was able to dispatch Subcase 1. Legendre was a referee of the paper, and he was able, after reading Dirichlet's work, to treat Subcase 2. Thus a paper was published in 1825 that completely settled the case $n = 5$ of Fermat's last theorem. Dirichlet himself was subsequently able to develop his own proof of Subcase 2 using an extension of his techniques for Subcase 1. Later on Dirichlet was also able to treat the case $n = 14$.

General Foy died in November of 1825 and Dirichlet decided to return to Germany. However, in spite of support from Alexander von Humboldt, he could not assume a position in a German university since he had not submitted an Habilitation thesis.[3] Dirichlet's mathematical achievements were certainly adequate for such a thesis, but he was not allowed to submit one because **(i)** he did not hold a doctorate and **(ii)** he did not speak Latin.

The University of Cologne interceded and awarded Dirichlet an honorary doctorate. He submitted his Habilitation on polynomials with prime divisors and obtained a position at the University of Breslau. Dirichlet's appointment was still considered to be controversial, and there was much discussion among the faculty of the merits of the case.

Standards at the University of Breslau were still rather low, and Dirichlet was not satisfied with his position. He arranged to transfer, again with Humboldt's help, to the Military College in Berlin. He also had an agreement that he could teach at the University of Berlin, which was really one of the premiere institutions of the time. Eventually, in 1828, he obtained a regular professorship at the University of Berlin. He taught there until 1855. Since he retained his position at the Military College, he was saddled with an unusual amount of teaching and administrative duties.

Dirichlet also earned an appointment at the Berlin Academy in 1831. His improved financial circumstances then allowed him to marry Rebecca Mendelssohn, the sister of the noted composer Felix Mendelssohn. Dirichlet obtained an eighteen-month leave from the University of Berlin to spend time in Italy with Jacobi (who was there for reasons of his health).

Dirichlet returned to his duties at the University of Berlin and the Military College, He continued to find his duties at both schools to be a considerable burden, and complained to his student Kronecker. It was quite a relief when, on Gauss's death in 1855, Dirichlet was offered Gauss's distinguished chair at the University in Göttingen.

Dirichlet endeavored to use the new offer as leverage to obtain better conditions in Berlin. But that was not to be, and he moved to Göttingen directly. There he enjoyed a quieter life with some outstanding research students. Unfortunately the new conditions were not to be enjoyed for long. Dirichlet suffered a (serious but nonfatal) heart attack in 1858, and his wife died of a stroke shortly thereafter.

Dirichlet's contributions to mathematics were monumental. We have already described some of his work on Fermat's last problem. He also made contributions to the study of Gauss's quadratic reciprocity law. It can be said that Dirichlet was the father of the subject

[3] Typically a Habilitation is a postdoctoral degree involving further research accomplishments and possibly the publication of a book.

of analytic number theory. In particular, he proved foundational results about prime numbers occurring in arithmetic progression.

Dirichlet did further work on what was later to become (in the hands of Emmy Noether—see Chapter 18) the theory of ideals. He created *Dirichlet series*, which are today a powerful tool for analytic number theorists. And he laid some of the foundations for the theory of *class numbers* (later to be developed by Emil Artin).

Dirichlet is remembered for giving one of the first rigorous definitions of the idea of function. He was also the first to define precisely what it means for a series to *converge*. He is noted as one of the fathers of the theory of Fourier series.

Dirichlet had a number of historically important students, including Kronecker and Riemann (see the next chapter). Riemann went on to make seminal contributions to complex variables, Fourier series, and geometry.

Further Reading

Leigh Atkinson, Where do functions come from?, *The College Mathematics Journal* 33 (2002), 107–112.

David M. Bloom and Stephen M. Gagola, Jr., Dirichlet knew it all along: 10718, *The American Mathematical Monthly* 108 (2001), 174.

14.2 The Pigeonhole Principle

Today combinatorics and number theory and finite mathematics are thriving enterprises. Cryptography, coding theory, queuing theory, and theoretical computer science all make use of counting techniques. But the idea of "counting", as a science, is relatively new.

One of the first masters of the theory of counting was Dirichlet. And one of his principal counting techniques, the one for which he is most vividly remembered, is that which was originally called the "Dirichletscher Schubfachschluss" (Dirichlet's drawer-shutting principle). Today we call it the "pigeonhole principle". It is a remarkably simple idea that has profound consequences.

> Suppose that $(n+1)$ letters are put into n mailboxes. Then one mailbox must contain at least two letters.
>
> This paradigm is so significant that we shall provide a couple of different proofs. But, before we do, let us introduce some notation. Let

$$\sum_{j=1}^{N} a_j$$

mean

$$a_1 + a_2 + \cdots + a_N .$$

For instance,

$$\sum_{j=1}^{6} [j^2 + j] = [1^2 + 1] + [2^2 + 2] + [3^2 + 3] + [4^2 + 4] + [5^2 + 5] + [6^2 + 6] .$$

Another example is

$$\sum_{j=1}^{5} \sin j = \sin 1 + \sin 2 + \sin 3 + \sin 4 + \sin 5 + \sin 6 .$$

It is also possible to begin the summation at an index other than 1:

$$\sum_{j=3}^{8} \frac{j+1}{2} = \frac{4}{2} + \frac{5}{2} + \frac{6}{2} + \frac{7}{2} + \frac{8}{2} + \frac{9}{2} .$$

Now we turn to the pigeonhole principle:

First Proof of the Pigeonhole Principle. Suppose that the principle is not true. Then each mailbox contains either 0 or 1 letters.

Returning to the statement of the pigeonhole principle, let us suppose that the n mailboxes are numbered 1 through n. We now have

$$\begin{aligned}\text{(total number of letters)} &= \sum_{j=1}^{n} \text{(number of letters in mailbox \# j)} \\ &\leq \sum_{j=1}^{n} 1 = n .\end{aligned}$$

But this says there are at most n letters, and that is incorrect. So the hypothesis is false and some mailbox must contain at least two letters. □

Second Proof of the Pigeonhole Principle. We proceed by induction. Our statement $P(n)$ is "If $(n+1)$ letters are distributed to n mailboxes then some mailbox must contain at least two letters. Now $P(1)$ is clearly true: If 2 letters are distributed among 1 mailbox then some mailbox (indeed, the *only* mailbox) must contain two letters.

Now suppose that $P(n)$ is true. We need to prove $P(n+1)$. So we have $n+1$ mailboxes and $n+2 = (n+1)+1$ letters and we distribute them among the mailboxes. If the last mailbox contains 2 letters then we are done. If not, then it contains 0 or 1 letter. So there are at least $n+1$ letters remaining and these are distributed among the first n mailboxes. By the inductive hypothesis, one of these must therefore contain two letters. That establishes $P(n+1)$ (assuming $P(n)$) and completes the inductive argument. □

One of the earliest, and most dramatic, applications of Dirichlet's pigeonhole principle is to prove Dirichlet's theorem in number theory. This result has been extremely influential in the study of irrational and transcendental numbers (we shall say more about these types of numbers in Chapters 16 and 17). It considers how closely an irrational number may be approximated by a fraction of the form m/n. The result is this:

Theorem 14.1. *Let ξ be a real number. If $n > 0$ is an integer, then there are integers p, q such that $0 \leq q \leq n$ and*

$$\left| \frac{p}{q} - \xi \right| < \frac{1}{qn} . \tag{†}$$

Refer to [MOS] and [NIV] for further reading about theorems of this kind.

EXAMPLE 14.1. As an instance of the last theorem, let $\xi = \sqrt{2}$ and $n = 5$. Then the numbers $p = 7$ and $q = 5$ satisfy the conclusion of the result. For $p/q = 1.4$ and

$$\left|\frac{p}{q} - \xi\right| = |1.4 - 1.414\ldots| = |0.014\ldots| < 0.04 = \frac{1}{5^2}.$$

For You to Try Illustrate the conclusion of Theorem 14.1 when $\xi = \pi$ and $n = 6$.

What should be noticed here is that we are making an assertion about how rapidly ξ can be approximated by rational numbers—and the rate of approximation is expressed in terms of the rational number itself (i.e., its denominator). The theorem is of particular interest when ξ is an irrational number. Now we shall present a very classical proof (due to Dirichlet) of this result:

Remark. In the proof that we are about to present, we shall use the idea of "greatest integer" and "fractional part". If x is any real number, then the *greatest integer* in x (denoted $\lfloor x \rfloor$) is the greatest integer that is less than or equal to x. For example,

$$\lfloor 2.3 \rfloor = 2\,,\ \lfloor 4 \rfloor = 4\,,\ \lfloor -1.5 \rfloor = -2\,,\ \lfloor \pi \rfloor = 3\,.$$

The fractional part of the real number x is just $(x) = x - \lfloor x \rfloor$. Fractional part is a fascinating idea. For instance, Hermann Weyl noticed nearly 100 years ago, that if ξ is irrational, then the fractional parts

$$\lfloor \xi \rfloor\,,\ \lfloor 2\xi \rfloor\,,\ \lfloor 3\xi \rfloor\,,\ \ldots$$

will form a dense set in the unit interval [0, 1] (meaning that every point in the unit interval is the limit of some subsequence of these numbers).

Proof of Theorem 14.1. Let $\epsilon > 0$ and set $Q = \lfloor 1/\epsilon \rfloor + 1$. Take $n \geq Q$. As usual, $\lfloor x \rfloor$ denotes the greatest integer in x and (x) the fractional part of x.

Now consider the $Q + 1$ numbers

$$0, (\xi), (2\xi), \ldots, (Q\xi)\,. \tag{$\star$}$$

These are $Q + 1$ numbers in the interval [0, 1]. Now we divide the interval [0, 1] into the subintervals

$$\left[\frac{j}{Q}, \frac{j+1}{Q}\right)\,, \qquad j = 0, 1, 2, \ldots, Q - 1\,. \tag{$\star\star$}$$

The $Q + 1$ points listed in ($\star$) are distributed among the Q intervals (or "pigeonholes") in ($\star\star$). Thus one of these intervals must contain two of the points.

As a result, we find integers q_1, q_2 (both not greater than Q) such that

$$|(q_1\xi) - (q_2\xi)| < \frac{1}{Q}\,.$$

Assuming as we may that $q_1 < q_2$, and setting $q = q_2 - q_1$, we thus see that $0 < q \leq Q$ and $|\overline{q\xi}| < 1/Q$. [Here we use the overbar to denote distance to the nearest integer.] It follows that there is an integer p such that

$$|q\xi - p| < \frac{1}{Q}\,.$$

We conclude that there are integers p, q such that

$$q \leq \left\lfloor \frac{1}{\epsilon} \right\rfloor + 1 \qquad \text{and} \qquad \left| \frac{p}{q} - \xi \right| < \frac{1/Q}{q} < \frac{\epsilon}{q}.$$

In particular,

$$\left| \frac{p}{q} - \xi \right| < \frac{1}{qQ}.$$

□

It may be noted—we shall not prove the result here—that, when ξ is rational, then there are only finitely many p/q which satisfy (†). For ξ irrational it is the case that (†) holds for infinitely many p/q.

Further Reading

Joan P. Hutchinson and Paul B. Trow, Some pigeonhole principle results extended, *The American Mathematical Monthly* 87(1980), 648–651.

Kenneth R. Rebman, The pigeonhole principle (What it is, how it works, and how it applies to map coloring), *The Two-Year College Mathematics Journal* 10(1979), 3–13.

14.3 Ramsey Theory

Frank Plumpton Ramsey (1903–1930) was a notably original scholar and thinker. In addition to his mathematical contributions, he participated in supervising Ludwig Wittgenstein's[4] Ph.D. thesis, and played a part in getting the *Tractatus Logico Philosphicus* translated into English. He also contributed to mathematical economics and to the study of the decision problem in logic. Ramsey single-handedly invented the mathematical subject that we now call *Ramsey theory*. The essence of that discipline is to make order out of chaos. Ramsey's philosophy was that, if a set is big enough, then there will be some order in its structure. Ramsey lived a tragically short life, but he made major contributions to several disciplines.

In 1930, in a paper entitled *On a Problem in Formal Logic* [RAM], Ramsey proved a very general theorem (now known as Ramsey's theorem) which we shall illustrate with a special case. The idea is encapsulated in the following question:

> How many people need there be in a room so that we can be sure that there are (at least) three people who are mutually acquainted or three people who are mutually unacquainted?

Here three people are mutually acquainted if each two of them knows each other. Three people are mutually unacqainted if each two of them does not know each other.

The problem is charming, for it has many different interpretations. Consider N points in the plane. The *complete graph* on these N points consists of arcs that connect every point to every other. Say that each point, or node, represents a person. One can color an arc blue if

[4]Ludwig Wittgenstein (1889 C.E.–1951 C.E.) was one of the outstanding figures of twentieth century philosophy. A student of Bertrand Russell, Wittgenstein is noted for his work on the use of language in philosophy. His most famous work is *Tractatus Logico Philosophicus*.

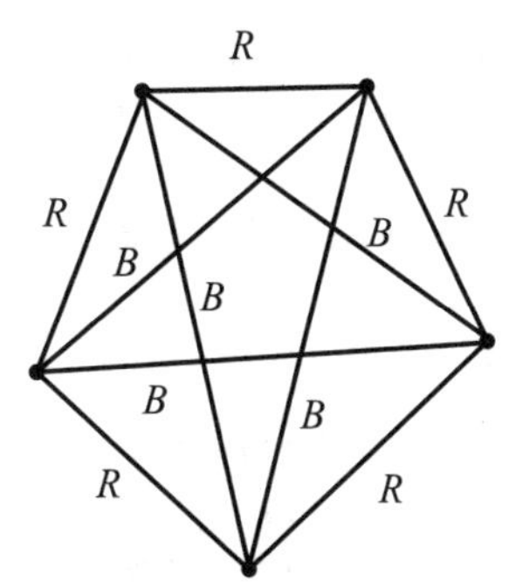

Figure 14.1. A complete graph on five vertices with no triangle that is all red or all blue.

the two corresponding people (represented by the two endpoints of the arc) are acquainted; and one can color the arc red if the two corresponding people are *not* acquainted. One may then ask whether there will always be a blue triangle or always be a red triangle? This corresponds (obviously) to three people being acquainted or three people being unacquainted, respectively.

Look at Figure 14.1. It shows the complete graph on five vertices. And it also indicates a coloring of the arcs in the graph so that there is no red triangle and no blue triangle. This shows that five people will *not* give an affirmative answer to the original question. A minimum of six is required.

In fact the magic number is six. To see this, imagine that one of the six people is named Bob. Of the five other people besides Bob, he is either acquainted with three of them or he is not. In case he is acquainted with three of them, either two of those are acquainted or not. If two are acquainted, then those two plus Bob make a mutually acquainted threesome. If no two are acquainted, then those three make a mutually unacquainted threesome.

If instead Bob only knows at most two of the others, then look at the other three. If they are mutually acquainted then we are done. If not, then two of them are unacquainted so those two plus Bob make a mutually unacquainted threesome.

In any circumstance, there will be three people who satisfy the conclusion of the theorem.

Let us use the notation k to denote the number of mutually acquainted or unacquainted people we seek, and let $N(k)$ denote the number of people required to guarantee k such acquainted or unacquainted people among them. So we have just seen that $N(3) = 6$. For $k = 4$, the answer is $N(4) = 18$ and that is quite difficult to see. For five people, it is known that $42 \leq N(5) \leq 55$. It has required 40 years to obtain these estimates. And we are far from knowing the exact answer. For $k = 6$ we know that $69 \leq N(6) \leq 102$, but determining the exact answer is considered to be slightly worse than hopeless. Erdős put the matter this way:

> If an evil spirit would appear and say, "Tell me the value of N when k equals 5, or I will exterminate the human race," it would be best to get all the computers in the world to try to solve the problem. But if the evil spirit would ask for the value when k equals 6, it would be best to try to exterminate the evil spirit.

He goes on to say, "And if we could get the right answer just by thinking, we wouldn't have to be afraid of him [the evil spirit], because we would be so clever that he couldn't do us any harm."

The Ramsey theorists have introduced the notation $R(s,t)$ to denote the number such that a complete graph on $R(s,t)$ vertices, with edges colored red and blue, will have either a complete subgraph on s vertices which is entirely red or a complete subgraph on t vertices which is entirely blue. We have already seen that $R(3,3) = 6$. Also $R(4,4) = 18$. And so forth. We call R a *Ramsey number.*

Ramsey's more general result is as follows:

Theorem 14.2. *Given any number of colors c, and any positive integers $n_1, n_2, \ldots, n_c$, there is a number $R(n_1, \ldots, n_c)$ such that, if the edges of a complete graph with $R(n_1, \ldots, n_c)$ vertices are colored with c different colors, then for some j between* 1 *and c there is a complete subgraph with n_j vertices whose edges are all colored j.*

Of course the result we have already discussed and proved is the special case $c = 2$, $n_1 = 3$, $n_2 = 3$. A *multicolor Ramsey number* is one for which c is at least 3. There is only one nontrivial multicolor Ramsey number whose exact value is known. To wit, $R(3,3,3) = 17$. To understand this assertion, we can reason as follows.

Let $\mathcal{G}$ be a complete graph on N vertices, some N. Assume that $\mathcal{G}$ is colored using three colors that we call red, yellow, and blue. Suppose further that this coloring has no monochromatic (i.e., all-one-color) triangle in it. Let $\mathbf{v}$ be a vertex in this graph. Let S be the set of all vertices (distinct from $\mathbf{v}$) which are connected to $\mathbf{v}$ by a blue edge. We call S the *blue neighborhood* of $\mathbf{v}$. If $\mathbf{w}_1$ and $\mathbf{w}_2$ are two points in S, distinct from $\mathbf{v}$, then $\mathbf{w}_1$ and $\mathbf{w}_2$ *cannot* be connected by a blue edge—for then $\mathbf{v}$, $\mathbf{w}_1$, and $\mathbf{w}_2$ would form a blue (monochromatic) triangle. Thus the elements of S can be connected with only red and yellow edges. Since we know that $R(3,3) = 6$, we must conclude that S has at most five elements—since there can be no monochromatic triangle. A similar argument shows that the red neighborhood and the yellow neighborhood of $\mathbf{v}$ will contain at most five elements. Since every vertex, except for $\mathbf{v}$ itself, lies in one of the blue, red, or yellow neighborhoods of $\mathbf{v}$, we must conclude that the entire graph can have at most $1+5+5+5 = 16$ vertices. Thus we see that $R(3,3,3) \leq 17$.

To see that $R(3,3,3) \geq 17$, we need to construct an edge coloring on the complete graph with 16 vertices that has no monochromatic triangles. We leave that as Exercise 5 for you.

Further Reading

Arie Bialostocki, An application of elementary group theory to central solitaire, *The College Mathematics Journal* 29(1998), 208–212.

T. C. Brown, Variations on van der Waerden's and Ramsey's theorems, *The American Mathematical Monthly* 82(1975), 993–995.

Exercises

1. Give an alternative derivation of the fact that $R(3,3) = 6$ by counting triangles in the complete graph on 6 vertices. There are 20 triangles, and at most 18 of these that are not all of one color.

2. If you have 23 people in the room, then the probability is greater than 0.5 that two of them were born on the same day of the year. Discuss this problem in class. Verify the

assertion.

3. Let $\xi = \sqrt{2}$ and $n = 10$. Find integers p and q that satisfy the conclusion of Dirichlet's Theorem 12.1.

4. Let $\xi = 3/2$ and $n = 7$. Find integers p and q that satisfy the conclusion of Dirichlet's Theorem 12.1.

5. Produce a coloring, with three colors, on the edges of the complete graph with 16 vertices, so that there is no monochromatic triangle.

6. Suppose that you deliver $n + 2$ letters to n mailboxes. Can you be sure that some mailbox contains three letters? Why or why not?

7. You deliver 10,000 letters to 9,000 mailboxes. What is the greatest number of mailboxes that could contain three letters? What is the greatest number of mailboxes that could contain two letters?

8. In a five-card poker hand, what is the probability that you hold a flush (i.e., all cards of the same suit)?

9. In poker, a royal flush is the $10 - J - Q - K - A$ of a single suit. How many different royal flush hands are there? What is the probability, with a standard deck of 52 cards, of being dealt a royal flush? What is the probability of being dealt a hand that differs from a royal flush by just one card?

10. A jar contains fifty black marbles and fifty white marbles. You choose three marbles at random. What is the probability that all three of them are white? What is the probability that at least two of them are white? Discuss this problem in class.

11. A woman distributes fifteen letters among ten mailboxes. She knows that two of the mailboxes each contain three letters. What can she say about the distribution of letters in the remaining mailboxes?

12. You have six white marbles and six black marbles and you will give three marbles to each of four children. How many different ways are there to perform this task? Discuss the problem in class.

13. In a five-card poker hand, what is the probability that all of the cards are of different denominations? [*Hint:* Ignore the suits. Just pay attention to the values of the cards.]

14. **Project:** The *Guinness Book of World's Records* contains a citation for the largest number (i.e., positive integer) ever created by a man. That man is the ineffable Ron Graham, and his number (now known as the "Graham number") is so large that it cannot be expressed with ordinary mathematical or scientific notation. Indeed, Donald Knuth has invented a new scientific argot for describing the Graham number.

    Let us first say what the number is, at least in intuitive terms. We seek a positive integer N with the following property. For a set S with N elements, consider all possible subsets (the power set). Now form a new set T consisting of all pairs of elements of the power set. Partition T into two subsets A and B. We seek the least N so that there are four subsets of S with the properties that **(i)** every pair of these four lies in A or every pair lies in B, **(ii)** every element x that lies in one of these four subsets actually lies in an even number of the subsets.

It was originally suspected that $N = 6$ will do. Now it is known, thanks to a recent result of Exoo [EXO], that N must be at least 11. Graham showed that there is some such number N, and that its value will not exceed the enormous number that we are about to describe.

We define a new type of exponentiation. Let $3 \uparrow k$ denote 3 to the power k. This is a familiar idea. Now let $3 \uparrow\uparrow 3$ denote $3 \uparrow (3 \uparrow 3)$. In general

$$3 \underbrace{\uparrow\uparrow \cdots \uparrow\uparrow}_{k \text{ arrows}} 3$$

denotes

$$3 \underbrace{\uparrow\uparrow \cdots \uparrow\uparrow}_{(k-1) \text{ arrows}} (3 \underbrace{\uparrow\uparrow \cdots \uparrow\uparrow}_{(k-1) \text{ arrows}} 3)\,.$$

Now let

$$M = 3 \underbrace{\uparrow\uparrow \cdots \uparrow\uparrow}_{3\uparrow\uparrow\uparrow\uparrow 3 \text{ arrows}} 3\,.$$

Next define

$$P = 3 \underbrace{\uparrow\uparrow \cdots \uparrow\uparrow}_{M \text{ arrows}} 3\,.$$

Now iterate this construction 61 more times. You will then obtain Graham's number G. Explain how to use Knuth's notation (described above) to specify the weight of the sun in ounces. Use this notation to specify the number of molecules in the universe. The reference [WIKIGN] has more information about the Graham number.

15. **Project:** Refer to Exercise 1. This is a simple instance of *Ramsey theory*, a subject invented by Frank Ramsey (1903–1930). Try to answer the question with 3 replaced by 4. You will find this to be quite difficult. For "five" the problem is intractable. The book [GRAH] has all the basics of Ramsey theory.

15

Bernhard Riemann and the Geometry of Surfaces

15.0 Introduction

Bernhard Riemann (1826–1866) was the son of Friedrich, a Lutheran minister. Bernhard was the second of six children. Their father taught all the children, and in particular Bernhard, until Bernhard was ten years old. At that point a local teacher named Schulz assisted in the education. Riemann died young, but his impact on mathematics was immense. The saga of his life appears in [LAU].

In 1840, at the age of fourteen, Bernhard Riemann entered the third class at the Lyceum in Hanover. He lived there with his grandmother. In 1842 the grandmother died, and Riemann transferred to the Johanneum Gymnasium in Lüneburg. Young Bernhard was a solid but not outstanding student who studied classical subjects like Hebrew and theology. He did show a particular interest in mathematics, and was allowed to borrow books from the Gymnasium director's personal library. One of these was Legendre's number theory, and he devoured it in six days.

In 1846 Riemann enrolled at the University in Göttingen. This institution was later, especially under the guidance of Hilbert and Felix Klein (1849–1925), to become the premiere mathematics institution in all of Germany. At the time when Riemann was a student it was not. Riemann's father had in mind for young Riemann to study theology, but Bernhard begged him to instead study mathematics. Fortunately for all of us, the elder Friedrich granted his permission. Riemann was fortunate to take courses from Moritz Stern and Carl Friedrich Gauss. Gauss was teaching only elementary courses in those days, and had no opportunity to observe Riemann's special talents. Stern, however, was quite impressed by Riemann. He said that Bernhard Riemann

> ...already sang like a canary.

Riemann moved from Göttingen to the University of Berlin in 1847. There he enjoyed classes from Steiner, Jacobi, Dirichlet, and Eisenstein. The chief influences on Riemann at this time were Eisenstein and especially Dirichlet. Riemann adopted Dirichlet's style of basing his mathematics on a strong intuitive foundation. At this time he worked out his theory of complex variables; this in turn formed the basis of some of his most important later work.

In 1849 Riemann returned to Göttingen. His thesis, under the supervision of Gauss, was submitted in 1851. Riemann was also strongly influenced by Weber and Listing, who were professors of physics. In particular, they gave him important ideas from topology. This would affect his later development of the theory of Riemann surfaces.

Riemann's doctoral dissertation was certainly one of the most remarkably original pieces of work ever to appear in a thesis. It contains foundational ideas of the geometric and topological theory of complex variables, including the basic ideas of Riemann surfaces. His ideas built on earlier work of Cauchy, Puiseaux, and of course Dirichlet. Even the rather austere and distant Gauss reported on the thesis that Riemann had

> ...a gloriously fertile originality.

Gauss recommended Riemann to a post in Göttingen, and Riemann was thereby able to work on his Habilitation. The subject of the Habilitation was trigonometric series and the functions that they represent. Here Riemann laid the foundations for what we now call the *Riemann integral*. He also developed special ideas about Fourier series that are today studied in their own right.

A seminal part of the Habilitation is a ceremonial lecture. Riemann's major professor, Gauss, was allowed to choose the topic (from among a predetermined list that included electricity and geometry). Riemann was surprised when Gauss asked to hear about geometry. On June 10, 1854, he delivered the now-famous lecture entitled "Über die Hypothesen welche der Geometrie zu Grunde liegen," which translates to "On the hypotheses that lie at the foundations of geometry." In this lecture Riemann completely reinvented how geometry should be conceived. In particular, he created a toolkit for producing a variety of important non-Euclidean geometries. One expert has described Riemann's delivery as

> It possesses shortest lines, now called geodesics, which resemble ordinary straight lines. In fact, at first approximation in a geodesic coordinate system such a metric is flat Euclidean, in the same way that a curved surface up to higher-order terms looks like its tangent plane. Beings living on the surface may discover the curvature of their world and compute it at any point as a consequence of observed deviations from Pythagoras' theorem.

It has further been written that

> Among Riemann's audience, only Gauss was able to appreciate the depth of Riemann's thoughts.... The lecture exceeded all his expectations and greatly surprised him. Returning to the faculty meeting, he spoke with the greatest praise and rare enthusiasm to Wilhelm Weber about the depth of the thoughts that Riemann had presented.

It must be noted, in fact, that Albert Einstein found in Riemann's work the mathematical framework to fit his ideas of general relativity. The basics of cosmology and cosmogony arise from Riemann's geometrical ideas. Riemann gave to physics the concept of a metric structure determined by data.

Riemann, with Gauss's help, obtained a post in Göttingen on the strength of his Habilitation work. Riemann continued to do brilliant work. We have noted that, on his death, Gauss's chair in Göttingen was filled by Dirichlet. There was a movement to obtain a second chair for Riemann, but this failed. However, two years later, Riemann was appointed to a Professorship in Göttingen. In 1857 Riemann published his pathbreaking work on abelian functions. This masterwork was based on lectures he had given to a small audience in 1855–1856. One of the three who listened to Riemann was Richard Dedekind. Dedekind in fact published an amplified version of Riemann's lectures after the latter's untimely demise.

Riemann's paper contained an overwhelming number of new ideas. This included ideas about Riemann surfaces, the Dirichlet principle, and conformal mapping. The paper appeared in Crelle's Journal, volume 54. In fact Weierstrass was so impressed by Riemann's work that he withdrew his own paper on the subject and published no more for the remainder of his life.

In 1859, Dirichlet died and Riemann was awarded his chair at Göttingen. A few days later he was elected to the Berlin Academy of Sciences. The nomination by Kummer, Borchardt, and Weierstrass read as follows:

> Prior to the appearance of his most recent work [Theory of abelian functions], Riemann was almost unknown to mathematicians. This circumstance excuses somewhat the necessity of a more detailed examination of his works as a basis of our presentation. We considered it our duty to turn the attention of the Academy to our colleague whom we recommend not as a young talent which gives great hope, but rather as a fully mature and independent investigator in our area of science, whose progress he in significant measure has promoted.

As part of his induction into the Berlin Academy, Riemann was required to give a lecture on his latest work. He spoke of the zeta function (now called the "Riemann zeta function") and the location of its zeros. He reported that it has infinitely many zeros and asserted that it is probable that they all lie on a particular vertical line in the right half of the complex plane. This last claim remains unproved to this day. It is an important cornerstone of analytic number theory known as the *Riemann hypothesis*.

In June of 1862 Riemann married Elise Koch, a friend of his sister. They had one daughter. Not long after the marriage Riemann contracted a serious cold which turned to tuberculosis (this was not an unusual turn of events, given the state of medicine in those days). Unfortunately his health suffered a serious decline after that, and it ultimately led to his death. Riemann had always suffered from poor health, and now things were catching up to him. He traveled several times to Italy hoping that the warmer climate would improve his health and disposition. But to no avail. His strength declined rapidly and he died in 1866.

Next we turn to the mathematical description of some of Bernhard Riemann's ideas.

Further Reading

James Pierpont, The geometry of Riemann and Einstein, *The American Mathematical Monthly* 30(1923), 425–438.

Stan Wagon, A Mathematical magic trick, *The College Mathematics Journal* 25(1994), 325–326.

15.1 How to Measure the Length of a Curve

We first begin by describing an important paradox about arc length (this idea was treated informally in Exercise 14 of Chapter 2). Consider the unit square depicted in Figure 15.1. We wish to determine the length of the diagonal.

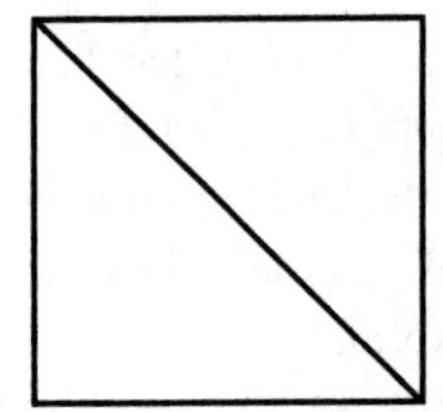

Figure 15.1. A paradox about arc length.

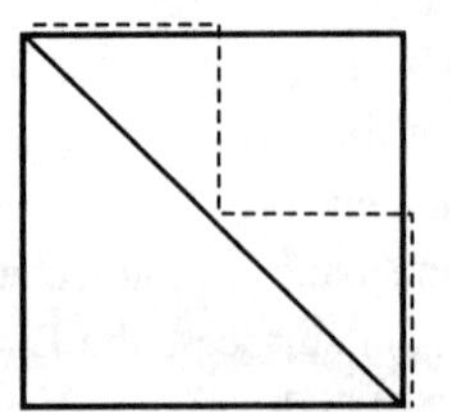

Figure 15.2. More on the paradox.

Now, in point of fact, we discussed this matter in Chapter 2, and we determined (using the Pythagorean theorem) that it is $\sqrt{2}$. But now let us set that result aside and take a new geometric approach to the question. We examine a blown-up picture of the square, and approximate the diagonal with a piecewise linear curve γ_1 (Figure 15.2). As you can see, this piecewise linear curve is composed of four segments, and each of these has length $1/2$. We conclude that γ_1 has length exactly 2.

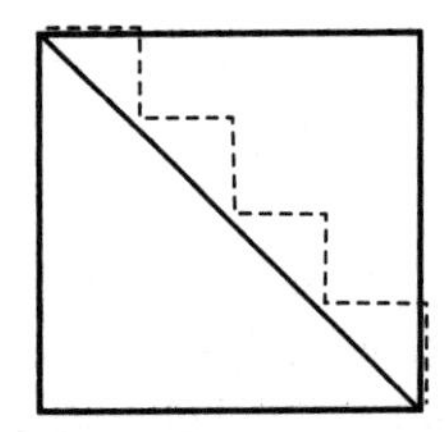

Figure 15.3. A piecewise linear curve.

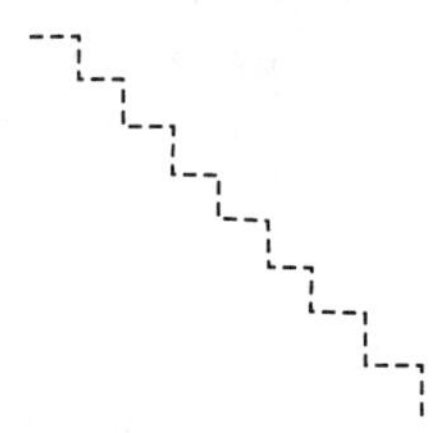

Figure 15.4. Enunciation of the paradox.

Now we consider a finer approximation γ_2 (Figure 15.3). As you can see, this piecewise linear curve is composed of eight segments, each of length $1/4$. We conclude that γ_2 has length exactly 2.

Now we can continue this procedure indefinitely. By using an approximating, piecewise linear curve with enough bends in it, we may approximate the diagonal very closely (Figure 15.4). Yet all these approximations have length 2. The conclusion, then, is that the diagonal itself must have length 2.

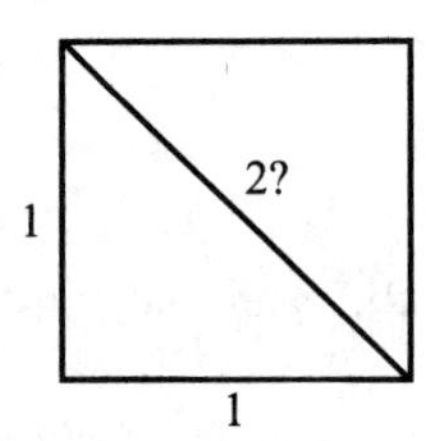

Figure 15.5. Two sides of the square.

But this result is inconsistent with our earlier discussions, and it also seems absurd on the face of it. After all, the length of two sides of the square is 2 (Figure 15.5). And surely *that* curve is longer than the diagonal. How could the diagonal itself have length 2?

What we have described here is a visual conundrum. The jagged, piecewise linear curve approximates the diagonal *visually*, that is to say, it gets *very close* to the diagonal in a

Figure 15.6. Approximation by secants.

reasonable sense. But the *length* of this approximating curve does not in fact approximate the *length* of the diagonal.

Now let us give a more effective, and in fact a correct, method for approximating the length of a curve. Examine Figure 15.6. What we do is to approximate the curve by *secants* to the curve. These segments are not only *visually* close to the curve in question, they are also (as can be proved using calculus) close in length.

Riemann's idea for creating new geometries (non-Euclidean geometries, in fact) was to approximate the length of a curve in a new way. He would use the approximation scheme in Figure 15.6, but he would assign *new lengths—not* Euclidean lengths—to each of the approximating secants.

The rigorous way to carry out Riemann's scheme is to use the integral—another idea from calculus. Since that technique is too advanced for the present text, we will use a more intuitive approach to the matter.

Further Reading

P. D. Barry, On arc length, *The College Mathematics Journal* 28(1997), 338–347.

James T. Sandefur, Using self-similarity to find length, area, and dimension, *The American Mathematical Monthly* 103(1996), 107–120.

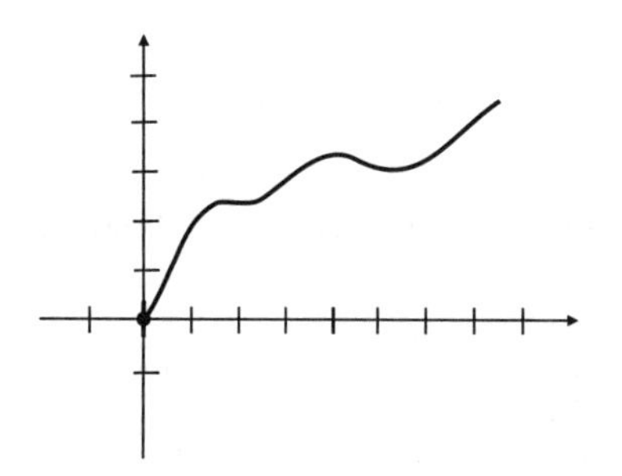

Figure 15.7. Riemann's method for measuring arc length.

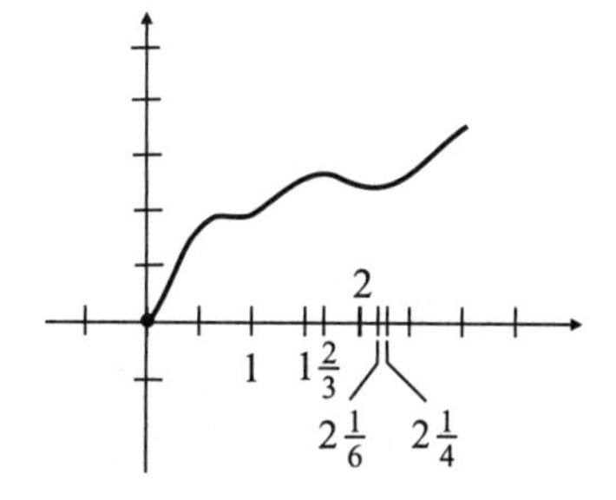

Figure 15.8. Dividing up the x-axis.

15.2 Riemann's Method for Measuring Arc Length

Consider the curve shown in Figure 15.7. We can estimate its arc length, following the model described in the last section, by dividing up the x-axis as indicated in Figure 15.8.

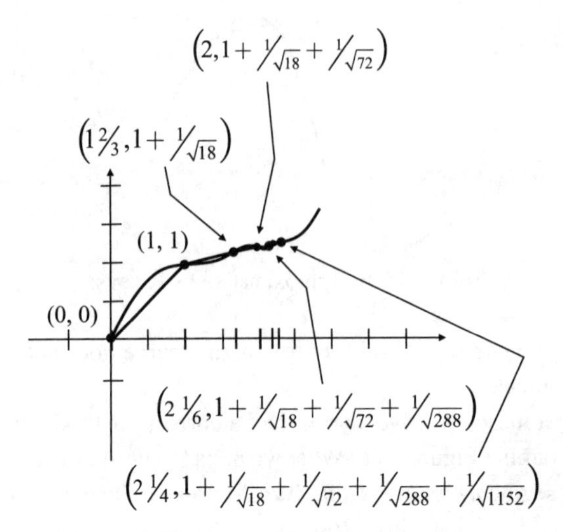

Figure 15.9. Approximating linear segments.

Then the approximating linear segments are shown in Figure 15.9. Let us analyze just one of those segments—see Figure 15.10.

We see that the length of the segment is determined by the Pythagorean theorem. It is

$$\ell = \sqrt{(\Delta x)^2 + (\Delta y)^2}\,.$$

Thus the first segment, connecting the points $(0, 0)$ and $(1, 1)$, has length $\sqrt{2}$. The second segment, connecting the points $(1, 1)$ and $(1 + \frac{2}{3}, 1 + \frac{1}{\sqrt{18}})$, has length $\sqrt{2}/2$. The third segment, connecting the points $(1 + \frac{2}{3}, 1 + \frac{1}{\sqrt{18}})$ and $(2, 1 + \frac{1}{\sqrt{18}} + \frac{1}{\sqrt{72}})$, has length $\sqrt{2}/4$. The fourth segment, connecting the points $(2, 1 + \frac{1}{\sqrt{18}} + \frac{1}{\sqrt{72}})$ and $(2 + \frac{1}{6}, 1 + \frac{1}{\sqrt{18}} + \frac{1}{\sqrt{72}} + \frac{1}{\sqrt{288}})$, has length $\sqrt{2}/8$. The fifth segment, connecting the points $(2 + \frac{1}{6}, 1 + \frac{1}{\sqrt{18}} + \frac{1}{\sqrt{72}} + \frac{1}{\sqrt{288}})$ and $(2 + \frac{1}{4}, 1 + \frac{1}{\sqrt{18}} + \frac{1}{\sqrt{72}} + \frac{1}{\sqrt{288}} + \frac{1}{\sqrt{1152}})$ has length $\sqrt{2}/16$.

Our choice of points here is a bit contrived, but we do things this way so that the calculations work out cleanly. In fact, with sufficient effort (and perhaps with the aid of a computer!) one could do the calculation for most any choice of points.

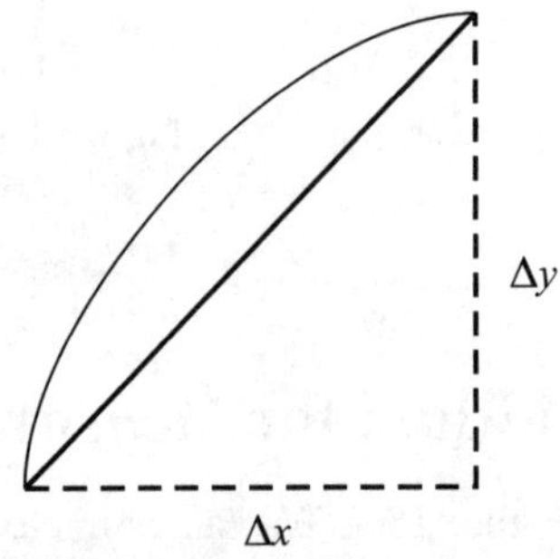

Figure 15.10. Analysis of a segment.

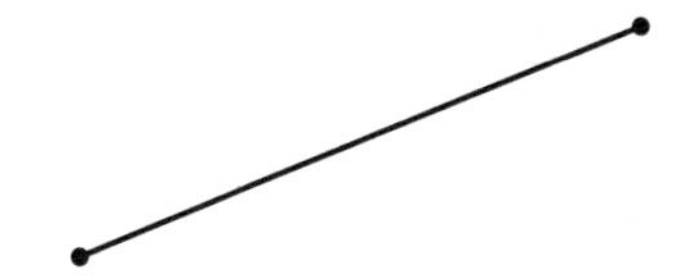

Riemmannian length is $2^{(j-1)}$ times Euclidean length.

Figure 15.11. Assigning a non-Euclidean length to each segment.

We see, then, that the total length of the curve, over the portion of the x-axis from 0 to 2.25, is about

$$\sqrt{2}\cdot\left(1+\frac{1}{2}+\frac{1}{4}+\frac{1}{8}+\frac{1}{16}\right). \qquad (*)$$

If we were to add pieces to the curve, then the length would be augmented in an obvious way by adding terms to this series. The sum tends to $\sqrt{2}\cdot 2$, even if infinitely many pieces are added. The curve terminates when $x = 7/3$.

Now Riemann's idea is to approximate the length of the curve by the sum of the lengths of the segments, but to reckon the length of a segment differently in different parts of space. Thus imagine viewing the segments through a strange telescope from the vantage point of the point $(1, 1)$ in the plane. The further a segment is from that base point, the longer it seems to be. As a result of these considerations, we assign to the first segment the length 1·(Euclidean length) or 1. We assign to the second segment the length 2·(Euclidean length). We assign to the third segment the length 4 · (Euclidean length). Refer to Figure 15.11. We assign to the fourth segment the length 8·(Euclidean length). We assign to the fifth segment the length 16 · (Euclidean length). Now the sum of the *Riemannian* lengths of all the little segments is

$$\sqrt{2}\cdot(1+1+1+1+1) = 5\sqrt{2}.$$

If we were to add pieces to the curve, then the length would be augmented in the obvious way by adding terms to this series. Each of those terms would be $\sqrt{2}$. Thus, even though the curve obviously has finite Euclidean length (as our earlier calculation shows), the Riemannian length is infinite.

Further Reading

Lawrence J. Wallen, Kepler, the taxicab metric, and beyond: An isoperimetric primer, *The College Mathematics Journal* 26(1995), 178–190.

Shiing-Shen Chern, What is geometry?, *The American Mathematical Monthly* 97(1990), 679–686.

15.3 The Hyperbolic Disc

One of the most famous, and most studied, examples of a Riemannian geometry is the hyperbolic disc. This is the unit disc

$$D = \{z \in \mathbb{C} : |z| = \sqrt{x^2+y^2} < 1\}.$$

Of course the ordinary Euclidean notion of distance makes perfectly good sense on D. But our goal now is to equip D with a special metric so that the distance from any point in D

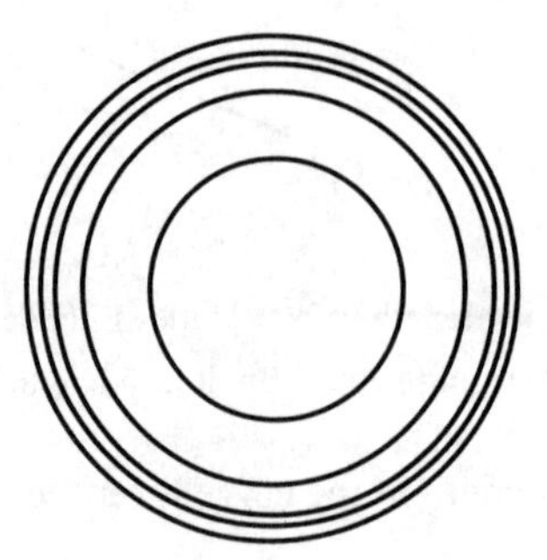

Figure 15.12. Breaking the disc up into annular pieces.

to the boundary is in fact infinite. This metric is of special interest because it is the unique metric on the disc that is invariant under certain special types of complex maps or functions (i.e., the *conformal mappings*). The metric is frequently termed the *Poincaré metric*.

We approach this metric, following the philosophy of Riemann, by assigning length in a special way. We proceed by breaking the disc up into annular pieces. Set

$$A_j = \{z \in D : 1 - 2^{-j+1} < |z| \leq 1 - 2^{-j}\}, \quad j = 1, 2, \ldots .$$

These are exhibited in Figure 15.12.

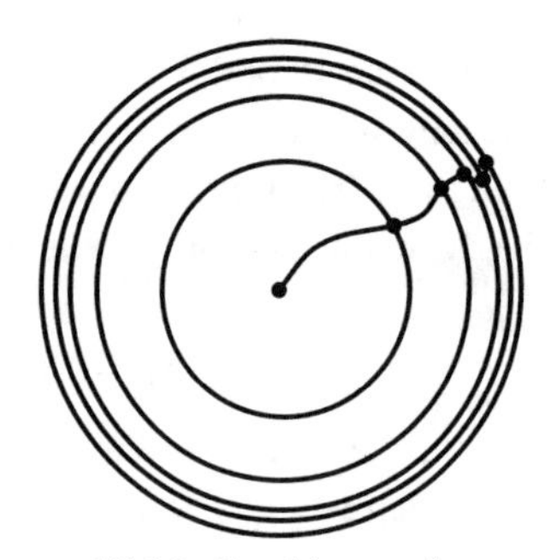

Figure 15.13. Breaking up the curve γ.

Now, if γ is a curve in the disc, then we break γ up into pieces:

$$\gamma = \bigcup_{j=0}^{\infty} \gamma_j ,$$

where

$$\gamma_j = \gamma \cap A_j$$

is the intersection of γ with A_j (refer to Figure 15.13). We define the *length* of γ_j to be

$$\ell(\gamma_j) = 2^j \cdot \|\gamma_j\| ,$$

where $\|\gamma_j\|$ denotes the ordinary Euclidean length of γ_j, as we discussed earlier in this chapter.[1] Finally, we set

$$\ell(\gamma) = \sum_{j=0}^{\infty} \ell(\gamma_j) .$$

[1] We could choose most any constants here. Certainly 2^j is *not* the only choice. We did it this way just for the convenience of the calculation.

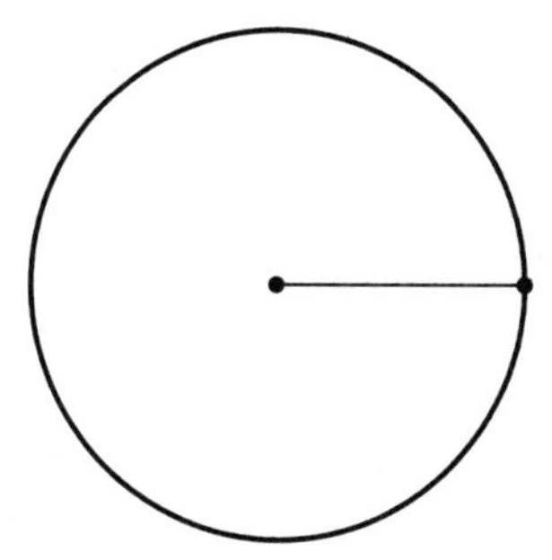

Figure 15.14. A straight line segment from the origin to the boundary point $(1, 0)$.

This is a somewhat convoluted definition, but it captures the spirit that we measure length according to how far we are from the boundary. Let us now calculate a specific example to show precisely what we have achieved:

EXAMPLE 15.1. Let γ be the straight line segment in D from the origin to the boundary point $(1, 0)$. See Figure 15.14. Calculate its length.

Solution. We break γ up into pieces as indicated in our paradigm for calculating length. The *Euclidean* length of the jth piece is 2^{-j}. And we are to multiply that by 2^j. So the contribution to length of the jth piece is 1. Thus we see that the overall Poincaré length of γ is $\sum_{j=0}^{\infty} 1 = +\infty$.

Although we shall not provide the details, it can be seen that the length of any curve stretching from an interior point of D to the boundary will be $+\infty$. The technical terminology for this situation is that *the disc is complete in the Poincaré metric.*

The point of interest here is that the boundary is infinitely far away from any point in the interior. At first this observation is counterintuitive. If instead we were living in the plane—that is if our domain of study were the plane—then it would make good sense that the boundary of the plane is infinitely far away. See Figure 15.15.

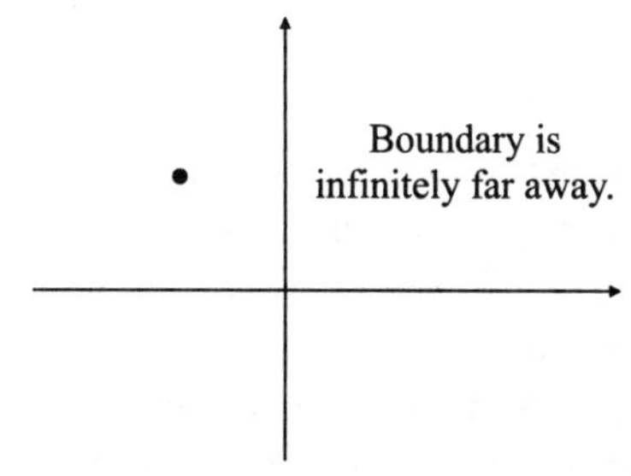

Figure 15.15. The boundary is infinitely far away and the region is complete.

Further Reading

H. S. Bear, Part metric and hyperbolic metric, *The American Mathematical Monthly* 98 (1991), 109–123.

Abraham A. Ungar, The hyperbolic Pythagorean theorem in the Poincaré disc model of hyperbolic geometry, *The American Mathematical Monthly* 106(1999), 759–763.

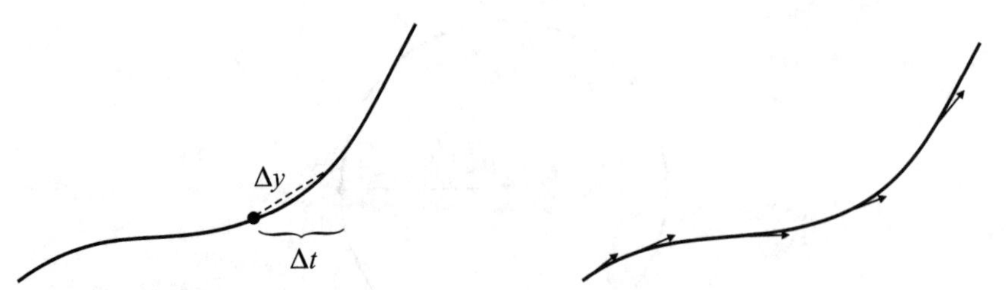

Figure 15.16. Increment of the length of a curve.

Figure 15.17. Length of the curve is the integral of the lengths of the tangent vectors.

15.4 The Use of the Integral

Since we introduced the concept of the integral in Chapter 8, it is appropriate here that we describe how to implement Riemann's ideas in that language.

Now suppose that M is a region in space or a surface, and we want to equip it with a (Riemannian) metric. Riemann's idea is that the nub of the matter is to specify a means for giving the length of a curve. So let $\gamma : [0, 1] \to M$ be a differentiable curve in M, and assume that the derivative of γ is continuous. We say in practice that γ is *continuously differentiable*.

It is useful to recall from calculus class that we learned to find the length of a curve γ by integrating

$$\int_0^1 |\gamma'(t)|\, dt\,.$$

See Figure 15.16. Thus the aggregate length of a curve is the integral of the moduli of the tangent vectors to that curve. Refer to Figure 15.17. In Riemannian geometry we let the modulus of the tangent vector vary from point to point. To put the idea in context, note that in our standard Euclidean model the inner product of two vectors $\nu = \langle n_1, n_2, n_3\rangle$ and $\mu = \langle m_1, m_2, m_3\rangle$ is

$$\nu \cdot \mu = n_1 m_1 + n_2 m_2 + n_3 m_3\,.$$

The base point—the actual *location* of the tangent vector—plays no role in this calculation.

What is going on here is that we are approximating a smooth curve by a piecewise linear curve, and then each linear piece is being approximated by a tangent vector. This is a familiar procedure from calculus (Figure 15.18).

Put in other words, we let there be given an inner product at each point $m \in M$, and we allow that inner product to vary in a smooth fashion from point to point. So if $m \in M$, then we have an inner product $\langle \eta, \mu\rangle_m$ on the collection of tangent vectors (i.e., the *tangent space*) at m (here both η and μ lie in the tangent space T_m to M at m). Then the *length of*

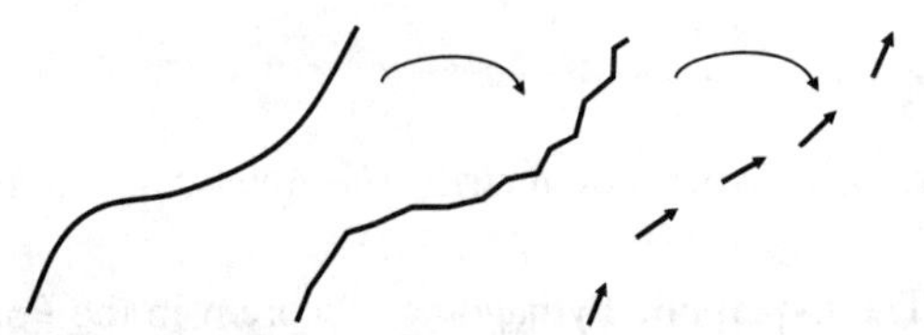

Figure 15.18. Approximation of a smooth curve.

a curve $\gamma : [0, 1] \to M$ is given, in the Riemannian model, by

$$\ell(\gamma) = \int_0^1 \left[\langle \gamma'(t), \gamma'(t) \rangle_{\gamma(t)}\right]^{1/2} dt\,.$$

It is convenient to let $\|\eta\|_z = \left[\langle \eta, \eta \rangle_z\right]^{1/2}$. Thus

$$\ell(\gamma) = \int_0^1 \|\gamma'(t)\|_{\gamma(t)}\, dt\,.$$

We define the *distance* between two points $P, Q \in M$ to be the infimum (like the minimum—see [KRA1] or [RUD]) of lengths of all possible continuously differentiable curves that connect P to Q.

EXAMPLE 15.2. It makes good sense that the first example we look at should be the Poincaré disc D. Define the inner product, at a point $z \in D$, to be

$$\langle \eta, \mu \rangle_z = \frac{\langle \eta, \mu \rangle_{\text{Euclid}}}{(1 - |z|^2)^2}\,.$$

We see that the numerator of the expression on the right is the usual Euclidean inner product, but we are scaling it with the factor of $(1 - |z|^2)$ in the denominator. Note that, as z approaches the boundary, the scaling blows up.

Now let $P = 0$, the origin, and let $Q = t + i0$, $0 < t < 1$. Let us consider the distance from P to Q. So we must examine the length of a curve γ connecting P to Q. It is convenient to consider a curve of the form $\gamma_t(s) = (s, g(s))$, $0 < t < 1$, and any curve *not* of this form will clearly be longer so is not of interest. We see that

$$\begin{aligned}
\ell(\gamma_t) &= \int_0^t \frac{|1 + |g'(s)|^2|^{1/2}}{1 - (s^2 + |g(t)|^2)}\, dt \\
&\geq \int_0^t \frac{1}{1 - s^2}\, dt \\
&= \frac{1}{2} \log\left(\frac{1+t}{1-t}\right)\, dt\,.
\end{aligned}$$

It is plain to see that

$$\lim_{t \to 1^-} \ell(\gamma_t) = +\infty\,.$$

This demonstrates that the distance of the origin to the boundary ∂D is infinity. Thus, in a palpable sense, the disc D is complete in the Poincaré metric.

A byproduct of the calculation in this last example is that the curve of least length (the geodesic) connecting the origin to a point of the form $t + i0$ is just a standard line segment. Now part of the interest of the Poincaré metric is that it is invariant under so-called conformal mappings of the disc D. These are mappings of a complex variable z that are one-to-one and onto from D to D. Certainly rotations are such mappings. Another type of such mapping, one that is of considerable analytical interest, is the Möbius transformation

$$\varphi_a : z \mapsto \frac{z - a}{1 - \overline{a}z}$$

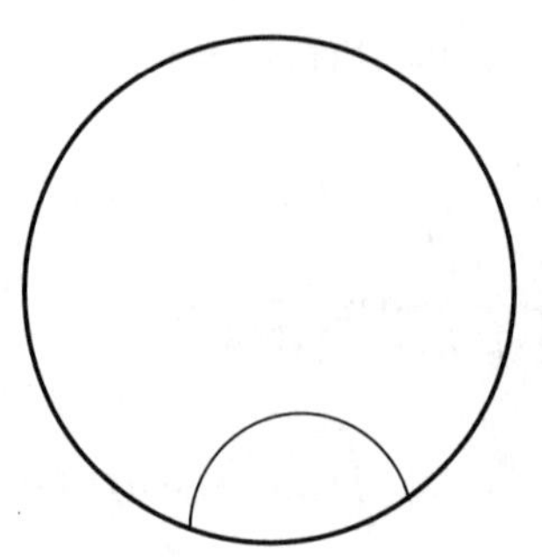

Figure 15.19. A geodesic in the Poincaré metric.

for a a fixed complex constant of modulus less than 1. You can check for yourself that φ_a maps D to D in a one-to-one and onto fashion. And the image of the segment-curve $t \mapsto t + i0$ under φ_a is the curve

$$t \longmapsto \frac{3t/4}{1+t^2/4} + i\frac{-1/2 - t^2/2}{1+t^2/4}. \qquad (\dagger)$$

One can check (just calculate the images of the points 0, 1/2, 1) that in fact this image curve is the arc of a circle.

What is the upshot of this reasoning? We began with the simply verified fact that the curve $t \mapsto t + i0$ is a length-minimizing curve or geodesic. We subjected that curve to a mapping φ_a that preserves the metric, hence it preserves geodesics. The result was the curve (†) that is also a geodesic. And what is this curve? It is the arc of a circle. See Figure 15.19. Thus we learn that geodesics in the Poincaré metric are line segments and arcs of circles.

Further Reading

Chaim Goodman-Strauss, Compass and straightedge in the Poincaré disk, *The American Mathematical Monthly* 108(2001), 38–49.

P. J. Kelly, Barbilian geometry and the Poincaré model, *The American Mathematical Monthly* 61(1954), 311–319.

Exercises

1. Imitate the construction of the Poincaré metric on the unit disc to produce a metric on the interval (0, 1) in the real line which makes this interval complete. This means that the distance from any interior point of the interval to the boundary is infinity.

2. Imitate the construction of the Poincaré metric on the unit disc to produce a metric on the unit ball

$$B(0,1) = \{(x_1, \ldots, x_N) \in \mathbb{R}^N : x_1^2 + x_2^2 + \cdots x_N^2 < 1\}.$$

in N-dimensional Euclidean space which makes this ball complete. This means that the distance from any interior point of the ball to the boundary is infinity.

3. Let $\gamma(t) = (t, 2t), 0 \le t \le 1/2$, be a curve in the unit disc. What is its Poincaré length?

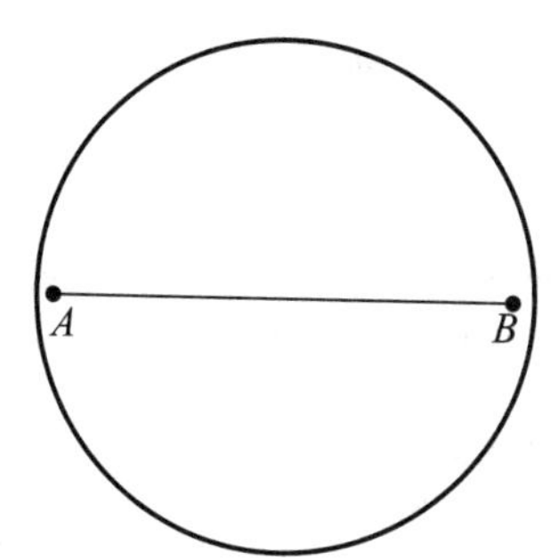

Figure 15.20. Curve of shortest length in the Poincaré metric.

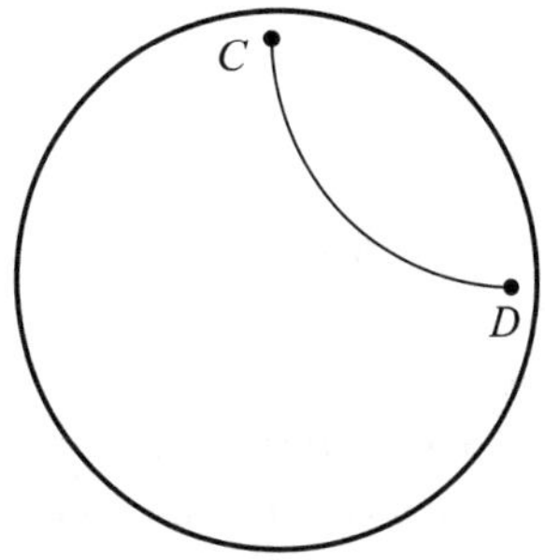

Figure 15.21. A geodesic in the Poincaré metric.

4. Show that, given two points in the disc, the curve of least Poincaré length connecting them is not necessarily unique. Thus Poincaré geometry is a non-Euclidean geometry. Read more about this topic in [KRA3] and [GRE].

5. Flesh out the discussion in the text: If A and B are diametrically opposite points in the disc D, then the *geodesic*, or curve of shortest length in the Poincaré metric connecting A and B, is a straight line segment (see Figure 15.20). Provide an argument that suggests why this is so.

But if C and D are points which are far apart in the Euclidean sense but both near the boundary, then the geodesic in the Poincaré metric that connects C and D is an arc of a circle (Figure 15.21). Provide an argument that suggests why this is so.

6. If we modify the definition of the Poincaré metric so that

$$\langle \mu, \eta \rangle_z = \frac{\langle \mu, \eta \rangle_{\text{Euclid}}}{1 - |z|^2},$$

then the metric is no longer complete. Explain why this is so.

7. Describe a metric on the annulus $A = \{(x, y) \in \mathbb{R}^2 : 1/4 < x^2 + y^2 < 4\}$ so that the curve $\gamma(t) = (t, 0)$, $1/2 < t < 2$, is of infinite length.

8. Refer to the metric on the annulus that you constructed in Exercise 7. What can you say about the lengths of the circles

- $\left\{(x, y) : x^2 + y^2 = \dfrac{1}{3}\right\}$;
- $\left\{(x, y) : x^2 + y^2 = \dfrac{1}{2}\right\}$;
- $\{(x, y) : x^2 + y^2 = 1\}$;
- $\{(x, y) : x^2 + y^2 = 2\}$;
- $\{(x, y) : x^2 + y^2 = 3\}$;
- $\left\{(x, y) : x^2 + y^2 = \dfrac{7}{2}\right\}$.

9. Project: Discuss all possible Riemannian metrics on the unit interval $(0, 1)$. Can you classify them (according to how they blow up at the endpoints)? Can you compare them? How many are there? See [ONE] for the basic ideas here.

10. Project: Any smooth surface has at least one Riemannian metric. Discuss why this is true. Is there only one? Are there finitely many? Are there infinitely many? Again see [ONE] or [DOC].

11. Project: Let S be a two-dimensional surface in Euclidean 3-space $\mathbb{R}^3$. We say that S is *totally geodesic* if, whenever $P, Q \in S$, then the geodesic path of least length connecting P to Q in $\mathbb{R}^3$ actually lies in S. For example, the sphere is *not* totally geodesic because the shortest path connecting two points of the sphere will be a segment, and a segment will not lie in the sphere. Provide the details of this assertion.

Give an example of a totally geodesic surface in $\mathbb{R}^3$. In fact show that the totally geodesic surface is unique in a certain sense. These ideas are considered in [DOC].

12. Project: If S and T are surfaces equipped with Riemannian metrics, then

$$S \times T = \{(s,t) : s \in S, t \in T\}$$

will also have a natural Riemannian metric (inherited from S and T). Explain why this is so. This idea is discussed in [KON].

16

Georg Cantor and the Orders of Infinity

16.1 Introductory Remarks

Georg Ferdinand Ludwig Philipp Cantor (1845–1918) was born to a merchant father in St. Petersburg and a talented violinist mother. He inherited considerable musical and artistic talent. For his first ten years young Cantor was educated at home by a private tutor. After that he attended primary school in St. Petersburg. In 1856 the family moved to Germany, and Cantor lived there for the rest of his life. He said that he never felt comfortable in Germany, and he remembered his early years in Russia with great nostalgia. The genius of Georg Cantor was not fully appreciated during his lifetime. He was one of the great mathematicians of all time, but he died an unhappy man. Details of his life appear in [DAU] and [PUR].

Cantor was one of the true geniuses of modern mathematics. Whereas most mathematical ideas, indeed most ideas in the world, can be said to have developed from earlier ones—from constructs that were already in the air—Cantor's theory of sets and the infinite seems to have been a wholly original creation. Even the ideas of calculus, for which Newton and Leibniz are justly revered and celebrated, can be said to have followed from earlier constructions of Archimedes, Descartes, and Fermat. Not so with Cantor's theory; it sprang in full blossom directly from the great man's cranium.

Yet poor Cantor suffered for his genius. The notion of infinity is one of those special ideas that occupy the collective unconscious of all human beings. Pretend to be some "super intellect" who actually understands infinity—and not a philosopher or a preacher but a lowly mathematician—and you court public criticism, damnation, and ostracism. Cantor spent a significant part of his adult life in sanitaria in his agonized attempts to deal with all the flak he got for his ideas.

In fact, just as thirty years later people from philosophy and theology and other nontechnical fields picked up on (and in some cases ran with) the Heisenberg uncertainty principle, so it happened to Cantor that various theologians attempted to show that his ideas provided a rock-solid proof that God exists. One such gentleman used the idea of cardinal numbers and set-theoretic isomorphism (see our discussion below) to give a proof of the existence of the holy trinity.

Cantor's father suffered from poor health, and the family moved to Germany seeking a climate that was warmer than St. Petersburg's harsh winters. Cantor studied at the Realschule in Darmstadt, the Höhere Gewerbeschule in Darmstadt, and finally the Polytechnic of Zurich. Although Cantor's father originally wanted him to study engineering, he ultimately consented to let Georg study mathematics. It is sad that his studies in Zurich were cut short by his father's untimely death in 1863.

Cantor moved to the University of Berlin, where he studied with Weierstrass, Kummer, and Kronecker. He was friends with fellow student Hermann Schwarz. He graduated with a doctorate in number theory in 1867 and went to work at a girls' school. In 1868 he joined the Schellbach Seminar for mathematics teachers. He was appointed to a position at the University of Halle in 1869, and immediately sought to do his Habilitation.[1] Cantor's director of research at this time was Heine, and his tastes were moving from number theory to analysis. Heine recognized Cantor's talent and challenged him to attack a famous unsolved problem: to prove the uniqueness of representation of a function by a trigonometric series. Cantor solved the problem in 1870. He published further papers in the subject between 1870 and 1872. Cantor was promoted to Extraordinary Professor at Halle in 1873.

At this time Cantor began his seminal research on infinity and the concept of cardinal number. Of all Cantor's ideas, this was the one for which he would be most remembered, and for which (during his lifetime) he would be most persecuted. During this period his communications with Richard Dedekind were of central importance.

In 1874 Cantor became engaged to Vally Guttmann, a friend of his sister. They married in the same year, and spent their honeymoon in Interlaken, Switzerland. Cantor spent some time even during his honeymoon on a sojourn discussing mathematics with Dedekind.

By 1877, thanks in part to ideas developed with Dedekind, Cantor fulfilled a long quest and proved that the points in an interval can be put in one-to-one correspondence with the points of a cube in any dimensional space. This is a profound and shocking result, with significant consequences for geometry, analysis, and the philosophy of mathematics. And it was at this time that Cantor's former teacher and mentor Leopold Kronecker began to attack Cantor. In fact Kronecker tried to block the publication of some of Cantor's work.

Cantor was in 1879 promoted to a full Professorship. But he longed for a chair at a more prestigious university. Beginning in 1879 Cantor's relations with several important mathematicians became strained, and ultimately ended. This included his friendly relations with Weber and Dedekind. At least part of the problem was that other mathematicians were unsure of the directions that Cantor's research was taking. On the positive side, Cantor began a correspondence with Mittag-Leffler, and began to publish in the latter's important journal *Acta Mathematica*.

In May of 1884 Cantor had his first known attack of clinical depression. He recovered in just a few weeks, but suffered a loss of confidence. Cantor took a vacation in the Harz mountains, and made an effort to mend his relationship with Kronecker. Here he enjoyed some success.

Cantor embarked on a bold new direction in his mathematics, attempting to prove what has become known as the "continuum hypothesis". The problem was to show that the order of infinity of the real numbers was the next one after the order of infinity of the natural

[1] Typically a Habilitation is a postdoctoral degree involving further research accomplishments and possibly the publication of a book.

numbers. This problem literally drove him crazy. One day he would think he had proved it true and the next he would think he had proved it false. At the same time, Cantor's friend Mittag-Leffler brought him up short by attempting to persuade Cantor to withdraw one of his papers—at the proof stage—from *Acta Mathematica*. Mittag-Leffler claimed that the paper was simply too far ahead of its time. Cantor later joked about the matter, but he was clearly very unhappy. Cantor ceased his correspondence with Mittag-Leffler at this time, and his flood of new ideas nearly stopped.

It should be stressed that Cantor certainly had his supporters. David Hilbert was a far-seeing mathematician who appreciated the importance and value of Cantor's ideas early on. He said

> No one shall expel us from the Paradise that Cantor has created.

And in fact these words were prophetic. The idea of cardinality is a principal tool in all of modern mathematics.

In 1886, Cantor bought a fine new house on Händelstrasse, a street named after the German composer Handel. By the end of the year the Cantors had a new son, completing the family to six children. Cantor's interests shifted at this time and he became involved with philosophical issues as well as the founding of the Deutsche Mathematiker-Vereinigung—a German mathematical society. Cantor invited his old teacher/nemesis Kronecker to address the first meeting of the society, but Kronecker was unable to accept because his wife became injured and then died in a mountain-climbing accident.

Cantor was elected president of the new mathematical society. He held the post until 1893. In 1897 Georg Cantor attended the first meeting of the International Congress of Mathematicians in Zurich. Hurwitz and Hadamard praised Cantor's work in their talks at that august meeting. One positive outcome of the meeting is that Cantor rekindled his friendship with Dedekind.

But Cantor's mathematics was taking some strange new turns. He continued to struggle with the continuum hypothesis (CH), and he would never resolve it.[2] Of course Cantor had no idea that it would take Kurt Gödel (1906 C.E.–1978 C.E.) and Paul Cohen (1934 C.E.–2007 C.E.), using powerful new ideas of abstract logic, to resolve this knotty problem. In fact they proved that the continuum hypothesis is independent of the axioms of set theory. He also began to learn of various paradoxes of set theory (such as Russell's Paradox). These really shook Cantor, for he felt that set theory was the fundament of his contribution to mathematics. These difficulties caused Cantor to turn away from mathematics and to instead concentrate on philosophy and Elizabethan literature. In fact one of Cantor's passions at that time was to prove that Francis Bacon had written the works of Shakespeare.

In 1911, after some years of mental turbulence, Cantor was delighted to be invited to the University of St. Andrews to be a distinguished foreign scholar at their 500th anniversary celebration. Unfortunately he was distracted by his own ill health and his son's failing health (his youngest son soon died). His behavior at the celebration was erratic.

Cantor retired in 1913 and spent his last years with little food because of World War I. He was ill for much of his final time. A major event was planned in 1915 to celebrate Cantor's 70th birthday; unfortunately the war prevented it being held. But a smaller event

[2] Cantor first began to think about CH in the mid-1880s. David Hilbert listed this question as the first of his famous 23 problems enunciated in his seminal address to the International Congress of Mathematicians in Paris in 1900.

was held near Cantor's home. In June of 1917 Cantor entered a sanitorium. This was not to his liking, and he continually wrote to his wife to be allowed to go home. He died there, of a heart attack, in 1918.

By the end of Kronecker's life, Cantor and Kronecker had finally mended some of their differences, and Cantor was receiving some of the honor and recognition he had so long deserved. In fact no less a light than David Hilbert, arguably one of the great spokesmen for twentieth century mathematics, described Cantor's transfinite arithmetics as "the most astonishing product of mathematical thought, one of the most beautiful realizations of human activity in the domain of the purely intelligible."

Further Reading

V. Wiktor Marek and Jan Mycielski, Foundations of mathematics in the twentieth century, *The American Mathematical Monthly* 108(2001), 449–468.

Raymond M. Smullyan, Satan, Cantor, and infinity, *The College Mathematics Journal* 16(1985), 118–121.

16.2 What is a Number?

In the 1960s there was a major educational movement in the United States called "The New Math". Formulated by prominent mathematical scholars at Stanford, Yale, and other distinguished universities, this paradigm for grade school education promulgated the idea that children should be taught mathematics axiomatically. That is to say, they should begin at age 6 learning the axioms of set theory and the construction of the number systems. They should learn first-order logic and methods of proof.

You can decide for yourself whether this program was a good idea. It was not very successful—in part because the teachers (who had been trained the old-fashioned way) could not understand it well enough to teach it and in part because the parents could not understand it well enough to be able to help their kids with their homework.

Be that as it may, we will take one of the famous questions from The New Math as the inspiration for our discussion of Cantor. Namely, the students were asked to muse about the difference between "number" and "numeral". In practice, the average person does not draw a formal distinction—at least could not precisely articulate a distinction—between the symbol 5 and the idea that it represents. But that is what the New Math question asks us to do. What is a cogent answer?

For this we need Cantor's idea of set and of set equivalence. A *set* is a collection of objects. For example,

$$\{2, 3, 5, 9, 13, 24, 48\}$$

is a set. The objects in the set are called *elements* of the set. Notice that we surround the elements with curly braces. We generally denote a set with a capital roman letter. So we write, for instance,

$$A = \{2, 3, 5, 9, 13, 24, 48\}$$

and refer to the set as A. Observe that $2, 3, 5, 9, 13, 24$, and 48 are all elements of the set A. We write $2 \in A$, $9 \in A$, and so forth.

If A and B are sets then we say that B is a subset of A, and we write $B \subset A$, if every element of B is also an element of A. As an example, let $A = \{1, 2, 3, 4\}$ and let $B = \{2, 3, 4\}$. Then B is a subset of A. There is a special set called *the empty set* that has no elements. We denote the empty set by $\emptyset$. The empty set is a subset of every set.

Now let A and B be sets. We say that A *is equivalent* to B (for the purposes of counting), and we write $A \cong B$, if the elements of A and of B can be matched up one-to-one. Let us look at an example to illustrate the point.

EXAMPLE 16.1. Let $A = \{1, 3, 5, 7, 9, 11\}$ and $B = \{\alpha, \beta, \gamma, \delta, \epsilon, \zeta\}$. Then the correspondence

$$\begin{aligned} 1 &\longleftrightarrow \alpha \\ 3 &\longleftrightarrow \beta \\ 5 &\longleftrightarrow \gamma \\ 7 &\longleftrightarrow \delta \\ 9 &\longleftrightarrow \epsilon \\ 11 &\longleftrightarrow \zeta \end{aligned}$$

shows that A and B are equivalent, i.e., $A \cong B$.

This notion of "equivalence" formalizes the idea of what it means for two sets to have the same number of elements. The sets do not have to be sets of numbers: they could be sets of fish, or sets of donuts, or sets of Michael Jackson CDs.

Now, to answer the "new math question": A *numeral* is a typographical symbol like 5. That symbol, 5, stands for the idea of a *number*. The number that it stands for is the collection of all sets that are equivalent to $\{1, 2, 3, 4, 5\}$. In English, the *numeral* 5 stands for the collection of all sets that are equivalent to $\{1, 2, 3, 4, 5\}$—in other words, for the collection of all sets having five elements. The collection of all sets having five elements is, by definition, the number 5.

So far this sounds like the pointless abstractification of fairly simple ideas that we have all known (in much simpler terms) since childhood. One point that should be noticed right away is this: If A is a set in the collection that is described by the numeral 4 and B is a set in the collection that is described by the numeral 5, then A is *not* equivalent to B. To see this, notice that A is in fact equivalent to a proper *subset* of B, consisting of the first four elements of B. And B certainly cannot be equivalent to a proper subset of itself. Of course the same comments apply to any other distinct pairs of numerals. Thus the different numerals denote completely distinct collections of sets.

All of this reasoning becomes much more interesting if we follow Georg Cantor's model and apply the ideas to infinite sets. Let

$$A = \{1, 2, 3, 4, \ldots\}$$

and

$$B = \{2, 4, 6, 8, \ldots\}.$$

Then of course $B \subset A$ and $B \neq A$. In other words, B is a proper subset of A. But the correspondence

$$A \ni n \longleftrightarrow 2n \in B$$

matches *all* the elements of A in a one-to-one fashion with *all* the elements of B. No element of A is omitted and no element of B is omitted. Thus A and B are equivalent, even though one is a proper subset of the other. This is quite astonishing!!

The set $\{1, 2, 3, 4, \ldots\}$ is commonly called *the natural numbers*, and is denoted by $\mathbb{N}$. Any set that is equivalent to the natural numbers we call *countable*. Thus, according to our last example, the set of positive, even numbers is countable.

When two sets A and B are equivalent, in the sense that we have just been discussing, then we say that A and B have the same cardinality. If A has the same cardinality as $\mathbb{N}$, then we say that A is countable.

For You to Try Demonstrate that the set

$$B = \{1, 3, 5, \ldots\}$$

of positive, odd integers has the same cardinality as the entire set $\mathbb{N}$ of positive integers. Thus the odd, positive numbers form a countable set.

For You to Try Demonstrate that the set

$$C = \{3, 7, 11, 15, \ldots\}$$

has the same cardinality as the set of natural numbers $\mathbb{N}$. Thus C is countable.

In fact there are many different countable sets.

EXAMPLE 16.2. Let us verify that the full set of integers has the same cardinality as the natural numbers. In other words, the set $\mathbb{Z}$ of integers is countable.

To see this, examine the correspondence

$$\begin{array}{cccccccccccc} \mathbb{N}: & \ldots & 9 & 7 & 5 & 3 & 1 & 2 & 4 & 6 & 8 & \ldots \\ \mathbb{Z}: & \ldots & -4 & -3 & -2 & -1 & 0 & 1 & 2 & 3 & 4 & \ldots \end{array}$$

You can see that the strategy is to bounce from left to right so that the positive and negative integers are systematically exhausted.

It is clear that all the integers, both plus and minus, are enumerated in this way. Thus the integers $\mathbb{Z}$ form a countable set.

EXAMPLE 16.3. Let $S = \mathbb{Q}^p$, the set of positive rational numbers. And let $T = \mathbb{N}$, the set of natural numbers (or positive integers). Then S and T have the same cardinality.

To see this, we lay out the rational numbers in a tableau (Figure 16.1).

$$\begin{array}{cccc} \frac{1}{1} & \frac{1}{2} & \frac{1}{3} & \frac{1}{4}\ldots \\ \frac{2}{1} & \frac{2}{2} & \frac{2}{3} & \frac{2}{4}\ldots \\ \frac{3}{1} & \frac{3}{2} & \frac{3}{3} & \frac{3}{4}\ldots \\ \frac{4}{1} & \frac{4}{2} & \frac{4}{3} & \frac{4}{4}\ldots \\ & \ldots & & \ldots \end{array}$$

Figure 16.1.

We now associate positive integers in a one-to-one fashion with the numbers in this tableau. We do so by beginning in the upper-left-hand corner and then proceeding along diagonals stretching from the lower left to the upper right (Figure 16.2):

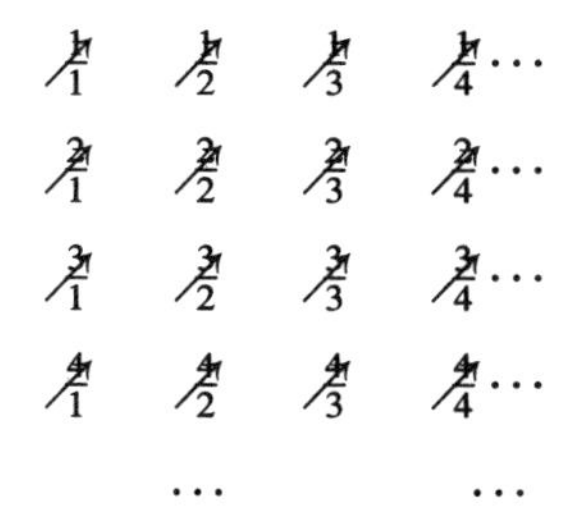

Figure 16.2.

This scheme clearly associates one positive integer to each fraction, and the association is one-to-one and onto:

$$\begin{array}{ccccccccccc} 1 & 2 & 3 & 4 & 5 & 6 & 7 & 8 & 9 & 10 & \cdots \\ \frac{1}{1} & \frac{2}{1} & \frac{1}{2} & \frac{3}{1} & \frac{2}{2} & \frac{1}{3} & \frac{4}{1} & \frac{3}{2} & \frac{2}{3} & \frac{1}{4} & \cdots \end{array}$$

Note that every fraction is counted multiple times, because the fraction $\frac{1}{2}$ also appears as $\frac{2}{4}$ and $\frac{3}{6}$ and so on. But we can skip the repeats, and the counting scheme still works. □

The preceding example is even more startling than the one before, because the positive integers form (apparently) a quite small subset of the positive rational numbers. Yet we are showing that the two sets have precisely the same number of elements. And we do so in a very graphic manner, exhibiting the correspondence quite explicitly. We commonly say that the argument in the last example *enumerates* the positive rational numbers. We have *enumerated* the positive rationals.

For You to Try Demonstrate that the set of *all* rational numbers—both positive *and* negative—has the same cardinality as the set of positive integers. In other words, enumerate all the rational numbers.

16.2.1 An Uncountable Set

It is natural to wonder whether every infinite set can be placed in one-to-one correspondence with the positive integers. The answer is "no." In fact, the set of all sequences of 0's and 1's forms a strictly larger infinite set, as the next example shows. In this example we shall be looking at objects like

$$0, 0, 1, 1, 1, 0, 1, 0, 1, 0, 0, 0, 0, 1, 1, 0, 1, \ldots$$

$$1, 0, 1, 0, 0, 1, 0, 1, 1, 1, 0, 1, 1, 0, 0, 0, 1, \ldots$$

$$0, 0, 0, 0, 1, 1, 0, 1, 1, 1, 1, 1, 0, 0, 1, 1, 1, \ldots$$

Each of these is a *sequence* of 0's and 1's. We shall consider the set of *all* such sequences.

EXAMPLE 16.4. Let S be the set of all sequences of 0's and 1's. We claim that S does *not* have the same cardinality as $\mathbb{N}$, the set of all positive integers.

The argument is by contradiction. Suppose, to the contrary, that there is a one-to-one correspondence between the set $\mathbb{N}$ of all positive integers and the set S of all sequences of 0's and 1's. Thus we can make a list:

$$\begin{array}{cccccc}
\textbf{(1)} & a_1^1 & a_2^1 & a_3^1 & a_4^1 & \ldots \\
\textbf{(2)} & a_1^2 & a_2^2 & a_3^2 & a_4^2 & \ldots \\
\textbf{(3)} & a_1^3 & a_2^3 & a_3^3 & a_4^3 & \ldots \\
\textbf{(4)} & a_1^4 & a_2^4 & a_3^4 & a_4^4 & \ldots \\
\ldots & & \ldots & & \ldots &
\end{array}$$

To be sure that this is clear, note that

$$a_1^1, a_2^1, a_3^1, a_4^1, a_5^1, \ldots$$

is a sequence of 0's and 1's, and

$$a_1^2, a_2^2, a_3^2, a_4^2, a_5^2, \ldots$$

is a sequence of 0's and 1's, and

$$a_1^3, a_2^3, a_3^3, a_4^3, a_5^3, \ldots$$

is a sequence of 0's and 1's, and so forth.

Thus each a_j^k is either a 0 or a 1. Thus the first row is the sequence of 0's and 1's (the element of S) corresponding to $\mathbf{1} \in \mathbb{N}$; the second row is the sequence of 0's and 1's (the element of S) corresponding to $\mathbf{2} \in \mathbb{N}$; the third row is the sequence of 0's and 1's (the element of S) corresponding to $\mathbf{3} \in \mathbb{N}$; and so forth. We claim to have explicitly exhibited a one-to-one correspondence between the set of positive integers and the collection S of all sequences of 0's and 1's.

But now we will find that this enumeration is in error. In fact, no matter how cleverly we think we have enumerated all the elements of S, there will always be a sequence of 0's and 1's that has been omitted from the list. That sequence is:

The first element is 0 if a_1^1 is 1 and is 1 if a_1^1 is 0.

The second element is 0 if a_2^2 is 1 and is 1 if a_2^2 is 0.

The third element is 0 if a_3^3 is 1 and is 1 if a_3^3 is 0.

The fourth element is 0 if a_4^4 is 1 and is 1 if a_4^4 is 0.

...And so forth...

In other words, we are constructing a new sequence that differs from the first in the list in the first entry, differs from the second in the list in the second entry, differs from the third in the list in the third entry, and so forth. Certainly the sequence we have now constructed cannot be on the list, so our claim to have enumerated all sequences of 0's and 1's cannot be true. That is a contradiction.

We conclude that the collection of all sequences of 0's and 1's cannot be enumerated. □

For You to Try Demonstrate that the set of all real numbers cannot be enumerated. [*Hint:* Consider only those real numbers with decimal expansions containing just the digits 0 and 1. Can you put those real numbers in one-to-one correspondence with the set of sequences considered in the last example?]

16.2.2 Countable and Uncountable

If a set S has the same cardinality as the set $\mathbb{N}$ of positive integers, then we say that S is *countable*. Thus it is immediate that the set $\mathbb{N}$ of positive integers is countable. A simple argument (see Example 16.2) shows that the set of all integers is countable, and the set of even integers is countable, and the set of odd integers is countable. The set of positive rational numbers is also countable (see Example 16.3). Example 16.4 shows that the set of sequences of 0's and 1's is not countable. It also follows that the set of real numbers is not countable, for any real number has a unique binary expansion (analogous to a decimal expansion but in base 2). And that is nothing other than a sequence of 0's and 1's.

If a set is infinite but is not countable, then we say it is *uncountable*.

EXAMPLE 16.5. Let S be the set of all subsets of the positive integers. Then S is uncountable.

To see this, observe that we can associate to any subset of the positive integers a sequence of 0's and 1's. We do so as follows. Let X be such a subset. If $1 \in X$, then the first element of the associated sequence is 1, otherwise it is 0. If $2 \in X$, then the second element of the associated sequence is 1, otherwise it is 0. If $3 \in X$, then the third element of the associated sequence is 1, otherwise it is 0, and so forth.

Just to be concrete, suppose that $X = \{1, 3, 5\}$. Then the sequence associated to this set is

$$1\,,\ 0\,,\ 1\,,\ 0\,,\ 1\,,\ 0\,,0\,,\ 0\,,\ \cdots.$$

If instead the set $X = \{2, 4, 6, 8, \ldots\}$ then the associated sequence is

$$0\,,\ 1\,,\ 0\,,\ 1\,,\ 0\,,\ 1\,,0\,,\ 1\,,\ \cdots.$$

In this way, every subset $X \subset S$ has associated to it a sequence of 0's and 1's (what amounts to the "indicator function" of the set) and vice versa. Since the set of all such sequences is uncountable, then so is the set of subsets of the positive integers uncountable.

□

For You to Try Consider the set of all real numbers of the form $j + k\sqrt{2}$ for $j, k \in \mathbb{N}$. Is this set countable or uncountable?

For You to Try Let S be the set of all polynomials whose coefficients are integer roots of positive integers (i.e., $\sqrt{2}$, $\sqrt[4]{5}$, $\sqrt[3]{4}$, etc.). Is this set countable or uncountable?

Proposition 16.1. *If S and T are each countable sets, then so is*

$$S \times T = \{(s, t) : s \in S, t \in T\}.$$

Proof. Since S is countable there is a bijection f from S to $\mathbb{N}$. Likewise there is a bijection g from T to $\mathbb{N}$. Therefore the function

$$(f \times g)(s,t) = (f(s), g(t))$$

is a bijection of $S \times T$ with $\mathbb{N} \times \mathbb{N}$, the set of ordered pairs of positive integers. But exactly the same argument as in Example 16.3 shows that the latter is a countable set. Hence so is $S \times T$. □

Proposition 16.2. *Let S b a countable set. If S' is a subset of S, then S' is either empty or finite or countable.*

Proof. This is really an exercise in logic.

If S' is neither empty nor finite then S' is infinite. Let $s_1, s_2, \ldots$ be an enumeration of the elements of S. Now let s_{j_1} be the first element of S that lies in S'. Let s_{j_2} with $j_2 > j_1$ be the second element of S that lies in S'. Continue in this manner. Since we are working with an enumeration of S, we shall certainly exhaust all the elements of S with our new counting process. Thus we shall also exhaust all the elements of S'. It follows that $\{s_{j_k}\}_{k=1}^{\infty}$ is an enumeration of S'. So S' is countable. □

Theorem 16.1. *Let S_1, S_2 be countable sets. Set $\mathcal{S} = S_1 \cup S_2$. Then $\mathcal{S}$ is countable.*

Proof. Let us write

$$S_1 = \{s_1^1, s_2^1, \ldots\},$$
$$S_2 = \{s_1^2, s_2^2, \ldots\}.$$

If S_1 and S_2 are disjoint, then the function

$$s_j^k \mapsto (j,k)$$

is a bijection of $\mathcal{S}$ with a subset of $\{(j,k) : j,k \in \mathbb{N}\}$. Then Proposition 16.2 shows that the set of ordered pairs of elements of $\mathbb{N}$ is countable. Thus $\mathcal{S}$ is equivalent to a subset of a countable set. It follows that $\mathcal{S}$ must be countable.

If there exist elements which are common to S_1, S_2, then discard any duplicates. The same argument shows that $\mathcal{S}$ is countable. □

Corollary 16.1. *If $S_1, S_2, \ldots, S_k$ are each countable sets, then so is the set*

$$S_1 \times S_2 \times \cdots \times S_k = \{(s_1, \ldots, s_k) : s_1 \in S_1, \ldots, s_k \in S_k\}$$

consisting of all ordered k-tuples $(s_1, s_2, \ldots, s_k)$ with $s_j \in S_j$.

Proof. We may think of $S_1 \times S_2 \times S_3$ as $(S_1 \times S_2) \times S_3$. Since $S_1 \times S_2$ is countable (by Proposition 16.1) and S_3 is countable, then so is $(S_1 \times S_2) \times S_3 = S_1 \times S_2 \times S_3$ countable. Continuing in this fashion (i.e., inductively), we can see that any finite product of countable sets is also a countable set. □

Corollary 16.2. *Let A_1, A_2, $A_3, \ldots$ each be countable sets. Let $\mathcal{A}$ be the union of all these sets:*

$$\mathcal{A} = \{a : a \in A_j \text{ for some } j\}.$$

Then $\mathcal{A}$ is countable.

Proof. Let $A_1, A_2, \ldots$ each be countable sets. If the elements of A_j are enumerated as $\{a_k^j\}_{k=1}^{\infty}$ and if the sets A_j are pairwise disjoint, then the correspondence

$$a_k^j \longleftrightarrow (j,k)$$

is one-to-one between the union $\mathcal{A}$ of the sets A_j and the countable set $\mathbb{N} \times \mathbb{N}$. By Proposition 16.1, this proves the result when the sets A_j have no common element. If some of the A_j have elements in common, then we discard duplicates and proceed as before. □

For You to Try Show that if B is countable and A is finite, then the set of all functions from A to B is countable. Discuss this problem in class.

Futher Reading

J. P. Jones, S. Toporowski, Irrational Numbers, *The American Mathematical Monthly* 80 (1973), 423–424.

Dan Kalman, The maximum and minimum of two numbers using the quadratic formula, *The College Mathematics Journal* 15(1984), 329–330.

16.3 The Existence of Transcendental Numbers

An *algebraic number* is a number that is the root of a polynomial equation with integer coefficients. For example, $\sqrt{2}$ is algebraic because it is the root of $x^2 - 2 = 0$. If a number is not algebraic then it is called *transcendental*.

EXAMPLE 16.6. Demonstrate that the number $\sqrt{2} + \sqrt{3}$ is algebraic.

Solution. Set $\alpha = \sqrt{2} + \sqrt{3}$. Then we may calculate that

$$\alpha^2 = 5 + 2\sqrt{6},$$

$$\alpha^3 = 9\sqrt{3} + 11\sqrt{2},$$

and

$$\alpha^4 = 49 + 20\sqrt{6}.$$

Thus it is natural to notice that

$$\alpha^4 - 10\alpha^2 = [49 + 20\sqrt{6}] - 10 \cdot [5 + 2\sqrt{6}] = -1.$$

We conclude that α satisfies the polynomial equation

$$x^4 - 10x^2 + 1 = 0.$$

It is extremely difficult to identify particular transcendental numbers. It can be proved that $\pi = 3.14159\ldots$, for example, is a transcendental number. But the proof is far beyond the scope of this text. Also the number $e = 2.71828\ldots$ is transcendental; again, the proof is extremely technical and difficult. In this section we shall use methods of Cantor to show that *most* real numbers are transcendental—without actually identifying any particular one of them.

Proposition 16.3. *The collection $\mathcal{P}$ of all polynomials $p(x)$ with integer coefficients is countable.*

Proof. Let $\mathcal{P}_k$ be the set of polynomials of degree k with integer coefficients. A polynomial p of degree k has the form

$$p(x) = p_0 + p_1 x + p_2 x^2 + \cdots + p_k x^k .$$

The identification

$$p(x) \longleftrightarrow (p_0, p_1, \ldots, p_k)$$

identifies the elements of $\mathcal{P}_k$ with the $(k+1)$-tuples of integers. By Corollary 16.1, it follows that $\mathcal{P}_k$ is countable. But then Corollary 16.2 implies that

$$\mathcal{P} = \bigcup_{j=0}^{\infty} \mathcal{P}_j$$

is countable. □

Proposition 16.4. *The set of all algebraic real numbers is countable. The set of all transcendental numbers is uncountable.*

Proof. Let $\mathcal{P}$ be the collection of all polynomials with integer coefficients. We have already noted in Proposition 16.3 that $\mathcal{P}$ is a countable set. If $p \in \mathcal{P}$ then let S_p denote the set of real roots of p. Of course S_p is finite, and the number of elements in S_p does not exceed the degree of p. Then the set A of algebraic real numbers may be written as

$$A = \cup_{p \in \mathcal{P}} S_p .$$

This is the countable union of finite sets so of course it is countable. We have demonstrated that the set of algebraic real numbers is countable.

Now that we know that the set of algebraic numbers is countable, we can notice that the set T of transcendental numbers must be uncountable. For $\mathbb{R} = A \cup T$. If T were countable then, since A is countable, it would follow that $\mathbb{R}$ is countable. But that is not so. □

For You to Try Take it for granted that the sum of two algebraic numbers is algebraic (it actually requires advanced ideas to prove this assertion). It is unknown whether $e + \pi$ is algebraic. It is also unknown whether $e - \pi$ is algebraic. But in fact one of them must be transcendental—we just do not know which one! Explain why.

For You to Try The square root of any positive integer is algebraic. In fact any integer root of any positive integer is algebraic. More subtle is that the *sum* of any two of these numbers is algebraic. Discuss in class why this claim is true.

Futher Reading

Robert Gray, Georg Cantor and transcendental numbers, *The American Mathematical Monthly* 101(1994), 819–832.

P. R. Halmos, Pure thought is better yet, *The College Mathematics Journal* 16(1985), 14–16.

R. W. Hamming, The transcendental character of $\cos X$, *The American Mathematical Monthly* 52(1945), 336–337.

Exercises

1. What is the cardinality of each of the following sets (i.e., is it countable or uncountable?)? Discuss these problems in class.

(a) $\mathbb{N} \times \mathbb{Q}$
(b) $\mathbb{N} \times \mathbb{N}$
(c) $\mathbb{R} \times \mathbb{Q}$
(d) $\mathcal{P}(\mathbb{Q})$ (i.e., the set of all subsets of $\mathbb{Q}$)
(e) $\mathbb{C}$
(f) $\mathbb{R} \setminus \mathbb{N}$ (i.e., the elements of $\mathbb{R}$ which are not natural numbers)
(g) $\mathbb{Q} \setminus \mathbb{N}$ (i.e., the elements of $\mathbb{Q}$ which are not natural numbers)
(h) The set of all decimal expansions, terminating or nonterminating, that include only the digits 3 and 7
(i) The set of all *terminating* decimal expansions that include only the digits 3 and 7
(j) The set of all solutions of all quadratic polynomials with integer coefficients
(k) The set of all solutions of all quadratic polynomials with real coefficients
(l) The set of all subsets of $\mathbb{N}$ that have at least three and not more than eight elements
(m) The set of all subsets of $\mathbb{Z}$ with at least six elements

2. Explain why every infinite set contains a countable subset. Discuss this problem in class.

3. Let S be a set. The *power set* of S is the collection of all subsets of S. For example, if $S = \{a, b, c\}$ then the power set of S is

$$\mathcal{P}(S) = \left\{ \{a\}, \{b\}, \{c\}, \{a, b\}, \{a, c\}, \{b, c\}, \{a, b, c\}, \emptyset \right\}.$$

Calculate the power set of $\{1, 2, 3, 4\}$.

4. Without attempting a rigorous proof, explain why if S is a finite set, then the power set of S (see Exercise 3 for terminology) will have more elements than S. Discuss this problem in class. Examine the question when $S = \{1, 2, 3\}$, when $S = \{1, 2, 3, 4, 5, 6\}$, and when $S = \mathbb{Z}$.

5. Refer to Exercise 3 for terminology. If S is a finite set with k elements then the power set of S has 2^k elements. Test out this statement for a set with two elements, and a set with three elements, and a set with four elements. Attempt an explanation for why this assertion is true in general.

6. Refer to Exercise 3 for terminology. Let S be the set of all positive integers. Consider the power set of S. Write down several elements of the power set. How many elements are in the power set? Is it countable or uncountable? Discuss this example in class.

7. Recall that $\emptyset$ is the set with no elements. If A is any other set, then confirm that $A \cup \emptyset = A$. Also show that $A \cap \emptyset = \emptyset$. Finally, check that $\emptyset \subseteq \emptyset$. Discuss these statements in class.

8. Let S and T be sets. Under what circumstances is it true that $S \cap T = \emptyset$? Under what circumstances is it true that $S \cup T = \emptyset$? Under what circumstances is it true that $S \times T = \emptyset$?

9. Let S be a set with k elements. Then how many elements are in $S \times \emptyset$?

10. Is it possible to write the set $\mathbb{R}$ of real numbers as the countable union of closed, bounded intervals $[a, b]$? Is it possible to write the set $\mathbb{R}$ of real numbers as the countable union of open, bounded intervals (a, b)? Is it possible to write the set $\mathbb{R}$ of real numbers as the countable union of countable sets?

11. Let S and T be uncountable sets. What can you say about the cardinality of $S \cup T$? About the cardinality of $S \cap T$? About the cardinality of $S \times T$?

12. Let

$$S_1 = \{1\},$$
$$S_2 = \{1, 2\},$$
$$S_3 = \{1, 2, 3\},$$
$$S_4 = \{1, 2, 3, 4\},$$

etc.. What can you say about the cardinality of $S_1 \times S_2 \times \cdots$?

13. **Project:** Let S be any set. Let $\mathcal{P}(S)$ denote the power set of S, that is the collection of all subsets of S. Then it is a fundamental result of Cantor that the cardinality of $\mathcal{P}(S)$ will always be greater than the cardinality of S itself. Prove this result. Consult [KRA6] for the details.

14. **Project:** Let S and T be sets. Suppose that there exists a function $f : S \to T$ that is one-to-one and there exists another function $g : T \to S$ that is one-to-one. Explain why S and T must have the same cardinality. This is a basic theorem to be found in [KRA6].

15. **Project:** Let S and T be sets. Supose that $f : S \to T$ and $g : T \to S$ are arbitrary functions (not necessarily one-to-one, not necessarily onto). Show that there is a subset $A \subseteq S$ and another subset $B \subseteq T$ such that $f(A) = B$ and $G(T \setminus B) = S \setminus A$. This result is treated in [HRJ].

17

The Number Systems

In everyday life, the numbers that we use most often are the whole numbers (the integers) and fractions (the rational numbers). If you go into a grocery store and request a quantity of chicken, or of carrots, or of flour, you express your needs in the form "Give me two and a half pounds of xyz." If you go to a lumber yard and place an order for wood, you ask for so many board feet of 12×1 pine. If you go to a bank for some money, you ask for a certain number of dollars and a certain number of cents—which of course is a *rational number* (of dollars).

The history of our number systems is a fascinating one. It is often said that primitive man counted, "One, two, three, many." Ancient cave paintings that depict the life of those times support this claim. In those days commerce was quite simple. Nobody owned more than a few pigs or cows or tunics. If a trade were to take place, it would most likely involve one or two or three items. On those rare occasions when a goodly number of pigs were involved, it was sufficient to say "many pigs". It was a long time later that man conceived of a need for numbers beyond three. And for a notation for writing those numbers down. And it was a long time after that before there was any notion that fractions were needed.

Much more fascinating is the history of zero, and of negative numbers. As you can imagine, it was impossible for people prior to 500 years ago—people whose lives were imbued with, and dominated by, religion—to consider zero and negative numbers in the absence of religious overtones. To talk about zero was to talk about *nothing*. And how could one do that? Was this not sacrilegious?

The beginnings of the idea of zero go back to the Sumerians of about 5,000 to 6,000 years ago and the Babylonians of about 4,000 years ago. The question of zero was intimately bound up with issues of place value. The Babylonian notation of 2000 B.C.E. did not distinguish between the integer 2106 and the integer 216. It was not until about 400 B.C.E. that a symbol was devised to mark a placeholder (where we would now put a zero).

In the Middle Ages, zero was disparaged as a mark of infidel sorcery, the sign of the Devil himself, the canceller of all meaning. For the Mayans, Zero was the Death God among their lords of the underworld, and men adopting the persona of Zero were ritualistically sacrificed in hopes of staving off the day of zero, the time when time itself would stop. Yet, over time, various unavoidable mathematical questions demanded that the idea of zero, and the idea of negative number, be addressed. In later years, zero was reinterpreted as a symbol of God's power to create a great deal out of naught.

For a long time, number systems were treated as languages that mathematical scientists just cooked up. But people eventually realized that this was not a dependable way to create mathematics. It could lead to paradoxes and contradictions. In the twentieth century we have realized that it is most rigorous, and minimizes the chance of error, to actually *construct* our number systems.

The purpose of the present chapter is to describe the twentieth-century methodology for creating and studying number systems. It is a fascinating journey, and will bring us into contact with a number of captivating people as well as ideas.

An important point to note, and we have discussed this idea elsewhere, is that the formalization of our number systems helped to remove them from religious considerations. The fact that we have an abstract construction for the integers means in particular that we need no longer worry that zero is a construct of the devil. We may now safely feel that the number systems are constructs of man; they are part of our mathematical machinery, that *we ourselves* have created to treat problems at hand.

Further Reading

W. C. Eels, Number systems of the North American indians, *The American Mathematical Monthly* 20(1913), 293–299.

Edwin G. Landauer, Methods of random number generation, *The Two-Year College Mathematics Journal* 8(1977), 296–303.

17.1 The Natural Numbers

17.1.1 Introductory Remarks

It is in fact quite difficult to construct—from first principles—a system of natural numbers in which the arithmetic operations are workable. Multiplication is particularly troublesome. In many treatments, an extra axiom is added in order to make multiplication have the properties that we want it to have (see [SUP, p. 136]). In other treatments, the natural numbers are taken as undefinables. We will take a third approach, which adheres more closely to the philosophy of ordinal numbers.

17.1.2 Construction of the Natural Numbers

Recall that $\emptyset$ is the *empty set*, that is the set with no elements (see Section 16.2). We inductively construct numbers as follows

$$\begin{aligned}
\mathbf{0} &= \emptyset \\
\mathbf{1} &= \{\emptyset\} \\
\mathbf{2} &= \{\emptyset, \{\emptyset\}\} \\
&\cdots \\
\boldsymbol{n} + \mathbf{1} &= n \cup \{n\} \\
&\cdots.
\end{aligned}$$

These are the natural numbers. The set of natural numbers is denoted by $\mathbb{N}$. We commonly enumerate them as $0, 1, 2, \ldots$.

What is important about the natural numbers, indeed about any number system, is its closure under certain arithmetic operations. Our particular construction of the natural numbers lends itself well to verifying this property for addition. We define addition inductively:

$$\begin{aligned} n + 1 &= n \cup \{n\}; \\ n + 2 &= (n+1) + 1 \\ &\text{etc.} \end{aligned}$$

For example,

$$\begin{aligned} 2 + 2 = 2 + (1+1) = (2+1) + 1 &= \left[\{\emptyset, \{\emptyset\}\} \cup \left\{\{\emptyset, \{\emptyset\}\}\right\}\right] + 1 \\ &= \left\{\emptyset, \{\emptyset\}, \{\emptyset, \{\emptyset\}\}\right\} \cup \left\{\left\{\emptyset, \{\emptyset\}, \{\emptyset, \{\emptyset\}\}\right\}\right\} \\ &= \left\{\emptyset, \{\emptyset\}, \{\emptyset, \{\emptyset\}\}, \{\emptyset, \{\emptyset\}, \{\emptyset, \{\emptyset\}\}\}\right\} \\ &= 4. \end{aligned}$$

It is convenient in the logical construction of the natural numbers to include zero as a natural number. In particular, our definition makes the additive law

$$n + 0 = n \cup 0 = n \cup \emptyset = n$$

very natural. But the reader should be warned that, in standard mathematical discourse, the name "natural numbers" and the symbol $\mathbb{N}$ are generally reserved for the set $\{1, 2, 3, \ldots\}$ of *positive* integers. In common mathematical discourse, the set $\{0, 1, 2, \ldots\}$ is generally denoted by $\mathbb{Z}^+$ and is called "the nonnegative integers."

17.1.3 Axiomatic Treatment of the Natural Numbers

In practice, it is most convenient to treat the natural numbers axiomatically. Guiseppe Peano (1858–1932) formulated the axiomatic theory of the natural numbers that we still use today. His axioms are these:

1. Each natural number has a unique successor.
2. There is a natural number 1 that is not the successor of any natural number.
3. Two distinct integers cannot have the same successor.
4. If M is a set of natural numbers such that $1 \in M$ and such that if a natural number n is in M then its successor is also in M, then every natural number is in M.

Observe that the first axiom guarantees that there are infinitely many natural numbers, and they are the the ones we expect. Axiom 2 guarantees that there is a first natural number (which is of course 1). Axiom 3 guarantees that the natural numbers are linearly ordered. Axiom 4 amounts to the principle of mathematical induction.

Peano's axiomatic system is simple and complete. But it does *not* provide the machinery for addition or multiplication. In fact it is an as yet unresolved problem to determine definitively how to perform the usual arithmetic operations in Peano's system. The customary method for handling addition and multiplication is to add some other axioms. We will not explore the details here, but refer the reader to [SUP, p. 136].

Further Reading

M. O. Gonzalez and J. D. Mancill, Remarks on natural numbers, *The American Mathematical Monthly* 58(1951), 186–188.

Robert E. Kennedy and Curtis N. Cooper, On the natural density of the Niven numbers, *The College Mathematics Journal* 15(1984), 309–312.

Myren Krom, On sums of powers of natural numbers, *The Two-Year College Mathematics Journal* 14(1983), 349–351.

17.2 The Integers

17.2.1 Lack of Closure in the Natural Numbers

The natural numbers are closed under addition and multiplication. If you add or multiply any two natural numbers then you will certainly obtain another natural number as your answer. They are *not* closed under subtraction (for example, $3 - 5$ is not an element of the natural numbers). To achieve closure under that new operation, we must expand the number system as follows.

Let

$$X = \{(m, n) : m, n \in \mathbb{N}\} \equiv \mathbb{N} \times \mathbb{N}.$$

We define a *relation* on X by

$$(m, n) \sim (m', n') \qquad \text{if and only if} \qquad m + n' = m' + n.$$

It turns out that this relation partitions X into disjoint subcollections. For example

$$S_{(1,1)} = \{(m, n) : (m, n) \sim (1, 1)\}$$

is one such subcollection. These are all the ordered pairs of natural numbers that are related to $(1, 1)$. They include $(2, 2)$, $(3, 3)$, $(4, 4)$, and in fact all the ordered pairs (k, k) for $k \in \mathbb{N}$. Another such subcollection is

$$S_{(1,2)} = \{(m, n) : (m, n) \sim (1, 2)\}.$$

These are all the ordered pairs of natural numbers that are related to $(1, 2)$. They include $(2, 3)$, $(3, 4)$, $(4, 5)$, and in fact all the pairs $(k, k + 1)$ for $k \in \mathbb{N}$.

We call each of these subcollections an "equivalence class". The entire set X is the disjoint union of these equivalence classes.

17.2.2 The Integers as a Set of Equivalence Classes

Now the set of equivalence classes of X (we denote the set of equivalence classes by $X/\sim$), under this equivalence relation, is the new number system that we will call the *integers* (denoted by $\mathbb{Z}$, from the German word *Zahl* for number). In fact, we think of the integer that we commonly denote by k (for $k \geq 0$) as the equivalence class $\{(m,n) : m \in \mathbb{N}, n \in \mathbb{N}, m+k = n\}$. For $k < 0$ (assuming that the reader is familiar with the ordinary arithmetic of negative integers), we think of k as the equivalence class $\{(m,n) : m \in \mathbb{N}, n \in \mathbb{N}, m+k = n\}$. Our rules of arithmetic in this new number system are

- $[(m,n)] + [(k,\ell)] = [(m+k, n+\ell)]$;
- $[(m,n)] - [(k,\ell)] = [(m+\ell, n+k)]$;
- $[(m,n)] \cdot [(k,\ell)] = [(m\ell + nk, n\ell + mk)]$.

17.2.3 Examples of Integer Arithmetic

These definitions are best understood by way of some examples:

$$3 + (-5) = [(1,4)] + [(9,4)] = [(1+9, 4+4)] = [(10,8)] = -2\,,$$

$$4 - 8 = [(2,6)] - [(1,9)] = [(2+9, 6+1)] = [(11,7)] = -4\,,$$

$$\begin{aligned} 3 \cdot (-6) &= [(2,5)] \cdot [(10,4)] = [(2\cdot 4 + 5\cdot 10, 2\cdot 10 + 5\cdot 4)] \\ &= [(58,40)] = -18. \end{aligned}$$

The definition of multiplication used here may seem unnecessarily complicated, or perhaps unnatural. But, in the backs of our minds, we are thinking of the pair (m,n) as representing the difference $(n-m)$ and we are thinking of the pair (k,ℓ) as representing the difference $(\ell - k)$. So our "product rule" derives from the ordinary algebraic product of these two expressions.

It is important to notice in each of these examples that we need not engage in any hocus pocus to explain the arithmetic of negative numbers. Their properties are *built in* to the number system and its operations. In particular, $3 \cdot (-6) = -18$ because *that is the way things are*. That is how we have defined integer multiplication.

17.2.4 Arithmetic Properties of the Integers

The satisfying thing about the construction given here is that the standard arithmetic properties of negative numbers are automatic—they are built into the way we have defined our new number system. Observe that the additive identity is $0 = [(1,1)]$ and, indeed, $n+0 = n$ for any integer n. We may check this claim in detail:

$$n + 0 = [(1, n+1)] + [(1,1)] = [(2, n+2)] = n\,.$$

Also the multiplicative identity is $1 = [(1,2)]$, and one may check that $1 \cdot n = n$ for any n. In point of fact,

$$1 \cdot n = [(1,2)] \cdot [(1,n+1)] = [(1\cdot(n+1) + 2\cdot 1, 2\cdot(n+1) + 1\cdot 1)] = [(n+3, 2n+3)] = n\,.$$

Further Reading

Roy Dubisch, Representation of the integers by positive integers, *The American Mathematical Monthly* 58(1951), 615–616.

Robert G. Stein, Exploring the Gaussian integers, *The Two-Year College Mathematics Journal* 7(1976), 4–10.

17.3 The Rational Numbers

17.3.1 Lack of Closure in the Integers

The number system $\mathbb{Z}$ of integers is closed under addition, subtraction, and multiplication; but it is *not* closed under division. For example, $5 \div 7$ makes no sense in the integers. In order to achieve the desired closure property, we enlarge the system of integers as follows:

Let

$$Y = \{(p,q) : p, q \in \mathbb{Z}, q \neq 0\}.$$

We define a relation on Y by

$$(p,q) \sim (p',q') \qquad \text{if and only if} \qquad p \cdot q' = p' \cdot q.$$

17.3.2 The Rational Numbers as a Set of Equivalence Classes

Now the set of equivalence classes of Y (we denote this set by $Y/\sim$), under this equivalence relation, is the new number system that we will call the *rational numbers* (denoted by $\mathbb{Q}$, where $\mathbb{Q}$ should be considered to be an abbreviation for "quotient"). In fact, we think of the rational number that we commonly denote by p/q (in lowest terms) as the equivalence class $\{(pk, qk) : k \in \mathbb{Z}, k \neq 0\}$. Our rules of arithmetic in this new number system are:

- $[(p,q)] + [(r,s)] = [(ps + qr, qs)]$;

- $[(p,q)] \cdot [(r,s)] = [(pr, qs)]$.

Notice that the rule for addition may seem counterintuitive. But, in the backs of our minds, we are thinking of adding p/q to r/s and we are following the ordinary rubric for putting the fractions over a common denominator and then adding. Multiplication is of course more straightforward: one simply multiplies the numerators together and then multiplies the denominators together.

17.3.3 Examples of Rational Arithmetic

These definitions are best understood by way of some examples.

$$\frac{3}{5}+\frac{2}{7}=[(3,5)]+[(2,7)]=[(3\cdot 7+5\cdot 2, 5\cdot 7)]$$
$$=[(31,35)]=\frac{31}{35},$$

$$\frac{-4}{9}+\frac{2}{5}=[(-4,9)]+[(2,5)]=[((-4)\cdot 5+9\cdot 2, 9\cdot 5)]$$
$$=[(-2,45)]=\frac{-2}{45},$$

$$\frac{-3}{11}\cdot\frac{5}{13}=[(-3,11)]\cdot[(5,13)]=[(-15,143)]$$
$$=\frac{-15}{143}.$$

17.3.4 Subtraction and Division of Rational Numbers

The operations of subtraction and division on the rationals are already implicit in addition and multiplication. For completeness, however, we record them here:

- $[(p,q)]-[(r,s)]=[(ps-qr,qs)]$;
- $[(p,q)]\div[(r,s)]=[(ps,qr)]$, provided $r\neq 0$.

EXAMPLE 17.1. Let x be the rational number $[(2,3)]$ and y be the rational number $[(5,7)]$. (We think of these as $2/3$ and $5/7$ respectively.) Then the difference of these two numbers is

$$y-x=[(5,7)]-[(2,3)]=[(5\cdot 3-7\cdot 2, 7\cdot 3)]=[(1,21)].$$

The quotient of these two numbers is

$$y\div x=[(5,7)]\div[(2,3)]=[(5\cdot 3, 7\cdot 2)]=[(15,14)].$$

Further Reading

Larry Cannon, Geometry of the rational plane, *The College Mathematics Journal* 17(1986), 392–402.

F. Cunningham, Jr., A construction of the rational numbers, *The American Mathematical Monthly* 66(1959), 769–777.

Emma Lehmer, Rational reciprocity laws, *The American Mathematical Monthly* 85(1978), 467–472.

17.4 The Real Numbers

17.4.1 Lack of Closure in the Rational Numbers

The set $\mathbb{Q}$ of rational numbers is closed under the standard arithmetic operations of $+$, $-$, $\times$, $\div$. In fact, $\mathbb{Q}$ is what we call a *field*. From a strictly algebraic perspective, this number system is completely satisfactory for elementary purposes, and in fact it is the rational numbers that are used in everyday commerce, engineering, and science.

However, from a more advanced point of view, the rational numbers are not completely satisfactory. This assertion was first discovered by the Pythagoreans more than 2000 years ago (Subsection 1.1.1). Indeed, they determined that there is no rational number whose square is 2. More generally, a positive integer has a rational square root if and only if it has an integer square root. Thus we find that a broader class of numbers is more desirable.

The modern point of view is that the rational numbers are deficient from the point of view of metric topology. More precisely, the sequence of rational numbers given (for instance) by

$$3, 3.1, 3.14, 3.141, 3.1415, 3.14159, \ldots$$

gives better and better approximations to the ratio of the circumference of a circle to its diameter. These numbers are becoming and staying closer and closer together, and appear to converge to some value. But, as it turns out, that value cannot be rational. In fact, the value is π, and it is known that π is not rational. By the same token, the numbers

$$1, 1.4, 1.41, 1.414, 1.4142, \ldots$$

give better and better approximations to the length of the diagonal of a square of side 1, and that number is known to be irrational. In summary, we require a system of numbers that is still closed under the basic arithmetic operations, but is also closed under the limiting processes just described. [We in fact discussed these ideas in some detail in Section 12.3.]

17.4.2 Axiomatic Treatment of the Real Numbers

In fact, the system of *real numbers* will fill the need just described. It is rather complicated to give a formal construction of the real numbers $\mathbb{R}$, and we refer the reader to [KRA1], [RUD], and [STR] for details. We content ourselves here with enunciating an axiom system for the reals (these are taken from [STR]). We state once and for all that it is possible to present an explicit model for a number system that satisfies these axioms (constructed, for example, by the method of Dedekind cuts—see [KRA1]). In fact we have already given a particular, hands-on construction of the real numbers in Section 12.3. We shall take it for granted that a model exists, and we shall have no compunctions about using the real numbers elsewhere in this book.

The real numbers are a system or a *field* of numbers equipped with a notion of distance that makes the field operations (addition $+$ and multiplication $\cdot$) continuous. With this notion of distance (or *metric*), the real numbers are *complete*. The detailed axioms are these:

Axiom 1 (Commutative Laws) For all $x, y \in \mathbb{R}$,

$$x + y = y + x \qquad \text{and} \qquad x \cdot y = y \cdot x.$$

Axiom 2 (Associative Laws) For all $x, y, z \in \mathbb{R}$,

$$x + (y + z) = (x + y) + z \qquad \text{and} \qquad x \cdot (y \cdot z) = (x \cdot y) \cdot z.$$

Axiom 3 (Distributive Law) For all $x, y, z \in \mathbb{R}$,

$$x \cdot (y + z) = x \cdot y + x \cdot z.$$

Axiom 4 (Identity Elements) There exist two distinct elements 0 and 1 in $\mathbb{R}$ such that, for all $x \in \mathbb{R}$,

$$0 + x = x \qquad \text{and} \qquad 1 \cdot x = x.$$

Axiom 5 (Inverse Elements) If $x \in \mathbb{R}$, then there exists a unique $-x \in \mathbb{R}$ such that

$$x + (-x) = 0.$$

If $x \in \mathbb{R}$ and $x \neq 0$, then there is a unique element $x^{-1} \in \mathbb{R}$ such that

$$x \cdot x^{-1} = 1.$$

Axiom 6 (Positive Numbers) The real number system $\mathbb{R}$ has a distinguished subset **P** (the positive numbers) that induces an ordering on $\mathbb{R}$. The three sets **P**, $\{0\}$, and $-\mathbf{P} = \{-x : x \in \mathbf{P}\}$ are pairwise disjoint and their union is all of $\mathbb{R}$. We write $a < b$ in case $b - a \in \mathbf{P}$.

Axiom 7 (Closure Properties of P) If $x, y \in \mathbf{P}$, then $x + y \in \mathbf{P}$ and $x \cdot y \in \mathbf{P}$.

Axiom 8 (Dedekind Completeness) Let A and B be subsets of $\mathbb{R}$ such that

(i) $A \neq \emptyset$ and $B \neq \emptyset$;

(ii) $A \cup B = \mathbb{R}$;

(iii) $a \in A$ and $b \in B$ imply that $a < b$.

Then there exists exactly one element $x \in \mathbb{R}$ such that

(iv) If $u \in \mathbb{R}$ and $u < x$, then $u \in A$;

(v) If $v \in \mathbb{R}$ and $x < v$, then $v \in B$.

Plainly, the number x described in **Axiom 8** must be either in A or in B but not in both. Thus $B = \mathbb{R} \setminus A$ and either $A = \{u \in \mathbb{R} : u \leq x\}$ or $A = \{u \in \mathbb{R} : u < x\}$.

The first seven axioms of the real numbers are also satisfied by the rational numbers. It is these first seven axioms that constitute the postulates for a field. (In some treatments, **Axioms 1, 2, 4,** and **5** are each split into two; so it is common to say that there are eleven field axioms.)

It is Axiom 8 that makes the real numbers special. It says, in effect, that the real numbers have no gaps or holes in them. In other word, the reals are *complete*.

We invite the reader at this time to review Chapter 12, especially Section 12.4, to see all the special properties that the real numbers enjoy because of their completeness. It is the real numbers that are the foundation for all scientific computing and reasoning, and for all theoretical mathematics in the real world. Other number systems, such as the complex numbers, are built on the reals. See the next section.

Further Reading

Paul Fjelstad and Peter Hammer, A picture for real arithmetic, *The College Mathematics Journal* 31(2000), 56–60.

Donald E. Knuth, Vaughan Pratt, and O. P. Lossers, The real numbers, algebraically: 10689, *The American Mathematical Monthly* 107(2000), 755.

P. M. Rice, A topological characterization of the real numbers, *The American Mathematical Monthly* 76(1969), 184–185.

17.5 The Complex Numbers

17.5.1 Intuitive View of the Complex Numbers

Intuitive treatments of the complex numbers are unsatisfactory because they posit (without substantiation) the existence of a number that plays the role of the square root of -1. With the formalism developed thus far in this book, we were actually able to construct the complex numbers in Section 9.1. Here we shall review some of their key properties as part of our general consideration of number systems.

17.5.2 Definition of the Complex Numbers

We let

$$\mathbb{C} = \{(x, y) : x \in \mathbb{R}, y \in \mathbb{R}\}.$$

We equip $\mathbb{C}$ with the following operations:

$$\begin{aligned}(x, y) + (x', y') &= (x + x', y + y') \\ (x, y) \cdot (x', y') &= (xx' - yy', xy' + x'y).\end{aligned}$$

Notice that $\mathbb{C}$ is not a set of equivalence classes; it is merely a set of ordered pairs. The rule for multiplication may seem artificial, but it is the rule that is needed to turn $\mathbb{C}$ into a field.

17.5.3 The Distinguished Complex Numbers 1 and i

We denote the complex number $(1, 0)$ by 1. Notice that if (x, y) is any other complex number, then

$$(1, 0) \cdot (x, y) = (1 \cdot x - 0 \cdot y, 1 \cdot y + 0 \cdot x) = (x, y).$$

Thus $1 \equiv (1, 0)$ is the multiplicative identity. We commonly denote the complex number $(0, 1)$ by i. Observe that

$$i \cdot i = (0, 1) \cdot (0, 1) = (0 \cdot 0 - 1 \cdot 1, 0 \cdot 1 + 1 \cdot 0) = (-1, 0) = -(1, 0) = -1.$$

Thus i is a bona fide square root of -1, but this property is *built into* the arithmetic of $\mathbb{C}$; it is not achieved by fiat.

It is common to write the complex number (x, y) as $x \cdot 1 + y \cdot i = x + iy$.

17.5.4 Algebraic Closure of the Complex Numbers

The most important property of the complex numbers is that of *algebraic closure*: any polynomial $p(z) = a_0 + a_1 z + \cdots + a_{n-1} z^{n-1} + a_n z^n$ with complex coefficients has precisely n complex roots (counting multiplicities). This profound fact is due to Gauss, and he produced five distinct proofs. Today there are several dozen proofs of this, the Fundamental Theorem of Algebra. We provided our own proof of the Fundamental Theorem in Section 9.4.

Further Reading

D. H. Potts, Axioms for complex numbers, *The American Mathematical Monthly* 70 (1963), 974–975.

D. E. Richmond, Complex numbers and vector algebra, *The American Mathematical Monthly* 58 (1951), 622–628.

Rosemary Schmalz, Complex numbers as residue classes of polynomials mod $(x^2 + 1)$, *The Two-Year College Mathematics Journal* 3 (1972), 78–80.

Exercises

1. Use our definition of addition in the integers to verify that $m + n = n + m$ for any integers m and n.

2. Use our definition of multiplication in the integers to verify that $m \cdot n = n \cdot m$ for any integers m and n.

3. Explain why, if n is an integer, if k is any nonzero integer, and if $n \cdot k = 0$, then $n = 0$.

4. Suppose that q is a rational number, n is an integer, and $q + n$ is an integer. What can you conclude about q?

5. Suppose that q is a rational number, n is a nonzero integer, and $q \cdot n$ is an integer. What can you conclude about q? [*Hint:* Be careful. The correct answer is *not* that q is an integer.]

6. Suppose that q and r are rational numbers. Give a precise explanation of what a *lowest common denominator* for q and r would be. Describe a method for finding the lowest common denominator. Discuss this problem in class.

7. If k and n are natural numbers, then say precisely what the *greatest common divisor* of k and n is. Describe a method for finding the greatest common divisor. Discuss this problem in class.

8. If r is a real number and s is another real number, then offer an explanation of what r^s should mean. Discuss this problem in class.

9. Find a square root for the complex number $1 + i$ by solving the equation $(x + iy)^2 = 1 + i$.

10. Sometimes in high school we "discover" the arithmetic laws for the integers with the following sort of reasoning:

Say that we want to understand what value $-3 + 1$ must have. Consider the expression

$$6 + (-3 + 1) = (6 + (-3)) + 1 = 3 + 1 = 4\,.$$

The only possible conclusion is that $-3 + 1 = -2$.

With similar reasoning, suppose that we want to understand the value of $-2 \cdot 4$. Consider the expression

$$(-2 + 5) \cdot 4 = [(-2) \cdot 4] + [5 \cdot 4]\,.$$

We may simplify the expression on the left and the second expression on the right. The result is

$$3 \cdot 4 = [(-2) \cdot 4] + 20$$

or

$$12 = [(-2) \cdot 4] + 20 .$$

We conclude that

$$-[(-2) \cdot 4] = 8$$

or

$$(-2) \cdot 4 = -8 .$$

The arguments that we have just presented are not incorrect, but they ignore a fundamental issue. What is the gap?

11. Explain why $\sqrt{2}$ exists as a real number but not as a rational number. Discuss in class.

12. Explain why π exists as a real number but not as a rational number. Discuss in class.

13. **Project:** Imagine $\mathbb{R}^4 \equiv \mathbb{R} \times \mathbb{R} \times \mathbb{R} \times \mathbb{R}$ equipped with the following operations: set $\mathbf{i} = (0, 1, 0, 0)$, $\mathbf{j} = (0, 0, 1, 0)$, $\mathbf{k} = (0, 0, 0, 1)$. Denote the 4-tuple $(1, 0, 0, 0)$ by $\mathbf{1}$. Define the multiplication laws

$$\mathbf{i} \cdot \mathbf{i} = -\mathbf{1} \quad , \quad \mathbf{j} \cdot \mathbf{j} = -\mathbf{1} \quad , \quad \mathbf{k} \cdot \mathbf{k} = -\mathbf{1}$$

and

$$\mathbf{i} \cdot \mathbf{j} = \mathbf{k} \quad , \quad \mathbf{j} \cdot \mathbf{k} = \mathbf{i} \quad , \quad \mathbf{k} \cdot \mathbf{i} = \mathbf{j}$$

and

$$\mathbf{j} \cdot \mathbf{i} = -\mathbf{k} \quad , \quad \mathbf{k} \cdot \mathbf{j} = -\mathbf{i} \quad , \quad \mathbf{i} \cdot \mathbf{k} = -\mathbf{j}.$$

Of course the element $\mathbf{1}$ multiplied times any 4-tuple z is declared to be equal to z. In particular, $\mathbf{1} \cdot \mathbf{1} = \mathbf{1}$.

Finally, if $z = (z_1, z_2, z_3, z_4)$ and $w = (w_1, w_2, w_3, w_4)$ are 4-tuples, then we write

$$z = z_1 \cdot \mathbf{1} + z_2\mathbf{i} + z_3\mathbf{j} + z_4\mathbf{k}$$

and

$$w = w_1 \cdot \mathbf{1} + w_2\mathbf{i} + w_3\mathbf{j} + w_4\mathbf{k}.$$

Then $z \cdot w$ is defined by using the (obvious) distributive law and the rules already specified. For example,

$$\begin{aligned}
(2, 0, 1, 3) \cdot (-4, 1, 0, 1) &= [2 \cdot \mathbf{1} + \mathbf{j} + 3\mathbf{k}] \cdot [-4 \cdot \mathbf{1} + \mathbf{i} + \mathbf{k}] \\
&= (2 \cdot (-4)) \cdot \mathbf{1} + (2\mathbf{i}) + (2\mathbf{k}) \\
&\quad + (\mathbf{j} \cdot (-4)) + (\mathbf{j} \cdot \mathbf{i}) + (\mathbf{j} \cdot \mathbf{k}) \\
&\quad + (3\mathbf{k} \cdot (-4)) + (3\mathbf{k} \cdot \mathbf{i}) + (3\mathbf{k} \cdot \mathbf{k}) \\
&= -8 \cdot \mathbf{1} + 2\mathbf{i} + 2\mathbf{k} - 4\mathbf{j} - \mathbf{k} + \mathbf{i} - 12\mathbf{k} + 3\mathbf{j} - 3 \cdot \mathbf{1} \\
&= -11 \cdot \mathbf{1} + 3\mathbf{i} - \mathbf{j} - 11\mathbf{k} \\
&= (-11, 3, -1, -11).
\end{aligned}$$

We call this number system the *quaternions*.

Addition of two quaternions is simply performed componentwise: if $z = (z_1, z_2, z_3, z_4)$ and $w = (w_1, w_2, w_3, w_4)$ then

$$z + w = (z_1 + w_1, z_2 + w_2, z_3 + w_3, z_4 + w_4).$$

Verify for yourself that the additive identity in the quaternions is $(0, 0, 0, 0)$. The multiplicative identity is $\mathbf{1} = (1, 0, 0, 0)$.

In fact it can be checked that each nonzero element of the quaternions has a unique two-sided multiplicative inverse. However, since multiplication is not commutative, the quaternions do not form a field; instead the algebraic structure is called a *division ring*.

Let $z = 2 \cdot \mathbf{1} + 3\mathbf{i} - 5\mathbf{j} + 9\mathbf{k}$ and $w = -4 \cdot \mathbf{1} - 6\mathbf{i} + 2\mathbf{j} + 8\mathbf{k}$. Then calculate $z \cdot w^{-1}$ and $w^{-1} \cdot z$. The text [COSM] contains a wealth of information about the quaternions and the closely related octonions.

14. Project: It is also possible to give $\mathbb{R}^8$ an additive and a multiplicative structure. The multiplication operation is both noncommutative and nonassociative. The resulting eight-dimensional algebraic object is called the *Cayley numbers*. We shall not present the details here. It is one of the great theorems of twentieth century mathematics (see [ADA], [BOM]) that $\mathbb{R}^1$, $\mathbb{R}^2$, $\mathbb{R}^4$, and $\mathbb{R}^8$ are the only Euclidean spaces that can be equipped with compatible addition and multiplication operations in a natural way (so that the algebraic operations are smooth functions of the coordinates).

In fact the complex numbers are a subfield of the Cayley numbers in a natural way, and the quaternions are a natural sub-division-ring of the Cayley numbers in a natural way. Look up the Cayley numbers on `Wikipedia` and verify these assertions. Again see [COSM].

15. Project: Various number systems have been invented which extend the reals in interesting and provocative ways. In 1960 Abraham Robinson created the *nonstandard real number system*, which contains the standard real numbers as well as infinitesimals (positive numbers that are smaller than any real) and infinitary numbers (positive numbers that are larger than any real). You may become acquainted with Robinson's ideas through Jerome Keisler's remarkable calculus book [KEI] modeled on nonstandard analysis.

More recently, John Horton Conway invented the *surreal numbers*. These were first exposited in Donald Knuth's fictional book *Surreal Numbers: How Two Ex-Students Turned on to Pure Mathematics and Found Total Happiness* [KNU]. It is a fact that the surreal numbers contain the nonstandard reals and some other new numbers as well (relating to Georg Cantor's theory of cardinals—see our Chapter 16). Consult Knuth's book, as well as Keisler's and explain how these two new number systems are related.

18

Henri Poincaré, Child Phenomenon

18.1 Introductory Remarks

Jules Henri Poincaré (1854–1912) is considered to have been one of the outstanding talents of twentieth century mathematics. Even while he was a child his special gifts were recognized, and the entire country of France watched in awe as he grew up to be a brilliant and creative man of science. There were other distinguished individuals in Henri Poincaré's family. Poincaré's father's brother's son Raymond was prime minister of France several times and president of the French Republic during World War I. The second son of that same uncle was a distinguished and high-ranking university administrator.

Young Henri Poincaré was so gifted that he was a hero in all of France. From a physical point of view, he was described in this way:

> ...ambidextrous and was nearsighted; during his childhood he had poor muscular coordination and was seriously ill for a time with diphtheria. He received special instruction from his gifted mother and excelled in written composition while still in elementary school.

Poincaré studied at the Lycée—today called the Lycée Henri Poincaré. He was the top student in every subject that he undertook. One of his instructors called him a "monster of mathematics". Poincaré was certainly one of the leaders of twentieth century mathematics. His ideas live on today. More on his life may be found in [APP] and [TOU].

Young Henri enrolled at the École Polytechnique in 1873 and graduated in 1875. He was vastly ahead of all the other students in mathematics. But in other subject areas, such as athletics and art, he did poorly. His deleterious physical coordination held him back in activities which were not cerebral. In fact his eyesight was so poor that he could not see what his teachers wrote on the blackboard. This failure helped him to develop his visual imagination.

After the École Polytechnique, Poincaré spent some time as a mining engineer at the École des Mines. At the same time he studied mathematics under the direction of Charles Hermite. He earned his doctorate in 1879. The examiners were not entirely happy with the thesis, for they found the presentation obscure and the organization confusing. Yet they acknowledged that this was a difficult subject area, and that the candidate had demonstrated great talent. So he was awarded the degree.

Poincaré's first academic position was teaching mathematics at the University of Caen. His lectures were criticized for their lack of organization. He remained at Caen for only two years, and then he moved to a chair at the Faculty of Sciences in Paris in 1881. In 1886 Poincaré was nominated, with the support of Hermite, to the chair of mathematical physics and probability at the Sorbonne. Hermite also promoted Poincaré for a chair at the École Polytechnique. These were the two most prestigious professorships in all of France, and so a measure of Poincaré's prestige and recognition. He was to remain in Paris for the remainder of his career, and he lectured on a different subject every year until his untimely death at the age of 58.

Poincaré was remarkable for his work habits. He engaged in mathematical research each day from 10:00A.M.until noon and from 5:00P.M.until 7:00P.M.. He would read mathematical papers in the evening. Rather than build new ideas on earlier work, Poincaré preferred always to work from first principles. He operated in this fashion both in his lectures and in his writing. One expert described Poincaré's method for organizing a paper as follows:

> ...does not make an overall plan when he writes a paper. He will normally start without knowing where it will end.... Starting is usually easy. Then the work seems to lead him on without him making a wilful effort. At that stage it is difficult to distract him. When he searches, he often writes a formula automatically to awaken some association of ideas. If beginning is painful, Poincaré does not persist but abandons the work.

Poincaré also believed that his best ideas would come when he stopped concentrating on a problem, when he was actually at rest:

> Poincaré proceeds by sudden blows, taking up and abandoning a subject. During intervals he assumes ... that his unconscious continues the work of reflection. Stopping the work is difficult if there is not a sufficiently strong distraction, especially when he judges that it is not complete ... For this reason Poincaré never does any important work in the evening in order not to trouble his sleep.

In 1894 Poincaré published his important *Analysis Situs*. This seminal work laid the foundations for topology, especially algebraic topology. He defined the fundamental group—which is an important device for detecting holes of different dimensions in surfaces and other geometric objects. He proved the foundational result that a 2-dimensional surface having the same fundamental group as the sphere is in fact topologically equivalent to the sphere. He conjectured that a similar result is true in 3 dimensions, and ultimately in all dimensions. This question has become known as the *Poincaré Conjecture*, and it is one of the most important questions of twentieth century mathematics. It is a curious fact that the Poincaré Conjecture was proved for dimensions 5 and higher by Stephen Smale in 1961 (see [SMA1]) and in dimension 4 by Michael Freedman in 1982 (see [FRE]). But the fundamental question of dimension 3 still remained open. In the year 2003 Grigori Perelman of St. Petersburg, Russia distributed three papers which appear to finally prove the Pioncaré conjecture in dimension 3. Notable in this work is that Perelman does *not* confine himself to methods of topology. He uses partial differential equations and differential geometry in powerful new ways. In 2006 Perelman was awarded the Fields Medal, the highest honor in mathematics, for this work. Sadly, he declined the honor (the remarkable article [NAS] provides some of the details of this incident).

Poincaré is also remembered as the founder of the analytic theory of functions of several complex variables, and he did important work in number theory. Poincaré maintained his interest in physics, and made contributions to optics, electricity, telegraphy, capillarity, elasticity, thermodynamics, potential theory, quantum theory, relativity, and cosmology.

Of particular historical interest is Poincaré's participation in an 1887 competition that the King of Norway and Sweden initiated to celebrate his sixtieth birthday. Poincaré's paper on the 3-body problem[1] solved an important problem in celestial mechanics. Even though this paper was ultimately discovered to contain an error, it won the prize and has been very influential in twentieth century mathematics. In particular, the theory of dynamical systems (which ultimately led to fractal geometry and chaos) was founded in this paper. The paper was ultimately corrected and published in 1890.

Poincaré lived in a time when the general public was not particularly interested in science. Nonetheless, after he had established his pre-eminence as a research scientist, he turned his considerable talent to the writing of expository works for the general public. Among his major works describing science and mathematics for the general public were *Science and Hypothesis* (1901), *The Value of Science* (1905), and *Science and Method* (1908).

Poincaré believed passionately in the separate roles of intuition and rigor. The first was for *finding* ideas; the second was for *establishing* those ideas. In Poincaré's words:

It is by logic we prove, it is by intuition that we invent.

He went on, in a later article, to say that

Logic, therefore, remains barren unless fertilized by intuition.

Poincaré was prescient in many ways. He saw that non-Euclidean geometry would be the correct geometry to use to understand physical space (thus anticipating Einstein's general relativity). He claimed that mathematics could not be completely axiomatized, as Bertrand Russell and others had tried to do. In particular, Poincaré asserted that the method of mathematical induction could never be proved. He also claimed that arithmetic could never be proved to be consistent if it were defined by a system of axioms. These ideas were later fleshed out and proved to be correct by Kurt Gödel and other geniuses of twentieth century logic.

A curious feature of Poincaré's career is that he never founded his own school since he never had any students. Poincaré's contemporaries used his results, but not his techniques. He certainly had considerable influence over the mathematics of his day (and on into the present day). He achieved many honors during his lifetime. In 1887 he was elected to the Académie des Sciences and in 1906 was elected President of that Academy. His research covered such a broad scope that he was the only scientist ever elected to each one of the five sections of the Academy: geometry, mechanics, physics, geography, and navigation. In 1908 Poincaré was elected to the Académie Française and he was chosen to be director of that august body in the year of his death 1912. He was made Chevalier of the Légion d'Honneur, and he received honors from numerous learned societies around the world.

It is not generally well known that Poincaré discovered special relativity at just about the same time as Einstein. He gave a lecture on the subject at Washington University in

[1]The three-body problem asks about the long-term motion of objects (say the sun, the earth, and the earth's moon) acting on each other with the force of gravity. The French Academy first offered a prize question about the 3-body problem in the late eighteenth century, so the problem is well over 200 years old.

Today a great deal is known about what might happen to three planets in these circumstances, but the problem is far from being completely understood. The corresponding problem for n planets—called, appropriately enough, the "n-body problem"—is still open.

St. Louis, on the occasion of the 1904 World's Fair, a full year before Einstein's ideas appeared in print. In fact Poincaré and Einstein had a considerable rivalry in this matter, and they never acknowledged each other's work. Poincaré's ideas appear in a journal called *The Monist*, and they bear a remarkable similarity to textbook treatments of relativity that we see today.

Poincaré is arguably the father of topology (popularly known as "rubber sheet geometry") and also of the currently very active area of dynamical systems. He made decisive contributions to differential equations, to geometry, to complex analysis, and to many other central parts of mathematics.

We shall begin this chapter by exploring the subject of topology.

Further Reading

Michael Atiyah, Mathematics in the 20th century, *The American Mathematical Monthly* 108(2001), 654–666.

Herbert S. Wilf, Self-esteem in mathematicians, *The College Mathematics Journal* 21(1990), 274–277.

Figure 18.1. The regions S and T.

18.2 Rubber Sheet Geometry

A common joke among mathematicians is that a topologist is a person who does not know the difference between a coffee cup and a donut. What could this possibly mean? By the end of this section you should know the answer.

Let A and B be sets. In topology we say that A and B are *homeomorphic* if A can be continuously deformed into B. Put a bit more technically, A and B are homeomorphic if there is a function $f : A \to B$ that is a one-to-one correspondence and is continuous with a continuous inverse. We call the mapping f a *homeomorphism*.

EXAMPLE 18.1. The regions S and T exhibited in Figure 18.1 are homeomorphic. Figure 18.2 shows how S can be continuously deformed into T.

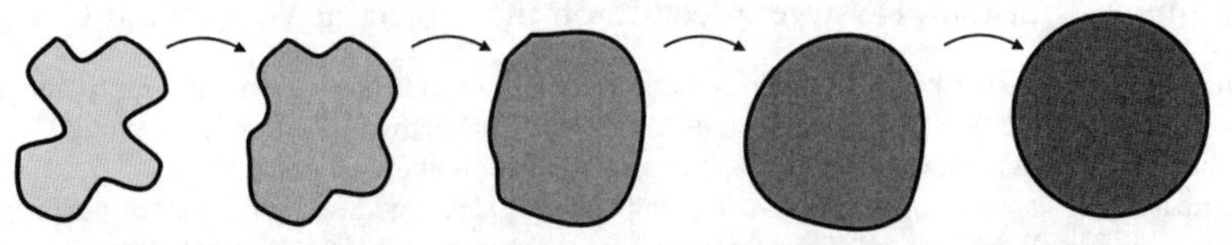

Figure 18.2. The region S can be continuously deformed into T.

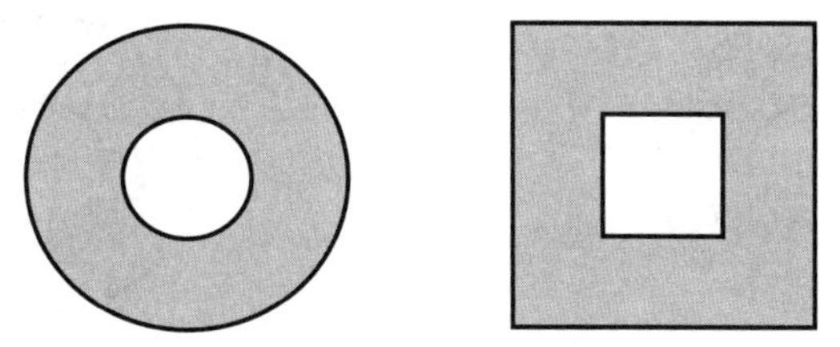

Figure 18.3. Homeomorphism of the annulus and the square frame.

One of the pleasures of this subject is that we frequently do not try to actually write down the function f that realizes the homeomorphism. In many instances, a picture will suffice.

The next example exhibits another two-dimensional instance of homeomorphism.

EXAMPLE 18.2. The annulus and the square frame in Figure 18.3 are homeomorphic.
Figure 18.4 shows how to deform one into the other.

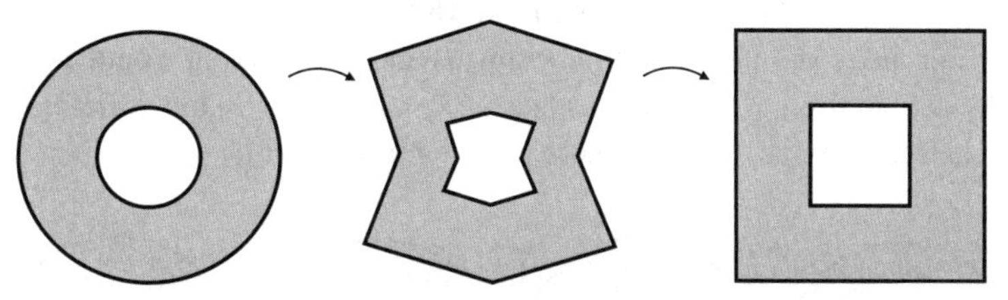

Figure 18.4. The deformation exhibited.

Further Reading

Beverly L. Brechner and John C. Mayer, Antoine's necklace or how to keep a necklace from falling apart, *The College Mathematics Journal* 19(1988), 306–320.

R. L. Wilder, Some unsolved problems of topology, *The American Mathematical Monthly* 44(1937), 61–70.

18.3 The Idea of Homotopy

Of course the notion of homeomorphism would not be interesting if just any old two geometric figures were homeomorphic. In point of fact, some pairs of objects are *not* homeomorphic, as the next example illustrates.

EXAMPLE 18.3. The two-dimensional geometric objects in Figure 18.5 are *not* homeomorphic.

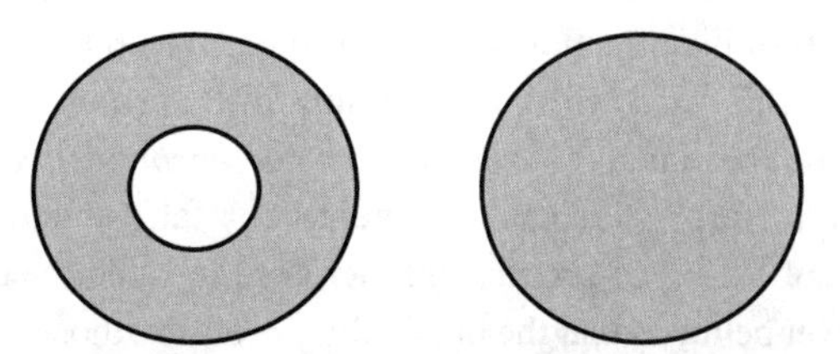

Figure 18.5. Two-dimensional geometric objects.

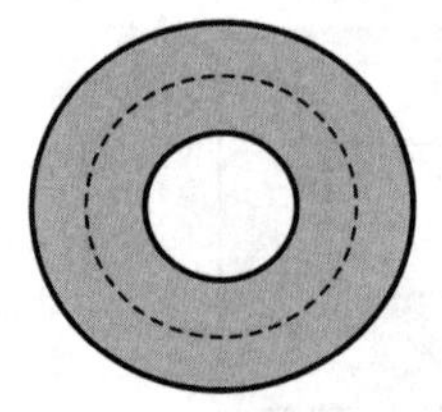

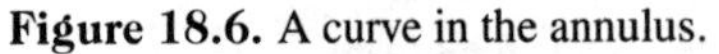

Figure 18.6. A curve in the annulus.

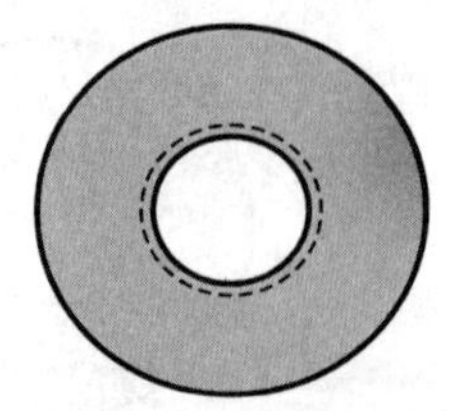

Figure 18.7. A curve that cannot be shrunk.

The justification for this assertion is very interesting, and is one of Henri Poincaré's great inventions: the idea of *homotopy*. Examine Figure 18.6. It shows a closed, dotted curve in the annulus. Now think of the annulus as rigidly fixed, and imagine moving the dotted curve around *inside the annulus*. Is it possible to transform the curve to a point?

The answer, of course, is "no"—just because the hole in the annulus prevents the curve from being shrunk any smaller than the hole itself—Figure 18.7.

On the other hand, a similar dotted curve in the disc on the right-hand side of Figure 18.5 is easily transformed to a point. See Figure 18.8.

Since the property of being able to transform a curve to a point would clearly be preserved under deformation of the domains, we see that it is not possible to deform the annulus to the disc (see Lemma 18.1 for a more precise formulation).

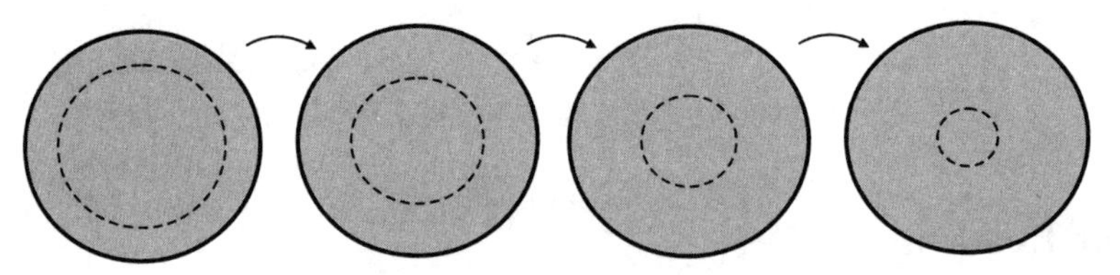

Figure 18.8. A curve that can be shrunk.

Further Reading

Raymond Redheffer, The homotopy theorems of function theory, *The American Mathematical Monthly* 76(1969), 778-787.

John W. Woll, Jr., One-sided surfaces and orientability, *The Two-Year College Mathematics Journal* 2(1971), 5–18.

18.4 The Brouwer Fixed Point Theorem

One of the most fascinating and important theorems of twentieth century mathematics is the Brouwer Fixed Point Theorem. Proving this theorem established Brouwer as one of the pre-eminent topologists of his day. But he refused to lecture on the subject, and in fact he ultimately rejected this (his own!) work. The reason for this strange behavior is that L. E. J. Brouwer (1882–1966) became a convert to *constructivism* or *intuitionism*. He rejected the Aristotelian dialectic (that a statement is either true or false and there is no alternative), at least in the context of existence proofs, and therefore rejected the concept of "proof by contradiction". Brouwer believed that the only valid proofs are those in which we *construct* the asserted objects being discussed.

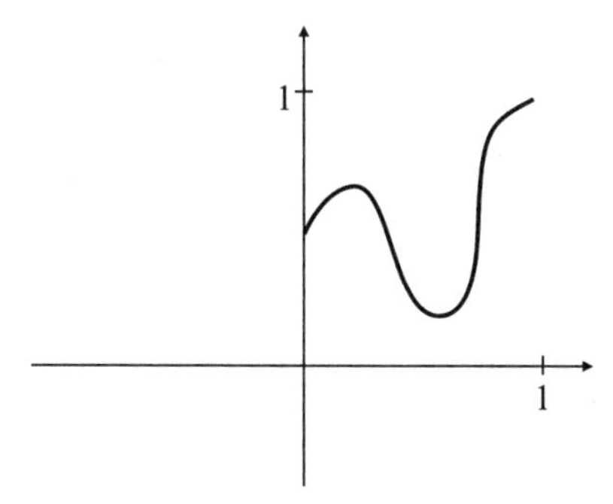

Figure 18.9. The graph of a function from $[0, 1]$ to $[0, 1]$.

As we shall see below, the Brouwer fixed point theorem asserts the existence of a "fixed point" for a continuous mapping. We demonstrate that the fixed point exists by assuming that it does not exist and deriving thereby a contradiction. This is Brouwer's original method of proof, but the methodology flies in the face of the intuitionism that he later adopted.

Let us begin by discussing the general idea of the Brouwer fixed point theorem. We proceed by considering a "toy" version of the question in one dimension. Consider a continuous function f from the interval $[0, 1]$ to $[0, 1]$. Figure 18.9 exhibits the graph of such a function.

Note here that the word "continuous" refers to a function that has no breaks in the graph. Some like to say that the graph of a continuous function "can be drawn without lifting the pencil from the paper." Although there are more mathematically rigorous definitions of continuity, this one will suffice for our purposes. The question is whether there is a point $p \in [0, 1]$ such that $f(p) = p$. Such a point p is called a *fixed point* for the function f. Figure 18.10 shows how complicated a continuous function from $[0, 1]$ to $[0, 1]$ can be. In each instance it is not completely obvious whether there is a fixed point or not. But in fact Figure 18.11 exhibits the fixed point in each case.

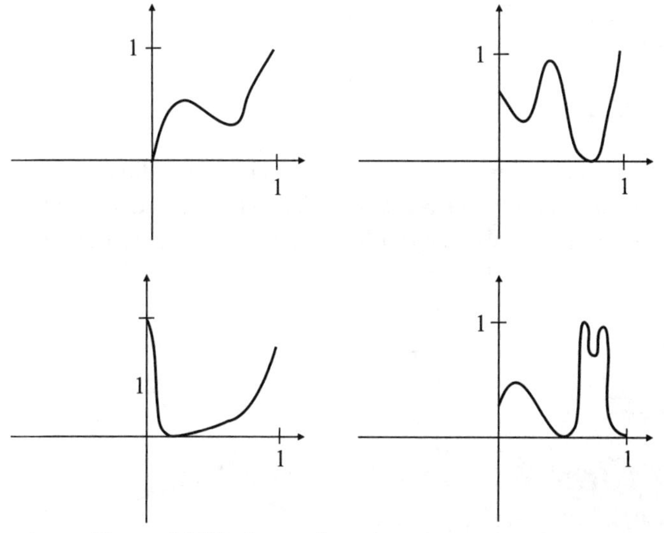

Figure 18.10. A complicated continuous function.

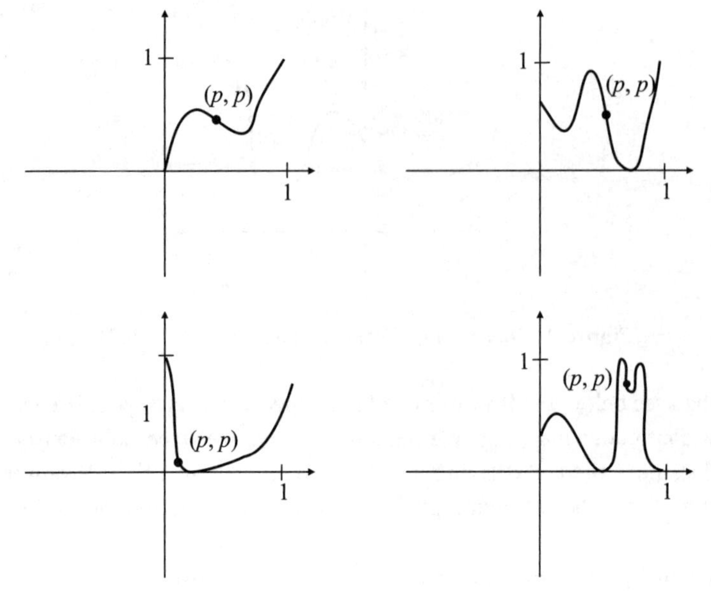

Figure 18.11. A fixed point.

Of course it is one thing to draw a few pictures and quite another to establish once and for all that, no matter what the choice of the continuous function $f : [0, 1] \to [0, 1]$, there is a fixed point p. What is required now is a *mathematical proof.* Now here is a formal enunciation and proof of our result:

Theorem 18.1. *Let $f : [0, 1] \to [0, 1]$ be a continuous function. Then there is a point $p \in [0, 1]$ such that $f(p) = p$.*

Proof. We may as well suppose that $f(0) \neq 0$ (otherwise 0 is our fixed point and we are done). Thus $f(0) > 0$. We also may as well suppose that $f(1) \neq 1$ (otherwise 1 is our fixed point and we are done). Thus $f(1) < 1$.

Consider the auxiliary function $g(x) = f(x) - x$. By the observations in the last paragraph, $g(0) > 0$ and $g(1) < 0$. Look at Figure 18.12. We see that a continuous function with these properties must have a point p in between 0 and 1 such that $g(p) = 0$. But this just says that $f(p) = p$. □

Now we turn to the higher-dimensional, particularly the 2-dimensional, version of the Brouwer fixed point theorem. This is the formulation that caused such interest and excitement when Brouwer first proved the result over eighty years ago. Before we proceed, we

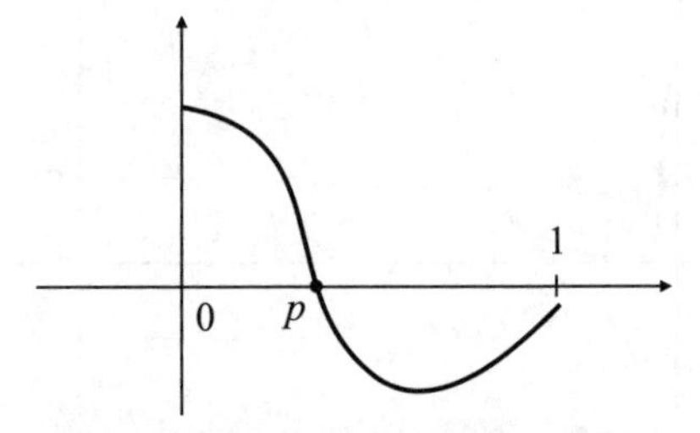

Figure 18.12. A point where g vanishes.

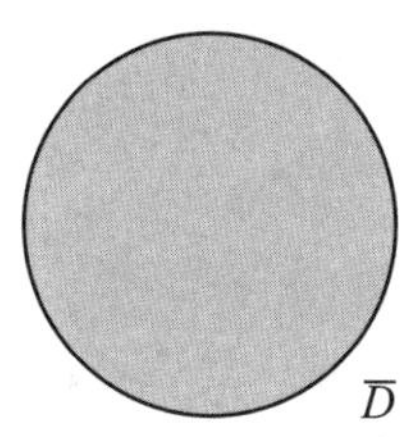

Figure 18.13. The closed unit disc.

must establish an auxiliary topological fact (Lemma 18.1). And it is for this purpose that we are going to use Poincaré's homotopy theory.

Lemma 18.1. *Let U, V be geometric figures and $g : U \to V$ be a continuous function. If $\gamma \subseteq U$ is a closed curve that can be continuously deformed to a point, then $f(\gamma) \subseteq V$ is also a closed curve that can be continuously deformed to a point.*

This statement makes good sense, but we do not prove it . Obviously a continuous function will not take a closed curve and open it up into an *un*closed curve; that is antithetical to the notion of continuity. And if we imagine a flow of curves, beginning with γ, that merge to a point in U, then of course their images under f will be a flow of curves in V that merge to a point in V.

Definition 1. Let $\overline{D}$ be the closed unit disc (i.e., the unit disc together with its boundary) as shown in Figure 18.13. Let C denote the boundary of $\overline{D}$. A continuous function $h : \overline{D} \to C$ that fixes each point of C is called a *retraction* of $\overline{D}$ onto C. See Figure 18.14.

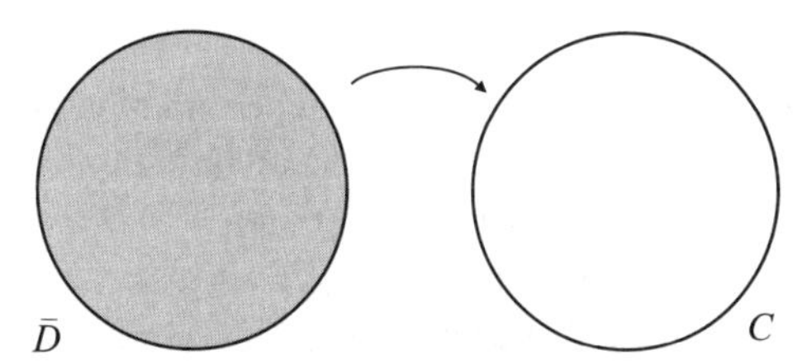

Figure 18.14. A retraction.

Proposition 18.1. *There does not exist any retraction from $\overline{D}$ to C.*

For the reasoning, consider Figure 18.15. We assume that there is a retraction $r : \overline{D} \to C$. The function u is just the inclusion map from C into $\overline{D}$. Now let γ be the curve in C that just wraps once around the circle—Figure 18.16. Then the composition $r \circ u$ obviously just takes γ to itself. On the other hand, u must take γ to a curve $u(\gamma) \subseteq \overline{D}$ that is shrinkable to a point—since all curves in $\overline{D}$ shrink to a point. And, by the lemma, r in turn must take $i(\gamma)$ to another curve that shrinks to a point.

But now we have a problem: On the one hand, $r \circ u$ takes γ to itself, and thus $r \circ u(\gamma)$ *cannot* be shrunk to a point. On the other hand, we just argued that $r \circ u(\gamma)$ *can be shrunk to a point*. It is impossible to have both statements be true. That is our contradiction. So the retraction cannot exist.

And now here is Brouwer's famous theorem:

Theorem 18.2. *Let $\overline{D}$ be the closed unit disc. Let $f : \overline{D} \to \overline{D}$ be a continuous function. Then there is a point $P \in \overline{D}$ such that $f(P) = P$.*

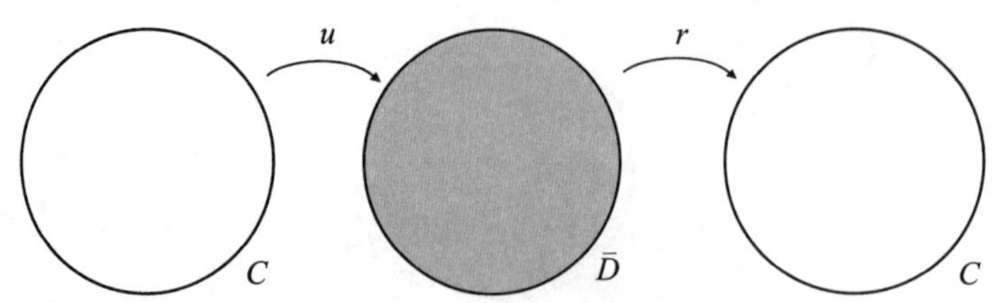

Figure 18.15. Nonexistence of a retraction.

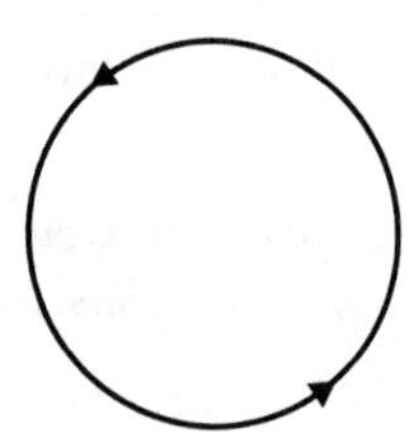

Figure 18.16. A curve that wraps once around.

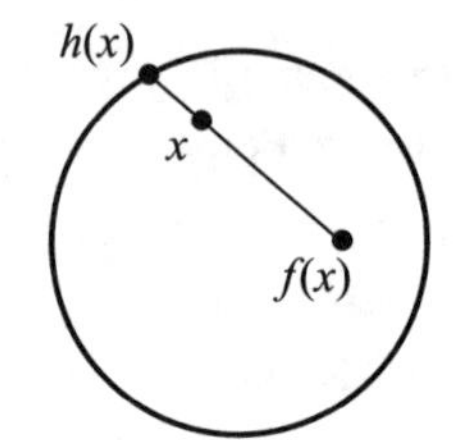

Figure 18.17. Existence of a retraction.

At the risk of offending Brouwer himself, we provide a proof by contradiction. Suppose that there is such a map f that does *not* possess a fixed point. Then, for each point $x \in \overline{D}$, $f(x) \neq x$. But then we can use f to construct a retraction of $\overline{D}$ to C as follows. Examine Figure 18.17. You can see that the segment that begins at $f(x)$, passes through x, and ends at a point $h(x)$ in C gives us a mapping

$$x \longmapsto h(x)$$

from $\overline{D}$ to C.

This mapping is evidently continuous, as a small perturbation in x will result in a small perturbation in $f(x)$ and hence a small perturbation in $h(x)$. Furthermore, each element of C is mapped, under h, to itself. So in fact *h is a retraction of $\overline{D}$ to C*. Note that the reason that we can construct this retraction is that $f(x) \neq x$; it is because of that inequality that we know how to draw the segment that defines $h(x)$. But we know, by the proposition, that that is impossible. Thus it cannot be that $f(x) \neq x$ for all x. As a result, some point P is fixed by f. And that is the end of our proof. We have established Brouwer's fixed point theorem using nonexistence of a retraction from $\overline{D}$ to C, which in turn uses Poincaré's homotopy theory. □

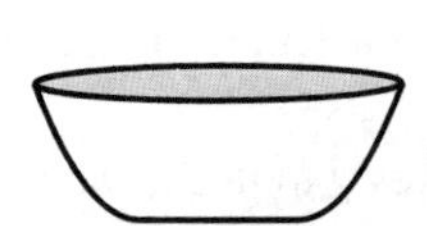

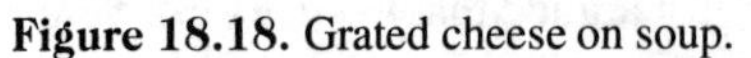

Figure 18.18. Grated cheese on soup.

Figure 18.19. More cheese.

In popular discussions of the Brouwer fixed point theorem, it is common for the teacher to suggest that we consider putting grated cheese across the top of a bowl of soup. See Figures 18.18 and 18.19. We are to imagine the entire surface of our creamed tomato soup covered by grains of ground parmesan. Then we stir up the soup (so that the cheese still remains on the surface) and the assertion is—can you anticipate?—that some grain of cheese ends up where it began! Refer to Figures 18.20 and 18.21.

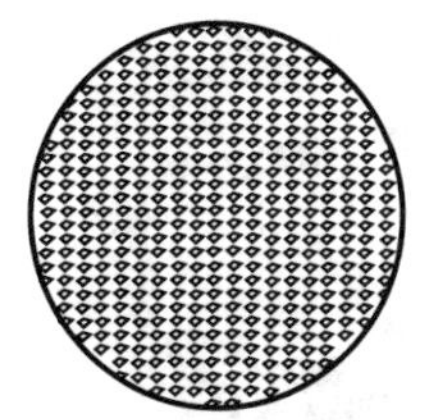

Figure 18.20. Distribution of the grated cheese.

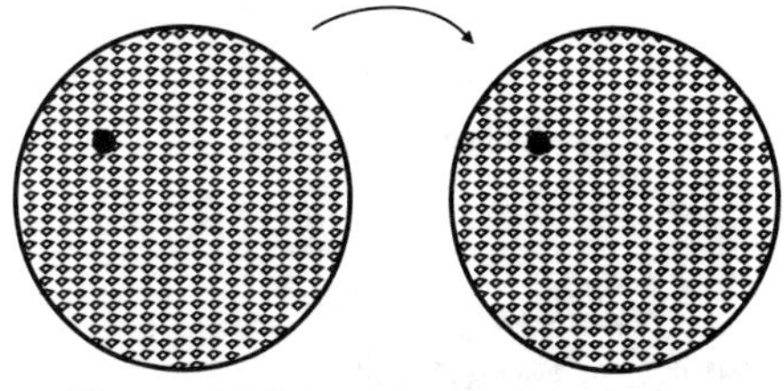

Figure 18.21. A fixed grain of cheese.

Further Reading

A. B. Brown, Extensions of the Brouwer fixed point theorem, *The American Mathematical Monthly* 69(1962), 643.

Bernhard Korte, Algorithmic mathematics versus dialectic mathematics, *The College Mathematics Journal* 16(1985), 6–8.

18.5 The Generalized Ham Sandwich Theorem

18.5.1 Classical Ham Sandwiches

In this section we are going to discuss a far-reaching generalization of the Brouwer Fixed Point Theorem. Our treatment will be almost entirely intuitive, as it must be. But it serves to show that mathematical ideas are not stagnant. Any good insight gives rise to further investigation and further discoveries. The "generalized ham sandwich theorem" is one of these.

First, let us define a classical ham sandwich. Such a sandwich consists of two square pieces of bread and a slice of ham (assuming that we are using packaged ham) and a slice of cheese (assuming that we are using packaged cheese). See Figure 18.22.

Now it is easy to see that, with a single planar slice of the knife, it is possible to cut the sandwich in such a way that

- the bread is sliced in half,
- the cheese is sliced in half,
- the ham is sliced in half.

Figure 18.23 illustrates one such cut. Figure 18.24 illustrates another.

Figure 18.22. A classical ham sandwich.

Figure 18.23. A bisecting cut.

Figure 18.24. Another bisecting cut.

In fact there are infinitely many ways to perform a planar cut of the classical ham sandwich that will bisect each of the bread, the cheese, and the ham.

In the next subsection we shall define a "generalized ham sandwich" and discuss an analogous but considerably more surprising result.

18.5.2 Generalized Ham Sandwiches

A generalized ham sandwich consists of *some ham*, *some cheese*, and *some bread*. But the ham could be in several pieces, and in quite arbitrary shapes. Similarly for the cheese and the bread. Figure 18.25 illustrates a generalized ham sandwich.

Please remember that these ham sandwiches live in 3-dimensional space. The generalized ham sandwich shown in Figure 18.25 is a 3-dimensional ham sandwich. Each of the ham, the cheese, and the bread is a solid, 3-dimensional, object.

Now we have the following astonishing theorem:

Theorem 18.3. *Let S be a generalized ham sandwich in 3-dimensional space. Then there is a single planar knife cut that*

- *bisects the bread,*
- *bisects the cheese,*
- *bisects the ham.*

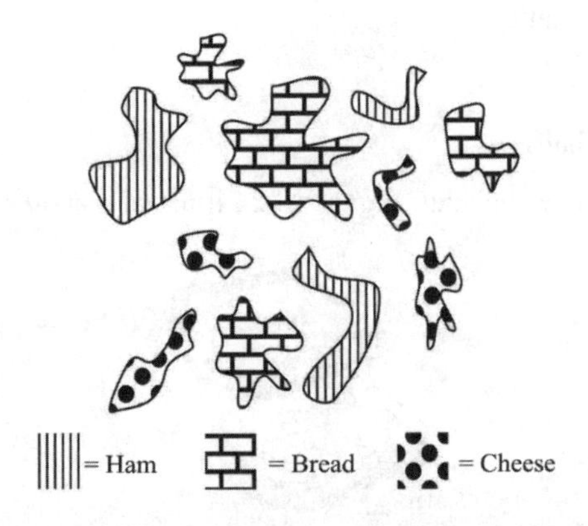

Figure 18.25. A generalized ham sandwich.

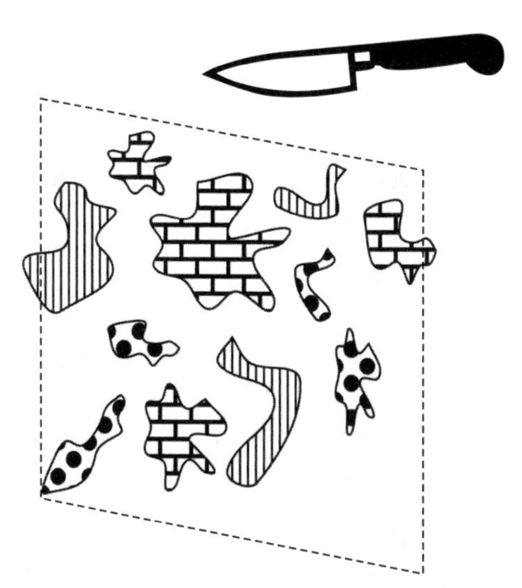

Figure 18.26. The generalized ham sandwich theorem.

See Figure 18.26. The proof, which is too complicated to present here, is a generalization of the intermediate value property that we used to prove the fixed point theorem in dimension 1.The ham sandwich problem was first posed by Stanislaw Ulam in 1930. The ham sandwich theorem as we have stated it here was proved by Arthur H. Stone and John W. Tukey in 1942—see [STT].

In fact it is worth pondering this matter a bit further. Let us consider the generalized ham sandwich theorem in dimension 2. In this situation we cannot allow the generalized ham sandwich to have three ingredients. In fact, in dimension 2, the generalized ham sandwich has only bread and ham. No cheese. Then the same result is true: a single linear cut will bisect the ham and bisect the bread. Examine Figure 18.27 and convince yourself that, with ham and cheese and bread configured as shown in dimension 2, there is no linear cut that will bisect all three quantities. But there is a certainly a linear cut that will bisect the ham and the cheese, or the bread and the cheese, or the ham and the bread.

In dimension 4, we can add a fourth ingredient to the generalized ham sandwich—such as turkey. And then there is a single hyper-planar slice that will bisect each of the four quantities: turkey, ham, cheese, and bread. This is all pretty abstract, and we cannot discuss the details here. But you should talk this up in class.

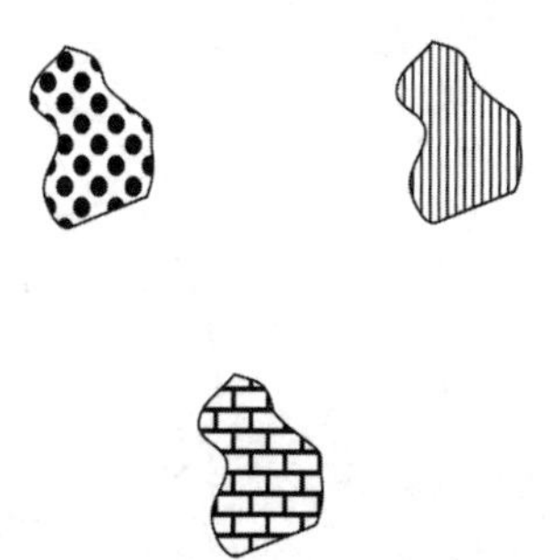

Figure 18.27. The generalized ham sandwich theorem in dimension 2.

Further Reading

W. A. Beyer and Andrew Zardecki, The early history of the ham sandwich theorem, *The American Mathematical Monthly* 111(2004), 58–61.

Theodore P. Hill, Common hyperplane medians for random vectors, *The American Mathematical Monthly* 95(1988), 437–441.

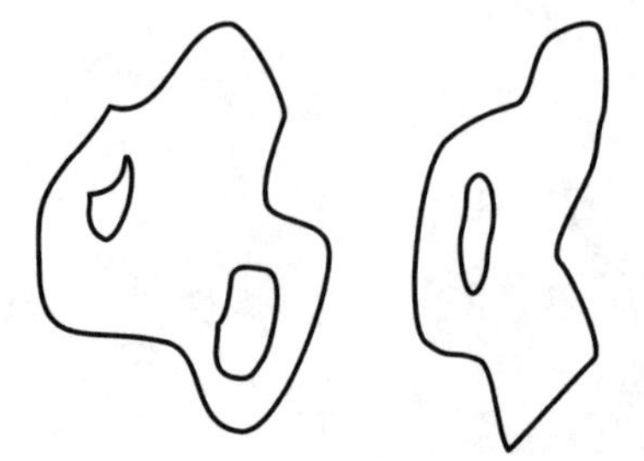

Figure 18.28. Two inequivalent domains.

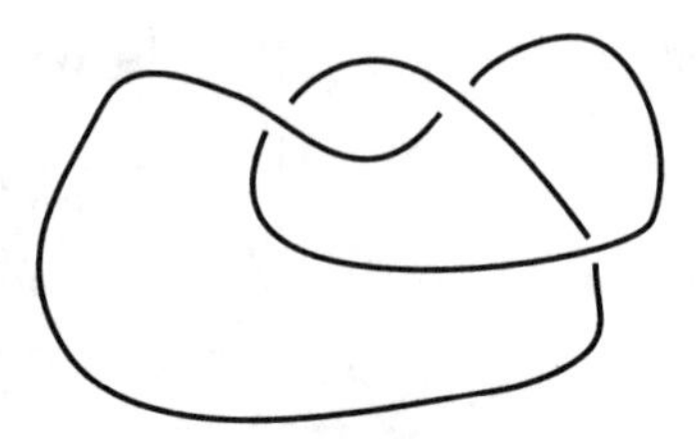

Figure 18.29. The trefoil knot.

Exercises

1. Compare the two domains shown in Figure 18.28. Use the idea of homotopy to establish that the two domains cannot be topologically equivalent. Discuss this question in class.

2. Examine the knot in Figure 18.29. This is a classic *trefoil knot*—the knot that most people know how to tie, and is used to tie one's shoe. Knot theory is an active area of mathematical investigation. Indeed, knot theory is used today in theoretical physics, mathematical analysis, and topology. A mathematician thinks of a knot as an exotic embedding of the circle (a simple loop) into space. See Figure 18.30. Explain why the knotted curve and the unknotted curve in Figure 18.30 are homeomorphic.

3. The function $f(x) = 1/x$ gives a homeomorphism between the bounded interval $(0, 1)$ and the unbounded interval $(1, \infty)$. Draw a picture to explain how this homeomorphism works. Discuss the question in class.

4. The circle (*not the disc*) and the (open) annulus have the same homotopy—see Figure 18.31. That is to say, there is a natural correspondence between loops in the circle and loops in the annulus. But the circle and the annulus are *not* homeomorphic. Explain these statements. Discuss the questions in class.

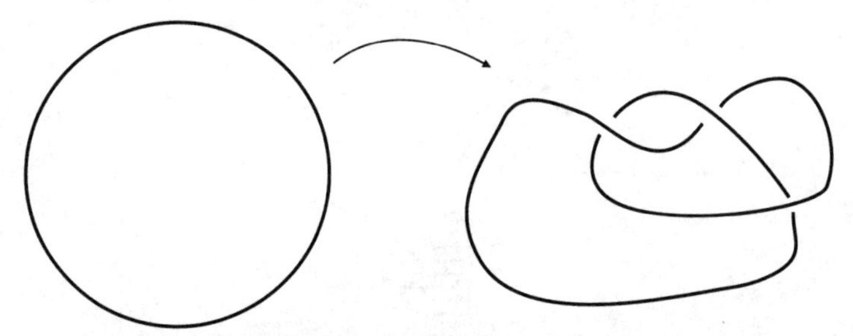

Figure 18.30. A knot is an embedding of the circle into space.

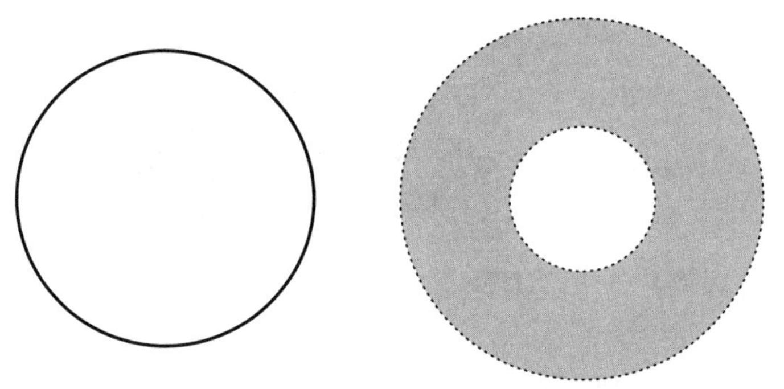

Figure 18.31. The circle and the annulus have the same homotopy.

5. The mapping

$$(x, y) \longmapsto \left(\frac{x}{x^2 + y^2}, \frac{y}{x^2 + y^2} \right)$$

gives a homeomorphism between the punctured disc

$$\widehat{D} = \{(x, y) : 0 < x^2 + y^2 < 1\}$$

and the set

$$U = \{(x, y) : 1 < x^2 + y^2 < \infty\}.$$

One of these sets is bounded and the other unbounded. Explain the nature of this homeomorphism. Draw some curves in $\widehat{D}$ and show to what curves they correspond in U.

6. Not all sets have the fixed point property. Consider the circle

$$S = \{(x, y) : x^2 + y^2 = 1\}.$$

The mapping $(x, y) \mapsto (-y, x)$ is a 90-degree rotation of this circle. It is certainly a continuous function of the circle to itself, and it is also one-to-one and onto. Yet it has no fixed points.

Now consider the unit sphere in 3-dimensional space. Describe a continuous function of the sphere to itself that has no fixed points.

7. Refer to Exercise 6. Describe a continuous function from the open unit interval $(0, 1)$ to itself that has no fixed points. Describe a continuous function from the entire real line to itself that has no fixed points.

8. Give an example of a continuous function from the closed unit interval $I = [0, 1]$ to itself that has two distinct fixed points.

9. Refer to Exercise 8. Give an example of a continuous function from the closed unit disc $\{(x, y) : x^2 + y^2 \leq 1\}$ to itself that has two distinct fixed points.

10. Examine the 2-dimensional generalized ham sandwich—containing only bread and ham—in Figure 18.32. Describe how to find a linear cut that will bisect the ham and bisect the bread.

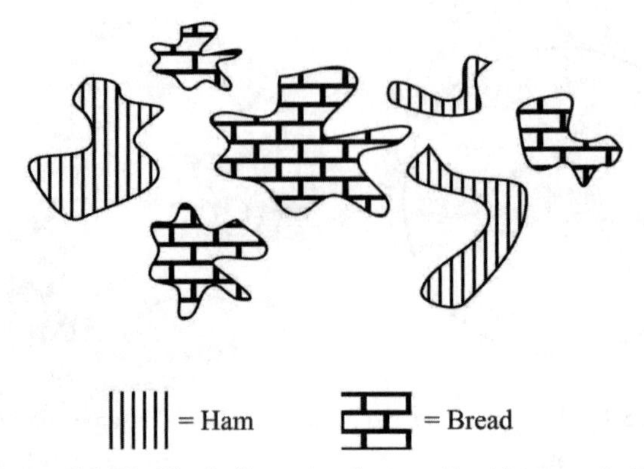

Figure 18.32. The 2-dimensional generalized ham sandwich.

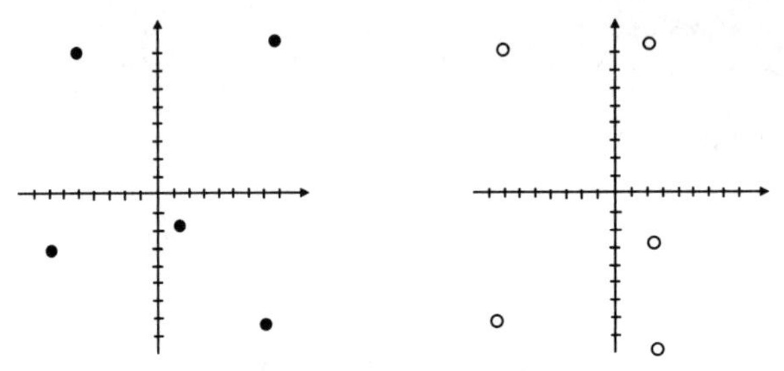

Figure 18.33. A homeomorphism preserving finite point sets.

11. Let A be the finite set of points exhibited in the left part of Figure 18.33 and B be the finite set of points exhibited in the right part of Figure 18.33. Describe a homeomorphism of the plane to itself that takes the points of A to the points of B (in a one-to-one, onto fashion, of course).

12. Project: Formulate a generalized ham sandwich theorem in four dimensions. Do not attempt a proof, but give an example to show that your enunciation is sharp. See [GOL] for further details.

13. Project: Calculate the fundamental group of the unit sphere in 3-dimensional space $\mathbb{R}^3$. See [GRH].

14. Project: Calculate the fundamental group (i.e., the first homotopy group) of the torus. Use this information to conclude that the torus and the sphere (refer to Exercise 13) are not topologically equivalent. Again see [GRH].

15. Project: For a continuous function $f : [0.1] \to [0, 1]$, the fixed point can be found by intersecting the graph with the diagonal line $y = x$. Draw some figures to illustrate this point, and explain the mathematics. What would be the analogous assertion in higher dimensions? See [KRA1].

19

Sonya Kovalevskaya and the Mathematics of Mechanics

19.1 The Life of Sonya Kovalevskaya

Sophie Vasilyevna Kovalevskaya (1850–1891) was the middle child in a family of minor nobility. The family name was Korvin-Krukovsky ("Kovalevskaya" was a married name that she adopted later). Her father was, among other things, an artillery general. Sophie was raised in plush surroundings by a strict governess who expended her efforts to turn little Sophie into a young lady. Sophie felt ignored in favor of her much-admired older sister Anyuta and her younger brother, who was of course the male heir. As a result, she was a nervous child having a withdrawn personality; these characteristics stuck with her throughout her life. Sophie was often called "Sonya" by her friends. Modern mathematicians know her by this name, so that is the name that we shall use in this discussion. Sonya Kovalevskaya was one of the most remarkable figures of modern mathematics. Her life is chronicled in the modern work [KOBA].

Sonya was educated by tutors and governesses. The family was well off and moved in high social circles. Among the esteemed members of her family's acquaintance was the author Dostoevsky.

Sonya Korvin-Krukovsky was attracted to mathematics at a very young age. In her autobiography/diary she wrote

> The meaning of these concepts I naturally could not yet grasp, but they acted on my imagination, instilling in me a reverence for mathematics as an exalted and mysterious science which opens up to its initiates a new world of wonders, inaccessible to ordinary mortals.

When Sonya's father retired from the military, he moved the family to an estate called Palibino near the Lithuanian border. When Sonya was only 11 years old, it was arranged for the walls of her bedroom to be papered with pages from Ostrogradski's lectures on differential and integral analysis (one explanation was that there was a shortage of wallpaper at the time). Unlike Sophie Germain (Chapter 11), little Sonya found her family to be most supportive of her quest to master mathematics. She learned calculus by studying her nursery wallpaper! She also studied her father's old calculus notes. Her uncle Peter (Fyodor) spent time talking to her about mathematics and explaining the subtleties of various abstractions.

Her uncle Vasily, though less well schooled, also made some effort to talk to little Sonya about squaring the circle and other topics of interest.

The Korvin-Krukovsky family tutor, Y. I. Malevich, mentored Sonya in her formal study of mathematics. She gives him much credit for her development, and indeed she wrote

> I began to feel an attraction for my mathematics so intense that I started to neglect my other studies.

It was when Sonya obtained a copy of Bourdeu's *Algebra*, and would study it late at night after the rest of the family had gone to bed, that her father began to object to her mathematical avocation. He ran into resistance not only from his daughter but from his acquaintance and neighbor Professor Tyrtov. Tyrtov presented the Korvin-Krukovsky family with a copy of his new, recently penned, physics textbook. Immediately Sonya began to read it. She struggled with the trigonometry in the text, and found that she had to work out many of the basic ideas from scratch herself. Tyrtov was impressed that her development of the sine function was quite similar to the way that the ideas had been developed historically, since the time of the Greeks. Tyrtov called her "a new Pascal". He argued vehemently with Mr. Korvin-Krukovsky that the bright young girl should be allowed to pursue her mathematical studies. It was only several years later that the rather conservative general finally acceded. At Tyrtov's insistence, it was agreed that Sonya would be allowed to go to St. Petersburg to continue her mathematical work under the tutelage of Professor Alexander Strannoliusky.

After St. Petersburg, Sonya Kovalvskaya determined to continue her studies at the university level. The closest universities that were open to women were in Switzerland. In the late nineteenth century, a Russian woman could not live apart from her family without the express permission of either her father or her husband. In order that she could go abroad to pursue a higher education, Sonya in 1868 entered into a nominal marriage with Vladimir Kovalevski (note that the man's name has the male ending while the woman's name has a different, feminine, ending). The marriage lasted fifteen years, but it was not a happy one. There were frequent quarrels and misunderstandings and the tension interfered greatly with Sonya's studies.

The Kovalevskis spent the first few months of their marriage in St. Petersburg. In 1869 Sonya Kovalevskaya traveled to Heidelberg to study mathematics and science. She was disappointed to learn that women were not allowed to matriculate at the university. She was eventually allowed to attend lectures on an unofficial basis, but she had to obtain the express permission of each individual lecturer. Sonya attended courses for three semesters, and made a strong impression on her professors. One of her fellow students noted that especially professors Königsberger (1837–1921) (mathematics) and Kirchhoff (1824–1887) (chemistry) were "ecstatic over their gifted student and spoke about her as an extraordinary phenomenon." It may not be out of place here to note that Sonya was an exceptionally beautiful woman. Her physical attraction no doubt contributed to her appeal.

In 1871 Sonya Kovalevskaya moved to Berlin to study with Karl Weierstrass (1815–1897), who had been Königsberger's teacher. Unfortunately, the university in Berlin simply would not allow her to attend classes—because she was a woman. It is ironic that this dreadfully unfair turn of events worked in Kovalevskaya's favor; for instead Weierstrass spent four years tutoring her privately.

By Spring of 1874 Kovalevskaya had completed three papers. One of these papers was about partial differential equations, one about Abelian integrals, and one about the rings of Saturn. Weierstrass was quite impressed by this work and recommended her for a doctorate. In fact in 1874 Sonya was granted her doctorate, *summa cum laude* (but in absentia), from the University of Göttingen. Sonya Kovalevskaya was the first woman to receive her doctorate in mathematics.

Unfortunately after that—as we have seen with other talented woman mathematicians—she was unable to obtain any academic position. Sonya took this so badly that, for a period of six years, she ceased doing research and would not respond to Weierstrass's letters. The one job that she did find was to teach arithmetic to school girls, and she reacted to this offer bitterly:

I was unfortunately weak in the multiplication table.

Sonya and her husband returned home to Palibino to be with her family. Shortly thereafter, her father died. During this period of sorrow, Sonya and her husband grew closer. In 1878 they produced a daughter. During this period, Sonya neglected her work in mathematics and instead developed her literary skills. She gave some effort to fiction, theater reviews, and science articles for a newspaper.

From 1880, however, Sonya returned to her mathematics. She began a study of the refraction of light, and wrote three articles on the subject. Years later, in 1916, Vito Volterra (1860–1940) discovered that Kovalevskaya had made a classical mistake originally due to Gabriel Lamé (1795–1870)—a scientist on whose work she had based her studies. Her work still had considerable value because it contained a thoroughgoing explanation of Weierstrass's theory of integrating certain partial differential equations.

In 1883, Sonya's husband Vladimir (from whom she had been separated for two years) committed suicide. His business ventures had all failed and he despaired. Sonya was deeply affected by this tragedy, and she attempted to deal with the situation by immersing herself in her mathematics. Gösta Mittag-Leffler (1846–1927), who had himself been a student of Weierstrass, championed her cause; he obtained for her a *privat docent* position in Stockholm.

Before Sonya moved to Sweden, she secretly visited Weierstrass in Berlin and rekindled her passion for mathematical research. Now that she was a single mother, she had to arrange to leave her daughter Sofya Vladimirovna in the care of friends so that she could pursue her newly reinvigorated mathematical career in Stockholm.

She began her lectures in Stockholm in 1884, and made such an impression that she was appointed to an *Extraordinary Professorship* in June of that year.[1] After Sonya was sure of her position at the university in Stockholm, she brought her daughter to Sweden to live with her.

In June of 1889 Sonya Kovalevskaya became the first woman since Laura Bassi (1711–1778, physics) and Maria Gaetana Agnesi (1718–1799, mathematics) to hold a chair (in Mechanics) at a European university. It was during this time that Sonya also co-authored the play *The Struggle for Happiness* with her friend Anna Leffler.

During her time in Stockholm, Sonya Kovalevskaya blossomed into a mathematician of international renown. She conducted the most important research of her career. She became

[1] In spite of the name, this position is only similar to a research assistantship—a very junior position, and not a prestigious one.

an editor of Mittag-Leffler's new journal *Acta Mathematica* (today one of the oldest and most prestigious mathematics journals). She organized international conferences. Sonya also had a great interest in literature, and she wrote some rather striking dramas and reminiscences at this time.

In fact it is not well known that Sonya Kovalevskaya was a talented writer of fiction and cultural works. Her book *Recollections of Childhood* is but a slight indication of her considerable skills. One appreciator of Sonya's writing said, "The Russian and Scandinavian literary critics have been unanimous in declaring that Sophie Kovalevskaya [Sonya Kovalevskaya] was the equal of the best writers of Russian literature, in style as well as in subject matter." Her early death cut short her plans for various literary projects. In particular, she had planned to write "The Razhevski Sisters during the Commune", a reminiscence of her trip to Paris in 1871. Kovalevskaya found writing to be a cathartic escape from her intense periods of work on mathematics. She wrote in her memoirs that, "At twelve years old I was thoroughly convinced I was born a poet." She also recalls rather fondly her uncle buying math books for her, and speaking to her about the quadrature of the circle.

In 1886 the prize topic for the Prix Bordin was announced to be significant contributions to the study of rigid bodies. Sonya Kovalevskaya entered her paper *Mémoire sur un cas particulier du problème de le rotation d'un corps pesant autour d'un point fixe, ou l'intégration s'effectue à l'aide des fonctions ultraelliptiques du temps.* Around this time Sonya also had a torrid affair with Russian lawyer Maxim Kovalevsky (unrelated to her late departed husband). It was a rocky relationship, evidently because each was too passionate about his/her work to give it up for the other. Maxim's work took him away from Stockholm, and he wanted her to give up her mathematical career to go with him and be his wife. Sonya flatly rejected the proposal, but was heartbroken as a result. She spent the summer in France with Maxim, but ended up falling into one of her frequent depressions. While in France, she returned to her writing and completed work on *Recollections of Childhood.*

In order to ensure fairness, the fifteen papers submitted for the Bordin Prize were submitted anonymously. Each author wrote a quote on their paper to be used for later identification (see Sonya's quote at the end of this article about her life). The judges chose Sonya Kovalevskaya's paper without knowing that it was written by a woman. The judges were so impressed by this work that they increased the prize amount from 3000 to 5000 francs. They were later much surprised to learn that their much-honored winner was a female. After receiving the prize, Sonya was feted and celebrated almost without cease. She was immensely happy, and finally felt that her work was properly appreciated.

In 1889 Sonya Kovalevskaya also won a prize from the Swedish Academy of Sciences. In the same year, on the recommendation of Chebychev, she was elected a corresponding member of the Russian Imperial Academy of Sciences. In fact the rules were changed to allow for Kovalevskaya's election.

Sonya developed a rather strong relationship with Mittag-Leffler during these years, and lived in his home for a time. That Kovalevskaya lived in Gösta Mittag-Leffler's home, and that they enjoyed an intense relationship (even though they were not married) is well known. An indication of the situation is given by Kovalevskaya herself: "Yesterday was a rough day for me, for big M [Mittag-Leffler] ... left in the evening. If he had stayed here I do not know how I would have been able to work. He is so tall, so powerfully built, that he manages to take up a great deal of room, not only on a sofa, but also in my thoughts and I would never have been able in his presence to think of anything but him."

Winning the Bordin Prize was a great triumph for Sonya Kovalevskaya. She was in Paris, and was honored and invited everywhere. She was in her glory, and completely happy. Mittag-Leffler came to join her. However, the good times came abruptly to an end. With her demands and her tyrannical and jealous love, she expected a great deal of Mittag-Leffler. She thought that his admiration for her did not measure up to her love for him. Further, she was unwilling to give up her mathematical career and become simply the wife of this man whom she so loved and admired. [Mittag-Leffler was of course married at the time, and the quality of his life depended on his wife's fortune.] Thus it was that they frequently separated; there was much bitterness and vituperation. Unable to live with him or without him, exhausted, torn by the incessant strife of their relationship, Kovalevskaya finally became ill with influenza complicated by pneumonia and died in 1891. She was at the height of her mathematical and scientific powers at the time, and enjoying an immense worldwide reputation. Her last paper, on properties of the potential function of a homogeneous body, was published right before her death.

Sonya Kovalevskaya's view of life is perhaps summarized by her remarkable statement

> Say what you know, do what you must, come what may.

She is remembered today in many ways. She was the first female Ph.D. and the first female professor anywhere. There is a lunar crater named after her, and also a minor planet (an asteroid) named after her.

Further Reading

Karen D. Rappaport, S. Kovalevsky: A mathematical lesson, *The American Mathematical Monthly* 88(1981), 564–574.

19.2 The Scientific Work of Sonya Kovalevskaya

19.2.1 Partial Differential Equations

Most of Kovalevskaya's scientific work is too technical to be discussed in any detail in these pages. We content ourselves with some informal descriptions of what she accomplished.

Perhaps the most widely cited result of Sonya Kovalevskaya is the so-called Cauchy-Kovalevskaya theorem in the theory of partial differential equations. A partial differential equation is an equation involving a function of several variables and its derivatives. Most of the laws of nature are formulated in terms of partial differential equations.

Kovalevskaya studied partial differential equations in which the coefficient functions are functions that have convergent power series expansions (that is to say, the coefficients are functions that can be written as infinite sums of powers of x and y). We shall say more about power series below. Thus—if we let $\partial/\partial x$ denote the derivative in the x-variable (holding the y-variable fixed) and also $\partial/\partial y$ the derivative in the y-variable (holding the x-variable fixed)—it holds that

$$a(x, y)\frac{\partial u}{\partial x} + b(x, y)\frac{\partial u}{\partial y} + c(x, y)\frac{\partial^2 u}{\partial x \partial y} = f(x, y)$$

is such an equation, provided that the functions a, b, c, f have convergent power series expansions. Her theorem (joint with A. Cauchy) is that there exists a solution of such a differential equation, and that solution will also have a convergent power series expansion.

The method of majorization that was originally used to prove the Cauchy-Kovalevskaya theorem is of wide utility in the mathematical sciences. Also, the result is frequently cited. It is the only truly general result about solvability in the entire literature of partial differential equations.

Further Reading

Debra Charpentier, Women mathematicians, *The Two-Year College Mathematics Journal* 8(1977), 73–79.

19.2.2 A Few Words About Power Series

We have already mentioned that Sonya Kovalevskaya made powerful contributions to the theory of power series in the context of differential equations. In fact power series are one of the oldest and most fundamental ideas in modern mathematics (going back at least to Issac Newton). In the present subsection we shall give a brief introduction to power series.

A power series is a sum—usually an infinite sum—of powers of x. We are allowed to put coefficients in front of the powers of x. Thus a power series is like a "generalized polynomial"—instead of having finitely many summands it has infinitely many summands.

Power series can be used to generate a vast and rich array of different functions.

EXAMPLE 19.1. Consider the series

$$f(x) = \sum_{j=0}^{\infty} x^j = 1 + x + x^2 + x^3 + \cdots .$$

We learned in Chapter 2 about series of this kind. This is a *geometric series*. The terms are all powers of the fixed number x. Thus we know from Sections 2.5, 2.6 that

$$f(x) = \frac{1}{1-x} .$$

Notice that this is a new concept. We have added up infinitely many monomials—1 and x and x^2 and x^3 and so forth—and produced a new function. That new function turns out to be $1/[1-x]$. Of course the summation process only makes sense when the series converges, and we learned in Chapter 2 that that is true precisely when $-1 < x < 1$. Put in other words, we have the series representation

$$\frac{1}{1-x} = 1 + x + x^2 + x^3 + \cdots ,$$

valid for $-1 < x < 1$.

EXAMPLE 19.2. Consider the power series

$$g(x) = \sum_{j=0}^{\infty} \frac{(-1)^j x^{2j+1}}{(2j+1)!} = x - \frac{x^3}{3!} + \frac{x^5}{5!} - \frac{x^7}{7!} + - \cdots .$$

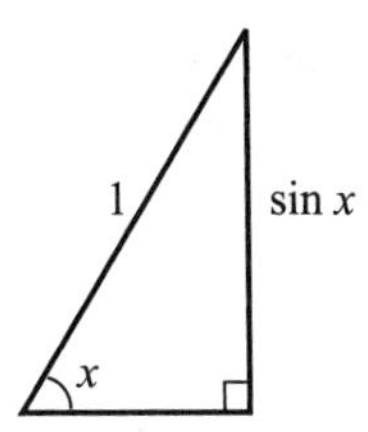

Figure 19.1. The sine function on a triangle.

It turns out—this involves advanced ideas that we cannot treat here—that this series converges for all values of x. And it defines the function $\sin x$.

You may have encountered the sine function in your earlier studies. Traditionally, we learn about sine in the context of a triangle (Figure 19.1).

The sine of the angle x is defined to be the ratio of the height of the triangle to the hypotenuse. It is a remarkable fact that this simple geometric quantity can be expressed in terms of a power series.

It is a deep theorem of mathematical analysis that a power series can be differentiated termwise. Thus

$$\sin x = x - \frac{x^3}{3!} + \frac{x^5}{5!} - \frac{x^7}{7!} + - \cdots$$

so

$$[\sin x]' = 1 - \frac{x^2}{2!} + \frac{x^4}{4!} - \frac{x^6}{6!} + - \cdots .$$

This last power series is known to represent the cosine function. So power series give us a way to re-discover the important fact that

$$[\sin x]' = \cos x .$$

For You to Try Verify that the function

$$y(x) = \sum_{j=0}^{\infty} (-1)^j \frac{x^{2j}}{(2j)!}$$

is a solution of the differential equation

$$y'' + y = 0 .$$

The power series that Sonya Kovalevskaya considered in her celebrated theorem with Cauchy (discussed above) were power series of *two* variables. Such a power series has the form

$$\sum_{j,k=0}^{\infty} a_{j,k} x^j y^k .$$

A concrete example is

$$f(x, y) = \sum_{j,k=0}^{\infty} (-1)^j \frac{2^{-(j+2k)} x^j y^{2k}}{j!(k!)^2} .$$

This is a subtle and quite general and flexible way to define new functions. The method of power series is in considerable use today for finding solutions of differential equations and for solving other physical problems. Sonya Kovalevskaya was one of the pioneers in this technique.

19.2.3 The Mechanics of a Spinning Gyroscope and the Influence of Gravity

Kovalevskaya studied the rotations of a rigid body about a fixed point. The foundational work in the subject was done by Leonhard Euler (1701–1783). He considered a freely tumbling body without the influence of gravity. He established that angular momentum and energy are conserved. Euler described the motion of the body by specifying the rotation matrix that changes spatial coordinates into body coordinates. He derived a set of nine differential equations. Of these nine, three turned out to be fundamental for the problem:

$$dx + \frac{c^2 - b^2}{a^2} yz\,dt = 0;$$

$$dy + \frac{a^2 - c^2}{b^2} zx\,dt = 0;$$

$$dz + \frac{b^2 - a^2}{c^2} xy\,dt = 0.$$

Here the quantities a^2, b^2, c^2 are the moments of inertia of the body about its principal axes. The variables x, y, z are the coordinates of a point on the body. Euler considered in detail the motion of a body on which no external torque is acting and showed that the entries of the rotation matrix can be expressed as elliptic integrals. He further showed that these entries reduced to elementary functions if two of the three principal moments of inertia are equal.

The second case, studied by J. L. Lagrange (1736–1813), considers a body with the symmetries of a spinning top. In particular, Lagrange studied the general case of motion of a body having equal moments of inertia about two of its principal axes. He noted that the center of gravity lay on the third principal axis provided the external torque acting on the body resulted purely from gravitational attraction. Lagrange's analysis is the one most commonly seen in the literature; in the nineteenth century it was referred to as the case of a "heavy body".

For a long time, the analyses of Euler and of Lagrange were the only two known situations in which the system was known to be "integrable", that is, that key physical constants are preserved.

Another piece of foundational work, which had considerable influence on Kovalevskaya, was that of Jacobi. For he showed how to integrate complicated mechanical systems using elliptic functions.

Kovalevskaya found a third system that is completely integrable. She realized that the general case of the motion of a heavy body about a fixed point was vastly too complicated for study. So she imposed some additional conditions. In her case, the moments of inertia of the body are related in a new and subtle way. Her method of analysis, using very delicate integrations of hyperelliptic functions, was a *tour de force* of mathematical analysis. It won her the Bordin Prize. It is worth quoting from the judges' report:

> The author has done more than merely adding a result of very high interest to those bequeathed to us by Euler and Lagrange; he has made a profound study of his result in which the resources of the modern theory of theta functions of two independent variables make it possible to give the solution in the most precise and elegant form. The result is a new and memorable example of a mechanical problem involving these transcendental functions, whose applications had previously been confined to pure analysis and geometry.

In a later paper Sonya Kovalevskaya studied the profile shape and equilibrium stability of a single gravitating ring around an attractive center. This work later proved to be decisive in her study of the rings of the planet Saturn.

19.2.4 The Rings of Saturn

In spite of some errors in her reasoning, Sonya Kovalevskaya's analysis of Saturn's rings, and of related ideas, turned out to be a foundational work of lasting importance. It helped to define the theory of rotating celestial body potentials. It is useful in the study of rotating fluid mass. Her method of integrating hyperelliptic functions has been particularly influential.

The early works by Galileo (1610–1616), Huygens (1655), Cassini (1675), Kant (1755), Laplace (1789–1798), and Maxwell (1859) set the stage for Kovalevskaya's discoveries. Sonya determined that Laplace's calculations were only accurate to first order; she was able to produce a more accurate mathematical model for the rings of Saturn.

In fact Kovalevskaya based her study on Laplace's work. Like Laplace, she assumed that the shape of the cross-section of the ring orbiting the planet was determined by assuming that a thin layer of liquid on the surface would be in equilibrium. Which is not to say that Kovalevskaya *assumed* that the ring was actually a liquid. The hypothesis was a way of saying that no shear stresses were acting on the ring as a result of the gravitational attraction of the planet. Laplace had assumed that the ring had an elliptical cross-section. Kovalevskaya took the cross section to satisfy the equation

$$r(\theta) = m_0 + m_1 \cos\theta + m_2 \cos 2\theta + \cdots .$$

Here θ is the angular variable and r the radial variable. The hypothesis of Laplace was that only the first two terms on the right were relevant. For Kovalevskaya, the higher order terms on the right were significant correction terms. Using numerical methods—a first for her—she was able to calculate the m_2 term and get a much more accurate idea of the shape of the ring.

19.2.5 The Lamé Equations

Right before her death, in collaboration with her teacher Weierstrass, Kovalevskaya studied Lamé's differential equations for the propagation of a wave in a solid medium. She rejected Lamé's unrealistic physical model for the problem, and succeeded in finding a solution for her more physically realistic model. Weierstrass was unwell at the time and delayed publication of the paper. He ultimately consented to the appearance of the work in *Acta Mathematica*, just so long as there were quotation marks around the portion of the work that was due to him. The paper was held in very high esteem at the time of its publication. Later, Vito Volterra (1860–1940) attempted to apply the mathematical model to a physical problem. He discovered errors, and was forced to publish a retraction to the Kovalevskaya paper in the same issue of *Acta* that contained an obituary of Kovalevskaya.

19.2.6 Bruns's Theorem

E. H. Bruns (1848–1919) was famous for having proved that the differential equations governing the three-body problem (e.g., the problem of analyzing the motion of the sun, the earth, and the earth's moon) have a ten-dimensional space of algebraic integrals. He also proved that the potential of a body bounded by a surface given by a power series is itself given by a power series at each regular point of the surface. Kovalevskaya was able to give a new derivation of these facts using the Cauchy-Kovalevskaya theorem. This is an impressive application of her earlier work.

Further Reading

R. P. Agnew, Summability of power series, *The American Mathematical Monthly* 53(1946), 251–259.

Ralph Boas, Power series for practical purposes, *The Two-Year College Mathematics Journal* 13(1982), 191–195.

G. Ervynck and P. Igodt, Power series and exponential generating functions, *The College Mathematics Journal* 20(1989), 411–415.

Garfield C. Schmidt, Uniqueness of power series representations, *The Two-Year College Mathematics Journal* 12(1981), 54–56.

R. Vyborny, Differentiation of power series, *The American Mathematical Monthly* 94(1987), 369–370.

19.3 Afterward on Sonya Kovalevskaya

Sonya Kovalevskaya is remembered as the first female Ph.D. and the first female professor since the Renaissance. She not only fought formidable social strictures against women in order to achieve her scientific goals; she also expended considerable effort to help other women. Mittag-Leffler, her friend, supporter, and confidante, remembers her as follows:

> ...it is perhaps neither as a mathematician nor writer that one should properly appreciate or judge this woman of so much spirit and originality. As a person she was even more remarkable than one would judge from her works. All those who knew her and were near to her, to whatever circle or part of the world they belonged, will remain forever under the lively and powerful impression which her personality produced.

Mittag-Leffler's house in Djurshom, Sweden lives on today in the mathematical community as the Mittag-Leffler Institute. It is a home for visiting mathematicians and a locale for mathematics conferences and gatherings. The house is much as Mittag-Leffler left it:

his books are still on the shelves, his correspondence still out in the open in his files, his scrapbooks still available for perusal, and his many boxes of personal photographs sit out in the library on the first floor. Among the personal photographs are dozens of pictures of Sonya Kovalevskaya—at parties, dressed in constumes, interacting with the family. She was a lovely person, and is remembered fondly for the many contributions she made during her short life.

Exercises

1. Consider the power series

$$1 + x^2 + x^4 + x^6 + \cdots .$$

For which x does the series converge? Can you write a closed form expression for the function defined by this series?

2. If

$$\sin x = x - \frac{x^3}{3!} + \frac{x^5}{5!} - \frac{x^7}{7!} + - \cdots$$

and

$$\cos x = 1 - \frac{x^2}{2!} + \frac{x^4}{4!} - \frac{x^6}{6!} + - \cdots ,$$

then can you write a power series expression for $\tan x = \sin x / \cos x$? [Hint: Use long division. Discuss this problem in class.]

3. Use the power series expansions for sine and cosine that we discussed in Exercise 2 to verify the formula

$$\sin^2 x + \cos^2 x = 1 .$$

4. Use the power series expansions for sine and cosine that we discussed in Exercise 2 to verify the formula

$$\sin 2x = 2 \sin x \cos x .$$

5. It is a fact—which we cannot derive here—that

$$\log(1 + x) = x - \frac{x^2}{2!} + \frac{x^3}{3!} - \frac{x^4}{3!} + - \cdots .$$

Use this power series expansion to obtain an approximation to the value of log 1.5. Use a calculator if you need to (but *do not* push the logarithm button!).

5. It is a fact—which we cannot derive here—that

$$(1 + x)^a = 1 + ax + \frac{a(a-1)}{2!}x^2 + \frac{a(a-1)(a-2)}{3!}x^3 + \cdots .$$

This is Newton's generalized binomial theorem. Use this formula to derive an approximate value for $\sqrt{2}$ (but *do not* push the square root button on your calculator!).

6. Why does a spinning top spin? Why does it remain upright? Discuss this question in class.

7. Suppose that a function $f(x) = \sum_{j=0}^{\infty} a_j x^j$ is one-to-one and onto, hence has an inverse. Then the inverse function f^{-1} should have a power series expansion. Explain why. What will be the coefficients of its power series (expressed in terms of the a_j)?

8. Suppose that you swing a weight on the end of a string. See Figure 19.2. At some moment you release the string, and the weight flys off. Explain in what direction, and along what sort of path, the weight will travel. What will be its initial velocity upon its release?

Figure 19.2. A weight on the end of a string.

9. Suppose that you shoot an arrow into the air—Figure 19.3. What will be shape of the path? Discuss this question in class. Perform some experiments. This is a curve that you know!

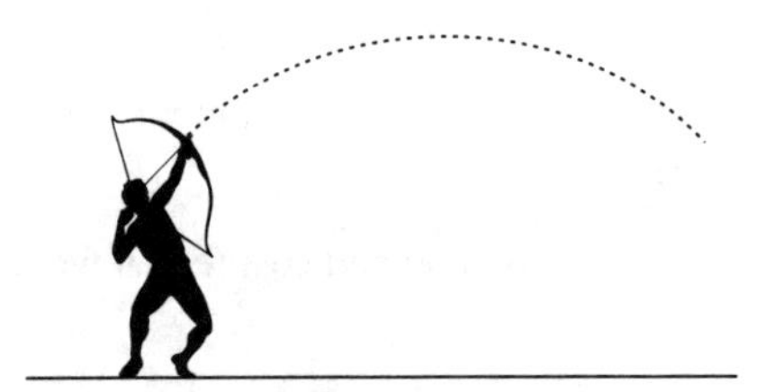

Figure 19.3. Shooting an arrow into the air.

10. Suppose that you suspend a heavy chain from two points. What shape will the chain assume? The answer is a *catenary*. See Figure 19.4. Read about the catenary on the Web site `http://mathworld.wolfram.com/Catenary.html`. Huygens (1629–1695) was the first to discuss the catenary—in correspondence with Leibniz (1646–1716). If you roll a parabola along a straight line, its focus traces out a catenary. See Figure 19.5. Try this last experiment yourself to generate the curve.[2]

Figure 19.4. The catenary.

[2]In fact the famous St. Louis arch is in the shape of a catenary.

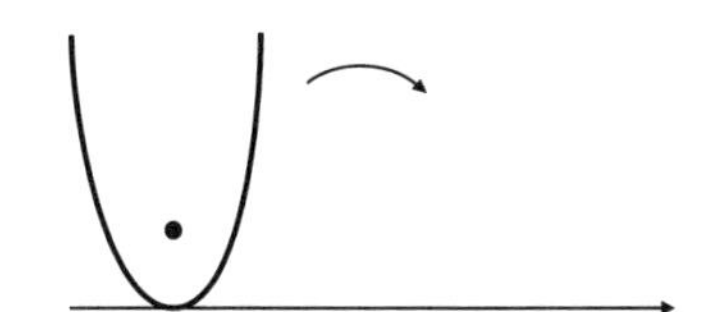

Figure 19.5. Rolling a parabola along a straight line.

11. **Project:** Find a solution to the differential equation

$$y' + 3y = x$$

by guessing a solution of the form $y = \sum_j a_j x^j$ and substituting into the equation. The book [SIK] has more on this technique.

12. **Project:** Probably the most famous and most-studied partial differential equation is Laplace's equation, or *the Laplacian*. In dimension 2, the equation takes the form

$$\Delta u = \left(\frac{\partial^2}{\partial x^2} + \frac{\partial^2}{\partial y^2} \right) u = f .$$

When the right-hand side of the equation is identically zero, the solutions of the equation are called *harmonic functions*. Harmonic functions are important for mathematical physics. For example, a steady-state heat distribution on the unit disk is given by a harmonic function.

Verify that the function

$$u(x, y) = x^2 - y^2$$

is harmonic. Verify that the function

$$v(x, y) = 2xy$$

is harmonic.

If f is any continuous function on the boundary of the unit disc (i.e., the circle), then there is a function u, continuous on the closure of the disc, so that u is harmonic on the interior and u agrees with f on the boundary. In the special case that $f(\cos t, \sin t) = \cos 2t$, say what this function u must be. Refer to [BEM] for background in this topic.

13. **Project:** Another famous and much-studied partial differential equation is the *heat equation*. It is given by

$$\frac{\partial}{\partial t} u = \frac{\partial^2}{\partial x^2} u .$$

Give an example of a solution of the heat equation on the upper halfplane $\mathcal{U} = \{(x, y) : y > 0\}$. Give an example of a solution of the heat equation on the upper halfplane that agrees with the function $f(x, 0) = x$ on the boundary of the upper halfplane. The book [BEM] is a good source for this topic as well. See [WID] for more advanced ideas.

20

Emmy Noether and the Birth of Modern Algebra

20.1 The Life of Emmy Noether

Emmy Amalie Noether (1882–1935) was the daughter of Max Noether, a distinguished German mathematician and professor at Erlangen. Emmy's mother was Ida Kaufmann, scion of a wealthy Cologne family. The family was Jewish, and Emmy had three male siblings. Her brother Fritz also made a career as a mathematician. The other two brothers died in childhood. The career of Emmy Noether suffered because of prejudices against women that were prevalent in her day. But the power of her ideas, and of her teaching, lives on. She was truly one of the outstanding woman mathematicians of all time. Her life story is told in [DIC] and [BSM].

Emmy Noether attended school in Erlangen, where she studied English, French, and arithmetic. She also took piano lessons. She took particular pleasure in dancing, and loved to attend parties with other university children. Her aim during her school days was to be a language teacher. Indeed, she ultimately took the examinations of the State of Bavaria and, in 1900, became a certified teacher of English and French for Bavarian girls' schools.

Emmy Noether never actually became a language teacher. She instead decided on an unusual and difficult course for a woman of her day: She decided to study university mathematics. In those days (the early twentieth century), women were allowed to attend the university unofficially. Professors had to give individual permission, on a case-by-case basis, for women to attend their courses. [We have seen similar phenomena in the context of Kovalevskaya's life.]

Emmy Noether attended courses at the University of Erlangen from 1900 to 1902. She passed the matriculation examination in Nürnberg in 1903 (she was allowed to matriculate in 1904), and then went on to the University of Göttingen—the premiere German institution for the study of mathematics. In the period 1903–1904, Emmy attended lectures in Göttingen by Blumenthal, Hilbert, Klein, and Minkowski. In 1907 Emmy Noether earned her doctorate under the direction of Paul Gordan. In her thesis she studied constructive approaches to the Hilbert basis theorem.[1]

[1] This is a theorem about generating sets for algebraic invariants. A great deal of mathematical effort was put into studying this problem in the late nineteenth century.

The ordinary career path for a *man* in Emmy Noether's position in 1907 would have been the *Habilitation* (which is a sort of apprenticeship under a full Professor, resulting in a second thesis). But this course was denied to a woman. So Emmy stayed in Erlangen assisting her father; because of his physical disabilities, Max Noether was grateful for her aid. But Emmy Noether was never paid for her work. Emmy continued to work on her own research, and was influenced by Professor Fischer, who was Paul Gordan's successor since 1911. Fischer succeeded in weaning Emmy away from Gordan's constructive approach to algebraic invariants, and instead instilled in her an appreciation for the new abstract approach.

When Emmy Noether's papers began to appear, her mathematical reputation was quickly established. In 1908 she was elected to the *Circolo Mathematico di Palermo*, and in 1909 she was invited to membership of the *Deutsche Mathematiker-Vereinigung*. Also in 1909 she was invited to address the annual meeting of the Scientific Society in Salzburg. In 1913 she gave an invited lecture in Vienna.

A real triumph for Emmy Noether was that, in 1915, David Hilbert and Felix Klein (1849–1925) invited her to return to Göttingen. They were working on developing some of Einstein's ideas, and felt that Noether's expertise in invariant theory would be valuable for their research program. Ultimately, her ideas about symmetry and conservation laws proved to be of seminal importance in the subject of relativity.

Hilbert and Klein convinced Noether to stay in Göttingen while they battled to have her officially on the faculty (for in those days women were not allowed). As late as 1919, Hilbert was arguing for Emmy's qualifications as a faculty member in Göttingen. His strongest opponents were the philologists and the historians. At one point he addressed the council of the university with these words: "I do not see why the sex of the candidate should be an argument against her appointment as Privatdozent; after all, we are not a bath-house ..."

Finally in 1919 Emmy Noether was allowed to obtain her *Habilitation* at the University of Göttingen. In the interim, Hilbert had allowed Emmy to lecture by advertising her courses under his own name. A typical example of Hilbert's subterfuge was a course that appeared in the 1916–1917 university catalog for the Winter semester as

> **Mathematical Physics Seminar:** Professor Hilbert, with the assistance of Dr. E. Noether, Mondays from 4:00 to 6:00, no tuition.

Eventually, Emmy Noether was granted permission to offer courses under her own name in Göttingen. At first she was not paid for her work, but at least she gained the recognition that her accomplishments merited. After several years, because of the efforts of Hilbert and Klein on her behalf, she finally received a small stipend.

Emmy Noether had quite a following among the students in Göttingen. Indeed, her crowd of acolytes became known as "Noether's boys". They traveled from as far away as Russia to study with her. Emmy was a warm and caring person who, unlike most German professors, was willing to listen to students' personal problems as well as their mathematical problems. She considered her students to be like her family, and treated them accordingly. She was particularly good at planting ideas in the minds of her eager young followers, and she produced a number of excellent Ph.D. students, among them the noted algebraist Ernst Witt (1911 C.E.–1991 C.E.).

The first piece of scientific work that Emmy Noether produced in Göttingen was a paper in 1915 that proved a relationship between symmetries in physics and certain conservation principles. Einstein himself praised this result, and in a letter to Hilbert spoke of Noether's "penetrating mathematical thinking". Apparently Emmy's expertise in invariants led her to formulate several important concepts for Einstein's general theory of relativity.

After 1919, Emmy Noether spent her time in Göttingen working on ring theory, particularly the theory of ideals in rings. She developed an abstract set of ideas that helped to make ring theory into a cornerstone of modern algebra. The "ascending chain" condition that she formulated has led to certain rings being called *Noetherian rings*. Her paper *Idealtheorie in Ringbereichen*, which appeared in 1921, gave a decomposition theorem for ideals that has proved influential to this day.

In 1924, the Dutch mathematician B. L. van der Waerden came to Göttingen to study with Noether. When he returned to Amsterdam he wrote his seminal two-volume book *Moderne Algebra*. This book has become one of the cornerstones of modern algebra, and is still used today. The second volume of that important work is based on ideas of Emmy Noether.

Emmy Noether had important collaborations with Helmut Hasse and Richard Brauer. She had considerable influence on her students and colleagues. Indeed, many of her important ideas appear in the work of others, without any attribution to her.

Emmy Noether was invited to address the International Congress of Mathematicians, both in 1928 and 1932. In 1932 she received, jointly with Emil Artin, the Alfred Ackermann-Teubner Memorial Prize for the Advancement of Mathematical Knowledge. In 1930 she was a visiting Professor both at the University of Moscow and the University of Frankfurt.

In 1933 the Nazis dismissed Emmy Noether from her post in Göttingen—just because she was Jewish. Her brother Fritz, who was also a professor at the time, moved to Siberia. Emmy could have moved to Moscow, but she decided to move to the United States. She was able to land a visiting faculty position at Bryn Mawr College outside Philadelphia, and she also spent time at the Institute for Advanced Study in Princeton.

Teaching at a women's college like Bryn Mawr was a new experience for Emmy Noether. For the first time she had *colleagues* who were women. The Head of the Mathematics Department at the time was Anna Pell Wheeler. She became a great friend of Emmy's. Wheeler had full knowledge, and truly understood, the struggles of a woman mathematician in the German system. She also empathized with the difficulties of being uprooted from her homeland. Even in her new environment at Bryn Mawr, Emmy Noether was a caring and compassionate teacher. Sometimes, when she had trouble getting her ideas across, she would lapse into German. But the students loved her.

Emmy Noether died young, of complications from uterine cancer. Only her good friend Anna Wheeler knew of the illness. Noether's ashes are buried near Bryn Mawr's library. She is memorialized by a coeducational high school in mathematics that is named after her.

Emmy Noether is remembered glowingly both for her deep mathematical insights and for her inspiring work with students and colleagues. She was particularly, and almost uniquely, comfortable with abstraction in algebra. Hermann Weyl said, in his Memorial Address:

> Her significance for algebra cannot be read entirely from her own papers, she had great stimulating power and many of her suggestions took shape only in the works of her pupils and co-workers.

B. L. van der Waerden wrote

> For Emmy Noether, relationships among numbers, functions, and operations became transparent, amenable to generalisation, and productive only after they have been dissociated from any particular objects and have been reduced to general conceptual relationships.

Further Reading

Clark H. Kimberling, Emmy Noether, *The American Mathematical Monthly* 79(1972), 136–149.

Edith H. Luchins, Sex differences in mathematics: How not to deal with them, *The American Mathematical Monthly* 86(1979), 161–168.

20.2 Emmy Noether and Abstract Algebra: Groups

The most basic algebraic structure is the *group*. A group is a collection of objects (i.e., a set) equipped with a binary operation that we usually think of as addition (denoted $+$) or multiplication (denoted $\cdot$). We require these properties of the operation:

(1) The operation is associative: $x \cdot (y \cdot z) = (x \cdot y) \cdot z$ (or $x + (y + z) = (x + y) + z$) for all x, y, z in the group.

(2) There is a *multiplicative/additive identity* element e (or 0) which satisfies $e \cdot x = x \cdot e = x$ (or $0 + x = x + 0 = x$) for every element x in the group.

(3) For each element x in the group there is a multiplicative/additive inverse x^{-1} (or $-x$) which satisfies $x \cdot x^{-1} = x^{-1} \cdot x = e$ (or $x + (-x) = (-x) + x = 0$).

An essential point here is that when we combine two group elements using the binary operation, then the result is also a group element. This property is called *closure* of the group under the binary operation.

We usually denote a group with the symbol G. As already noted, the binary operation is denoted either by $+$ (in case the group is commutative) or $\cdot$. The examples below will make this notation clear.

Groups were first invented by Evariste Galois (1811–1832) and Augustin Cauchy (1789–1857), but it was Noether and others who later developed the subject of group theory into full bloom.

An abstract concept like "group" is best understood by way of examples. We now provide several.

EXAMPLE 20.1. Let G be the set $\mathbb{Z}$ of all integers and let the binary operation be ordinary arithmetic addition, denoted by $+$. Certainly, as we know from our past experience, addition is associative. The additive identity is the number 0. For if $x \in G$ is any integer then $x + 0 = 0 + x = x$.

Finally, if x is any element of this group, then $-x$ is its additive inverse. That is to say, $x + (-x) = (-x) + x = 0$. For example, the number 5 is in our group. Its additive inverse is -5. Likewise, the number -3 is in our group. Its additive inverse is 3.

EXAMPLE 20.2. Let G be the collection of all positive rational numbers. Let the binary operation be multiplication, denoted by $\cdot$. Certainly multiplication is associative, so we shall say no more about that property. The multiplicative identity is the number 1. That is to say, if $x \in G$ then $1 \cdot x = x \cdot 1 = x$.

Finally, if x is any element of this group, then $1/x$ (the reciprocal) is the multiplicative inverse. That is to say, $x \cdot (1/x) = (1/x) \cdot x = 1$. For example, the number $3/4$ is in our group. Its multiplicative inverse is $4/3$. Likewise, the number $11/3$ is in our group. Its multiplicative inverse is $3/11$.

EXAMPLE 20.3. Let G be the collection of positive integers and let the binary operation be multiplication. Certainly this operation is associative. And the multiplicative identity is 1. However, this G is *not* a group. For example, the number $7 \in G$ does not have a multiplicative inverse (it ought to be $1/7$, but the number $1/7$ does not lie in G).

EXAMPLE 20.4. Consider the collection of 2×2 matrices. These are displays of the form

$$M = \begin{pmatrix} a & b \\ c & d \end{pmatrix}.$$

A matrix is said to be *nonsingular* if $ad - bc \neq 0$. We will consider how to make the collection of nonsingular 2×2 matrices into a group.

First, we multiply two such matrices

$$\begin{pmatrix} a & b \\ c & d \end{pmatrix} \cdot \begin{pmatrix} a' & b' \\ c' & d' \end{pmatrix}$$

by the following rule. The upper left entry of the product is obtained from the componentwise products of the elements of the first row in the first matrix and the first column in the second matrix:

$$\text{product} = \begin{pmatrix} aa' + bc' & * \\ * & * \end{pmatrix}.$$

The upper right entry of the product is obtained from the componentwise products of the elements of the first row in the first matrix and the second column in the second matrix:

$$\text{product} = \begin{pmatrix} * & ab' + bd' \\ * & * \end{pmatrix}.$$

The lower left entry of the product is obtained from the componentwise products of the elements of the second row in the first matrix and the first column in the second matrix:

$$\text{product} = \begin{pmatrix} * & * \\ ca' + dc' & * \end{pmatrix}.$$

Finally, the lower right entry of the product is obtained from the componentwise products of the elements of the second row in the first matrix and the second row in the second matrix:

$$\text{product} = \begin{pmatrix} * & * \\ * & cb' + dd' \end{pmatrix}.$$

As an example of matrix multiplication, we have

$$\begin{pmatrix} 3 & -5 \\ -2 & 1 \end{pmatrix} \cdot \begin{pmatrix} -4 & 6 \\ 5 & 9 \end{pmatrix} = \begin{pmatrix} 3 \cdot (-4) + (-5) \cdot 5 & 3 \cdot 6 + (-5) \cdot 9 \\ (-2) \cdot (-4) + 1 \cdot 5 & (-2) \cdot 6 + 1 \cdot 9 \end{pmatrix}$$

$$= \begin{pmatrix} -37 & -27 \\ 13 & -3 \end{pmatrix}.$$

Thus we have a binary operation. It is associative, as can be checked by a tedious calculation. The multiplicative identity is the matrix

$$I = \begin{pmatrix} 1 & 0 \\ 0 & 1 \end{pmatrix}.$$

In fact, for any matrix

$$M = \begin{pmatrix} a & b \\ c & d \end{pmatrix}$$

we have

$$I \cdot M = \begin{pmatrix} 1 & 0 \\ 0 & 1 \end{pmatrix} \cdot \begin{pmatrix} a & b \\ c & d \end{pmatrix}$$

$$= \begin{pmatrix} 1 \cdot a + 0 \cdot c & 1 \cdot b + 0 \cdot d \\ 0 \cdot a + 1 \cdot c & 0 \cdot b + 1 \cdot d \end{pmatrix}$$

$$= \begin{pmatrix} a & b \\ c & d \end{pmatrix}$$

$$= M .$$

The issue of multiplicative inverse is a bit more complex. If

$$M = \begin{pmatrix} a & b \\ c & d \end{pmatrix}$$

is a 2×2 matrix such that $D = ad - bc \neq 0$, then we set

$$M^{-1} = \begin{pmatrix} d/D & -b/D \\ -c/D & a/D \end{pmatrix}.$$

Notice that

$$(d/D) \cdot (a/D) - (-c/D) \cdot (-b/D) = [ad - bc]/D \neq 0 .$$

So the new matrix M^{-1} is still an element of the group. Furthermore, we invite the reader to check by hand that

$$M \cdot M^{-1} = M^{-1} \cdot M = I .$$

So, indeed, M^{-1} is a multiplicative inverse.

There are infinitely many examples of groups. An interesting feature of the last example (the 2×2 matrices) is that it is *not* commutative. That is to say, $M \cdot N \neq N \cdot M$ in general.

What is important about an abstract structure like "group" is that we can deal with all groups at once and therefore simultaneously establish properties for many different collections of objects. We now provide just one simple example:

Proposition 20.1. *If G is a group and $x, y \in G$, then*

$$(xy)^{-1} = y^{-1} \cdot x^{-1} .$$

Proof. What we are claiming is that the multiplicative inverse of xy is given by $y^{-1}x^{-1}$. We may verify this claim directly:

$$\begin{aligned}(y^{-1}x^{-1}) \cdot (xy) &= y^{-1} \cdot [x^{-1} \cdot (xy)] \\ &= y^{-1} \cdot [(x^{-1} \cdot x] \cdot y] \\ &= y^{-1} \cdot [e \cdot y] \\ &= y^{-1} \cdot y \\ &= e ,\end{aligned}$$

just as was claimed. Notice that, in the first two equalities, we used the associative property of group multiplication. In the next equality we used the definition of multiplicative inverse. In the next we used the defining property of the multiplicative identity. And in the last we used the definition of multiplicative inverse.

We leave it to the reader to check that $(xy) \cdot (y^{-1}x^{-1}) = e$. □

Further Reading

Jan F. Andrus and Alton T. Butson, Ordered groups, *The American Mathematical Monthly* 70(1963), 619–628.

Roy Dobyns, A condition equivalent to associativity for finite groups, *The Two-Year College Mathematics Journal* 3(1972), 10–13.

W. G. Leavitt, An algorithm and its connection with abelian groups, *The Two-Year College Mathematics Journal* 7(1976), 16–21.

Gary L. Walls, Automorphism groups, *The American Mathematical Monthly* 93(1986), 459–462.

20.3 Emmy Noether and Abstract Algebra: Rings

While Emmy Noether made contributions to all parts of modern algebra, she is particularly remembered for her ideas about ring theory. In the present section we shall discuss rings and give a number of examples.

A *ring* is a collection of objects (a set) with two binary operations: addition and multiplication. We require that the operation of addition induce a commutative group. Of course we require that the ring be closed under addition and multiplication. We require that multiplication be associative. Finally, we require that there be a distributive law:

$$x \cdot (y + z) = x \cdot y + x \cdot z .$$

Rather than engage in extensive discussion of the formalities of the definition of a ring, we instead concentrate on some examples. Note that we do *not* require that the ring have a commutative multiplication operation—see Example 20.7.

EXAMPLE 20.5. Consider R the set of all integers. The addition operation will be the usual arithmetic notion of addition, and likewise the multiplication operation will be the usual arithmetic operation of multiplication.

Certainly, as we have already noted in Example 20.1, the integers form a group under addition. Obviously multiplication is associative. And we have the distributive law

$$x \cdot (y + z) = x \cdot y + x \cdot z\,.$$

Thus R is a ring.

EXAMPLE 20.6. Consider R the set of all polynomials with real number coefficients. Such a polynomial has the form

$$p(x) = a_0 + a_1 x + a_2 x^2 + \cdots + a_k x^k\,.$$

Each coefficient a_j here is a real number.

The addition operation is the ordinary algebraic notion of addition of polynomials. For example,

$$(3 - 2x + x^2) + (5 + 6x^2 + 9x^3) = 8 - 2x + 7x^2 + 9x^3\,.$$

The multiplication operation is the ordinary algebraic operation of multiplication of polynomials. For example,

$$(3 - 2x + x^2) \cdot (5 + 6x^2 + 9x^3) = 9x^5 - 12x^4 + 15x^3 + 23x^2 - 10x + 15\,.$$

Clearly R is closed under these two operations. The distributive law holds. Thus R is a ring.

EXAMPLE 20.7. Consider R the set of all 2×2 matrices with real number entries. The addition operation is ordinary componentwise addition of matrices:

$$\begin{pmatrix} a & b \\ c & d \end{pmatrix} + \begin{pmatrix} a' & b' \\ c' & d' \end{pmatrix} = \begin{pmatrix} a + a' & b + b' \\ c + c' & d + d' \end{pmatrix}.$$

The multiplication operation is just as we defined it in Example 20.4. Then R forms a ring.

20.3.1 The Idea of an Ideal

Emmy Noether was particularly noted for her contributions to the theory of ideals. Let us say a few words about what an ideal is and how we can recognize an ideal.

Let R be a ring. An *ideal* is the collection of elements in R obtained by taking a subcollection $L \subseteq R$ and considering all expressions of the form

$$a_1 \ell_1 + a_2 \ell_2 + \cdots + a_k \ell_k\,,$$

where the ℓ_j are elements of L and the a_j are arbitrary elements of R. It is clear that an ideal is closed under addition and multiplication. It is also closed under multiplication on the left by any element of R. We call L the *generating set* for the ideal. We will frequently denote an ideal with the letter $\mathcal{I}$.

Put in slightly different terms, an ideal in R is a subset that is closed under addition and closed under multiplication on the left by *any* element of R.

EXAMPLE 20.8. Let R be the ring of all polynomials in the variable x, with coefficients taken from the real numbers. Then the set of all multiples of x^2 forms an ideal $\mathcal{I}$.

Observe that this ideal may be described as the set of all polynomials with no constant term and no linear term. In other words, it is the set of all polynomials of the form

$$p(x) = a_2x^2 + a_3x^3 + \cdots + a_kx^k .$$

This set $\mathcal{I}$ is obviously closed under addition and multiplication. It is also clear that if $p \in \mathcal{I}$ and $r \in R$, then $r \cdot p \in \mathcal{I}$. For example, let $p(x) = x^2 - x^3 + 4x^4$ and $r(x) = 3 + 5x - x^2$. Then $p \in \mathcal{I}$ and r is an arbitrary element of R. Thus

$$p(x) \cdot r(x) = 3x^2 + 2x^3 + 6x^4 + 21x^5 - 4x^6 .$$

Clearly this product is an element of $\mathcal{I}$.

EXAMPLE 20.9. Let R be the ring of all integers. Let $\mathcal{I}$ be the ideal generated by $L = \{6, 15\}$. Then the ideal consists of all integers of the form

$$m = 6 \cdot a + 15 \cdot b ,$$

where a and b are integers.

We notice that 3 divides both 6 and 15. More importantly, 3 is the greatest common divisor of 6 and 15, so we may express 3 in terms of 6 and 15:

$$3 = 1 \cdot 15 + (-2) \cdot 6 .$$

Thus $3 \in \mathcal{I}$. So in fact the ideal may be described more elegantly as the one that is generated by $\widetilde{L} = \{3\}$.

For You to Try The ring of integers is an important example of what is known as a *principal ideal domain*. This means that each ideal in $\mathbb{Z}$ is generated by just one element. As an exercise, the reader should determine the single generator for the ideal generated by the set $L = \{14, 21, 35\}$.

EXAMPLE 20.10. Let R be the set of polynomials in the two variables x, y with real variable coefficients. Consider the ideal $\mathcal{I}$ generated by $\{x, y\}$. This is simply the set of polynomials of the form

$$p(x, y) = a_{10}x + a_{01}y + a_{11}xy + a_{21}x^2y + a_{12}xy^2 + \cdots + a_{mn}x^my^n .$$

In other words, it is the set of polynomials with no constant term.

It is clear that $\mathcal{I}$ is closed under addition and multiplication. It is also obvious that if $p \in \mathcal{I}$ and $r \in R$, then $r \cdot p \in \mathcal{I}$. But notice that there is no single polynomial that will generate the entire ideal $\mathcal{I}$. It requires a minimum of two elements (such as x and y) to generate $\mathcal{I}$. Thus R is not a principal ideal domain.

The set of ideals in a ring tell us a great deal about the structure of that ring. The subject of algebraic geometry, which concerns itself with the zero sets of polynomials, is studied today largely with an algebraic language. That language centers about rings and ideals.

Particularly important among ideals are maximal ideals. An ideal $\mathcal{I}$ in a ring R is called *maximal* if it is *not* contained in any other proper ideal that is distinct from $\mathcal{I}$. It turns out—and this can be proved using Zorn's lemma—that any ideal is contained in some maximal ideal.

EXAMPLE 20.11. Let R be the ring of polynomials in the single variable x with real coefficients. Let $\mathcal{I}$ be the ideal generated by x^2. This $\mathcal{I}$ is *not* a maximal ideal, but it is contained in the maximal ideal generated by x.

In the ring of integers, the ideal generated by 4 and 6 is contained in the maximal ideal generated by 2—in fact it is the very same ideal.

Further Reading

Pasquale J. Arpaia, Rings, subrings, identities and homomorphisms, *The Two-Year College Mathematics Journal* 5(1974), 25–28.

Irl C. Bivens, Reward of the rings, *The College Mathematics Journal* 22(1991), 418–420.

P. M. Cohn, Rings of fractions, *The American Mathematical Monthly* 78(1971), 596–615.

Desmond MacHale, Commutativity in finite rings, *The American Mathematical Monthly* 83(1976), 30–32.

J. B. Wilker, Rings of sets are really rings, *The American Mathematical Monthly* 89(1982), 211.

Exercises

1. Consider the set of all 2×2 matrices with integer entries. Verify that this set forms a group under matrix addition.
2. Consider the set of all 2×2 matrices with integer entries. Verify that this set *does not* form a group under matrix multiplication. Which group property fails? Discuss this problem in class.
3. Consider the set of all 2×2 matrices with integer entries. We learned in Exercise 2 that this set does not form a group under matrix multiplication. But now restrict attention to those matrices which are nonsingular. Does this help? Now you can form the multiplicative inverse of any matrix. But something new goes wrong. What is it?
4. Refer to Exercise 3. Consider now the set of all 2×2 matrices with rational number entries and which are nonsingular. This *does* form a group under multiplication. Discuss this matter in class, and verify the assertion.
5. Consider the set of all 3×3 matrices of the form

$$\begin{pmatrix} 1 & x & z \\ 0 & 1 & y \\ 0 & 0 & 1 \end{pmatrix},$$

where x, y, and z are real numbers. Verify that this set is a group under multiplication.

6. Consider the ring of all 2×2 matrices with real coefficients. Verify that the set of all matrices of the form
$$\begin{pmatrix} 0 & x \\ 0 & 0 \end{pmatrix}$$
does *not* form an ideal. What goes wrong? Which property fails?

7. Consider the ring of all 2×2 matrices with integer coefficients. Verify that the set of all matrices of the form
$$\begin{pmatrix} 2m & 2n \\ 2k & 2\ell \end{pmatrix},$$
for m, n, k, ℓ integers, forms an ideal.

8. The collection of all polynomials with real coefficients in the single variable x forms a principal ideal domain. To illustrate this point, find the single generator for the ideal generated by $x + 1$, $x^2 - 4$, and $x^3 + x$. This ideal is special. Why is that?

9. Consider the ideal—in the ring $\mathbb{Z}$ of integers—of all integers that are divisible by both 2 and 3. Why is this an ideal? What single number will generate this ideal? Is the generator unique? Can you find more than one generator?

10. **Project:** Let G be the collection of 2×2 matrices with real entries, with rows that are unit vectors which are orthogonal, and having determinant 1. Show that G is a group when equipped with ordinary matrix multiplication as its binary operation. See [HER] for the fundamental ideas.

11. **Project:** Let R be the collection of 2×2 matrices with real entries, having a zero in the lower left corner. Let the binary operations be ordinary matrix addition and ordinary matrix multiplication. Show that this R is a ring. Give an example of a nontrivial ideal in this ring. Give an example of a maximal ideal. Again see [HER].

21

Methods of Proof

Part and parcel of the modern method of doing mathematics is *proofs*. It is quite amazing, really, because proofs are quite foreign to everyday discourse. Listen to politicians bickering, or to religious fanatics arguing, or even to discussions in your own family. There is more emotion and less reason than any of us would like to admit.

Mathematics is different. The entire subject hinges on logic. We cannot begin any discussion without first defining our terms. After we define the terms then we set up certain rules or "axioms". After that, we *prove* that certain relations or facts are true. This is how the discourse of mathematics proceeds. It is a rigorous discipline, and one that is rock-solid in its reliability and reproducibility. Throughout this book we have used proofs to establish the various mathematical truths that we have studied.

It is a fairly recent development—from the past two hundred years—that generally accepted methods for mathematical proof have been established. And the methodology and language for writing those proofs down has been rigorized and stabilized and generally accepted. A mathematical proof generated in France today will be just like a mathematical proof created in Japan or Italy or in the United States. This common language makes mathematics an international language, and a unifying enterprise for all peoples.

In fact it was the Pythagoreans, in the time of ancient Greece, who first insisted on the role of rigorous proof in mathematics. Prior to their time, mathematical assertions were "established" by way of plausibility arguments and examples and pictures. Euclidean geometry—with its axiomatic system, strict format for statements of theorems, and highly structured proofs—developed this important idea and created a template for how modern mathematics is studied. Euclid was the father of the axiomatic method and the systematic use of rigorous proof that follows a strict paradigm of formal logic.

In modern times—beginning in the seventeenth century let us say—the tradition of establishing mathematical facts by way of rigorous proof lived on. But there was little agreement on what constituted a rigorous proof, and there was no established format or mechanism for recording proofs.

At the beginning of the twentieth century it is safe to say that the heartbeat of mathematics was in France and Germany. In each of these countries, movements began to put mathematics on a more rigorous footing. In Germany, David Hilbert and his school observed that Euclidean geometry, number theory, and other parts of mathematics were a chaotic mish-mash: it was difficult to sort out the assumptions from the theorems. In France, under the guidance of André Weil and others, a different sort of movement arose.

In the mid-1930's, a cabal of French mathematicians was formed with the purpose of writing definitive texts in the basic subject areas of mathematics. They ultimately decided to publish their books under the *nom de plume* Nicolas Bourbaki. In fact the inspiration for their name was an obscure French general named Charles Denis Sauter Bourbaki. This general, so it is told, was once offered the chance to be King of Greece but (for unknown reasons) he declined the honor. Later, after suffering an embarrassing retreat in the Franco-Prussian War, Bourbaki tried to shoot himself in the head—but he missed. He was quite the buffoon, and the authors of these mathematical texts decided that he was the perfect foil for their purposes.

In fact the founding mathematicians in this group—André Weil (1906–1998), Jean Dieudonné (1906–1992), Jean Delsarte (1903–1968), Henri Cartan (1904–2009), Claude Chevalley (1909–1984), and some others—came from the tradition of the École Normale Supérieure. This is perhaps the most elite university in all of France, but it also has a long-standing tradition of practical joking. Weil himself tells of one particularly delightful story. In 1916, Paul Painlevé (1863–1933) was a young and extremely brilliant Professor at the Sorbonne. He was also an examiner for admission to the École Normale Supérieure. Each candidate for admission had to undergo a rigorous oral exam, and Painlevé was on the committee. So the candidates came early in the morning and stood around the hall outside the examination room awaiting their turn. On one particular day, some of the more advanced students of the École began to chat with the novices. They told the youngsters about the fine tradition of practical joking at the school. They said that one of the standard hoaxes was that some student would impersonate an examiner, and then ridicule and humiliate the student being examined. The students should be forewarned.

Armed with this information, one of the students went in to take the exam. He sat down before the extremely youthful-looking Painlevé and blurted out, "You can't put this over on me!" Painlevé, bewildered, replied, "What do you mean? What are you talking about?" So the candidate smirked and said, "Oh, I know the whole story, I understand the joke perfectly, you are an impostor." The student sat back with his arms folded and waited for a reply. And Painlevé said, "I'm Professor Painlevé, I'm the examiner, ..."

Things went from bad to worse. Finally Painlevé had to go ask the Director of the École Normale to come in and vouch for him.

When André Weil used to tell this story, he would virtually collapse in hysterics.

In any event, we have Hilbert and Bourbaki to thank for our modern notion of mathematical rigor, and for the modern paradigm of what a proof should be. Today there is little doubt of what constitutes a correct mathematical proof, or what are the proper modes of mathematical discourse.

In the present chapter we shall become acquainted with some of the most standard methods of mathematical proof. Along the way, we shall learn a number of interesting mathematical facts and some important mathematical techniques.

Further Reading

Kenneth Appel and Wolfgang Haken, The nature of proof: limits and opportunities, *The Two-Year College Mathematics Journal* 12(1981), 118–119.

David Perkins, A serendipitous proof, *The College Mathematics Journal* 34(2003), 359–361.

R. L. Wilder, The nature of mathematical proof, *The American Mathematical Monthly* 51(1944), 309–323.

21.1 Axiomatics

21.1.1 Undefinables

The basic elements of mathematics are "undefinables". Since every new piece of terminology is defined in terms of old pieces of terminology, we must begin with certain terms that have no definition. Most commonly, the terms "set" and "element of" are taken to be undefinables. We simply say that a set S is a collection of objects and x is an element of S if it is one of those objects. We denote a set with a capital roman letter like S or T or U and we write $x \in S$ to mean that x is an element of S.

21.1.2 Definitions

From this beginning, we formulate more complex definitions. For example, if A and B are sets, then we can define $A \times B$ to be all ordered pairs (a, b) such that $a \in A$ and $b \in B$. Of course, this presupposes that we have defined $\in$ ("element of") and "ordered pair." Then we can define a function from A to B to be a certain type of subset of $A \times B$. And so forth.

21.1.3 Axioms

Once a collection of definitions is put in place, then we can formulate axioms. An *axiom* is a statement whose truth we take as given. The axiom uses terminology that consists of undefinables plus terms introduced in the definitions. An axiom usually has a subject, a verb, and an object. For example, a famous axiom from Euclidean geometry[1] says

> For each line ℓ and each point $\mathbf{P}$ that does not lie on ℓ there is a unique line ℓ' through $\mathbf{P}$ such that ℓ' is parallel to ℓ.

Figure 21.1 illustrates this postulate (the famous *Parallel Postulate*, in Playfair's formulation) of classical geometry.

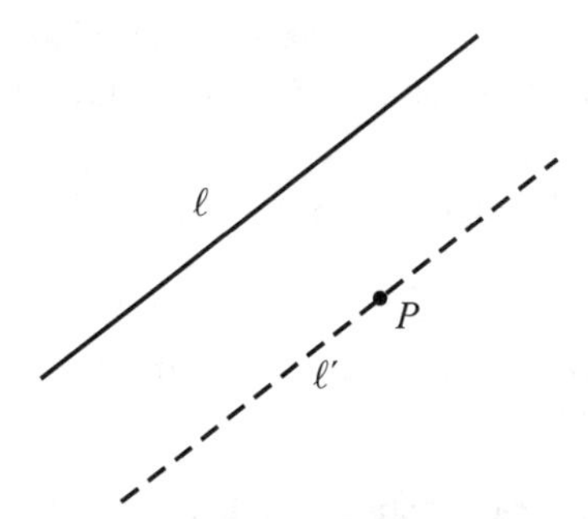

Figure 21.1. The Parallel Postulate.

[1] This axiom is known as the "Parallel Postulate", and was the subject of intense study for over 2000 years. We discussed the Parallel Postulate in Chapter 1.

Following the spirit of Occam's Razor,[2] we generally strive to have as few axioms as possible. Euclidean geometry has just five axioms. Group theory has three axioms, and field theory has eleven axioms. The natural numbers have five axioms. There are eight axioms for the real numbers.

21.1.4 Theorems, *Modus Ponendo Ponens, and Modus Tollens*

Next, we begin to formulate theorems. A theorem is a statement that we derive from the axioms using rules of logic. There is really only one fundamental rule of logic, and it is this:

modus ponendo ponens:

If **A** and (**A** $\Rightarrow$ **B**), then **B**.

Although the terminology is less frequently encountered in the literature, some books refer to a complementary rule called *modus tollendo tollens*. It is the contrapositive form of *modus ponendo ponens*:

modus tollendo tollens:

If **B** and ($\sim$ **A** $\Rightarrow \sim$ **B**), then **A**.

Even though there is only one rule of logic, there are several different proof strategies. The purpose of this chapter is to enunciate, discuss, and illustrate some of the most prominent and useful of these.

Further Reading

H. C. Kennedy, The origins of modern axiomatics: Pasch to Peano, *The American Mathematical Monthly* 79(1972), 133–136.

George Polya, Guessing and proving, *The Two-Year College Mathematics Journal* 9(1978), 21–27.

R. L. Wilder, The role of the axiomatic method, *The American Mathematical Monthly* 74(1967), 115–127.

21.2 Proof by Induction

The word "induction" is used in ordinary parlance to describe any method of inference. In mathematics it has a very specific meaning, which is summarized as follows.

21.2.1 Mathematical Induction

We remind the reader of the paradigm for mathematical induction:

Mathematical Induction: For each $n \in \mathbb{N}$, let $P(n)$ be a statement. If

(1) $P(1)$ is true;

(2) $P(j) \Rightarrow P(j+1)$ for every natural number j;

[2]This is an old tenet of philosophy posited in the fourteenth century (by William of Occam (1288 C.E.–1348 C.E.)). It asserts that we should, in the interest of elegance and precision, work with as few definitions and hypotheses as possible.

then $P(n)$ is true for every n.

The method of induction is best understood by way of several examples.

21.2.2 Examples of Inductive Proof

EXAMPLE 21.1. Prove that, if n is a positive integer, then

$$1 + 2 + \cdots + n = \frac{n(n+1)}{2}.$$

Proof. Let $P(n)$ be the statement

$$\mathbf{P(n):}\ 1 + 2 + \cdots + n = \frac{n(n+1)}{2}.$$

Then $P(1)$ is the simple equation

$$1 = \frac{1 \cdot 2}{2}.$$

This is certainly true, so we have established step **(1)** of the induction process.

The second step is the more subtle. We *assume* $P(j)$, which is

$$1 + 2 + \cdots + j = \frac{j(j+1)}{2}, \qquad (*_j)$$

and we use it to *prove* $P(j+1)$, which is

$$1 + 2 + \cdots + (j+1) = \frac{(j+1)(j+2)}{2}. \qquad (*_{j+1})$$

To accomplish this goal, we add the quantity $(j+1)$ to both sides of $(*_j)$. Thus we have

$$[1 + 2 + \cdots + j] + (j+1) = \frac{j(j+1)}{2} + (j+1). \qquad (\dagger)$$

The right-hand side can be simplified as follows:

$$\frac{j(j+1)}{2} + (j+1) = \frac{j^2 + j + (2j+2)}{2}$$
$$= \frac{j^2 + 3j + 2}{2} = \frac{(j+1)(j+2)}{2}.$$

As a result of this calculation, we may rewrite $(\dagger)$ as

$$1 + 2 + \cdots + (j+1) = \frac{(j+1)(j+2)}{2}.$$

But this is precisely $(*_{j+1})$.

We have assumed $P(j)$ and used it to prove $P(j+1)$. That is step **(2)** of the induction method. Our proof is complete. □

EXAMPLE 21.2. Prove that, for any positive integer n, the quantity $n^2 + 3n + 2$ is even.

Proof. Obviously the statement $P(n)$ must be

P(n): The quantity $n^2 + 3n + 2$ is even.

Observe that $P(1)$ is the assertion that $1^2 + 3 \cdot 1 + 2 = 6$ is even. That is obviously true.

Now assume $P(j)$ (i.e., that $j^2 + 3j + 2$ is even). We must use this hypothesis to prove $P(j+1)$ (i.e., that $(j+1)^2 + 3(j+1) + 2$ is even). Now

$$\begin{aligned}(j+1)^2 + 3(j+1) + 2 &= (j^2 + 2j + 1) + (3j + 3) + 2 \\ &= [j^2 + 3j + 2] + [2j + 4] \\ &= [j^2 + 3j + 2] + 2[j + 2].\end{aligned}$$

The number $j^2 + 3j + 2$ is even by the inductive hypothesis $P(j)$. And $2[j+2]$ is even since it is a multiple of 2. The sum of two even numbers is even, because each will be a multiple of 2. So we see that $(j+1)^2 + 3(j+1) + 2$ is even. That establishes $P(j+1)$, assuming $P(j)$. The inductive proof is complete. □

The next example illustrates a mathematical device, due to Peter Gustav Lejeune Dirichlet (1805–1859), known as the the pigeonhole principle. In early days it was known as the *Dirichletscher Schubfachschluss*. Refer to our discussion of this idea in Section 14.2.

EXAMPLE 21.3. Prove that if $n + 1$ letters are placed in n mailboxes, then some mailbox will contain (at least) two letters.

Proof. Let $P(n)$ be the statement

P(n): If $n + 1$ letters are put in n mailboxes, then some mailbox will contain (at least) two letters.

Then $P(1)$ is the simple assertion that if two letters are placed in one mailbox, then some mailbox contains at least two letters. This is trivial: there is just one mailbox and it indeed contains two letters.

Now we suppose that $P(j)$ is true and we use that statement to prove $P(j+1)$. Now suppose that $j + 2$ letters are placed into $j + 1$ mailboxes. There are three possibilities:

- If the last mailbox contains no letter, then all of the letters actually go into the first j mailboxes. And there are $j + 2$ such letters (even more than $j + 1$). So the inductive hypothesis $P(j)$ applies. Therefore some mailbox contains at least two letters.
- If the last mailbox contains only one letter, then $j + 1$ letters have gone into the first j mailboxes, and the inductive hypothesis $P(j)$ applies. So some mailbox contains at least two letters.
- If the last mailbox contains (at least) two letters, then we have identified a box with two letters.

Thus, by breaking the proof into three cases, we have established $P(j+1)$ (assuming $P(j)$). The proof is complete. □

If S is a set, then we call T a *subset* of S if T is also a set and each element of T is also an element of S. For example, if $S = \{1, 3, 5, 7\}$ then $T = \{3, 5\}$ is a subset of S.

EXAMPLE 21.4. Let us prove that a set with n elements has 2^n subsets.

Proof. Our inductive statement is

P(n): $\boldsymbol{P(n)}$ **:** A set with n elements has
2^n subsets.

Now $P(1)$ is clearly true: A set with one element has $2 = 2^1$ subsets, namely the empty set and the set itself.

Suppose inductively that $P(j)$ has been established. Consider a set $A = \{a_1, \ldots, a_{j+1}\}$ with $j + 1$ elements. Write $A = \{a_1, \ldots, a_j\} \cup \{a_{j+1}\} \equiv A' \cup \{a_{j+1}\}$. Now A' is a set with j elements, so the inductive hypothesis applies to it. The set A' therefore has 2^j subsets. These are also, of course, subsets of A. The additional subsets of A are obtained by adjoining the element a_{j+1} to each of the subsets of A'. That gives 2^j more subsets of A, for a total of $2^j + 2^j$ subsets.

We conclude that A has $2^{j+1} = 2^j + 2^j$ subsets. That completes the inductive step, and the proof. □

For You to Try Use induction to prove that the number of different orderings of n objects is $n!$

For You to Try Use induction to prove the Euclidean algorithm: That if n is a positive integer and $0 < k < n$ is another integer, then

$$n = k \cdot q + r ,$$

where q is the quotient and $r < n$ is the remainder.

Further Reading

Karl David, An odd induction proof, *The College Mathematics Journal* 15(1984), 251.
W. L. Duren, Jr., Mathematical induction in sets, *The American Mathematical Monthly* 64(1957), 19–22.
Leon Henkin, On mathematical induction, *The American Mathematical Monthly* 67(1960), 323–338.

21.3 Proof by Contradiction

Proof by contradiction is predicated on the classical "law of the excluded middle"—an idea that goes back to Aristotle.[3] The substance of the proof strategy is that an idea is either true or false. There is no "middle" status. With this premise in mind, we can prove that something is true by excluding the possibility that it is false. The way that we exclude the possibility that the assertion is false is to assume it is false and show that such an assumption leads to an untenable position (i.e., a contradiction). The only possible conclusion therefore is that the assertion is true. We now illustrate with some examples.

[3]The law of the excluded middle is sometimes referred to as *tertium non datur*.

21.3.1 Examples of Proof by Contradiction

We begin by revisiting the pigeonhole principle.

EXAMPLE 21.5. Prove that if $n + 1$ letters are placed in n mailboxes, then one mailbox must contain (at least) two letters.

Proof. Seeking a contradiction, we suppose the contrary. Thus we have a way to put $n + 1$ letters into n mailboxes so that each mailbox contains only 0 or 1 letter. Let m_j be the number of letters in the jth mailbox. Then

$$n + 1 = \sum_{j=1}^{n} m_j \leq \sum_{j=1}^{n} 1 = n.$$

Under our hypothesis, we have derived the absurd statement that $n + 1 \leq n$. That is a contradiction. As a result, our hypothesis must be false, and some mailbox must contain (at least) two letters. □

EXAMPLE 21.6. Prove that, if n is a positive integer, then $n^2 + 3n + 2$ is even.

Proof. If not, then n^2+3n+2 is odd for some n. Any odd number has the form $2m+1$ for some integer m. Hence

$$n^2 + 3n + 2 = 2m + 1.$$

But then

$$n^2 + 3n - 2m = -1,$$

or

$$n(n + 3) - 2m = -1.$$

Now, if n is even, then $n + 3$ is odd and if n is odd, then $n + 3$ is even. In either case, $n(n + 3)$ will be the product of an even and an odd number and will thus be even. So $n(n + 3) = 2k$ for some integer k. As a result we have

$$2k - 2m = -1$$

or

$$2(k - m) = -1.$$

But this shows that the number -1 is even, and that is impossible. We conclude that our initial hypothesis is false: $n^2 + 2n + 3$ cannot be odd; it must be even. □

We already saw the following result in Chapter 1, but it is well worth repeating at this time. We now examine a different proof of the fact.

EXAMPLE 21.7. **Theorem (Pythagoras)**: There is no rational number whose square is 2.

Proof. Assume to the contrary that there is a rational number α whose square is 2. Write $\alpha = p/q$, where p and q are integers having no common prime factor. Let the prime factorization of p be $p = p_1 \cdot p_2 \cdots p_k$ and the prime factorization of q be $q = q_1 \cdot q_2 \cdots q_m$. Note that if any prime factor is repeated we simply list it several times.

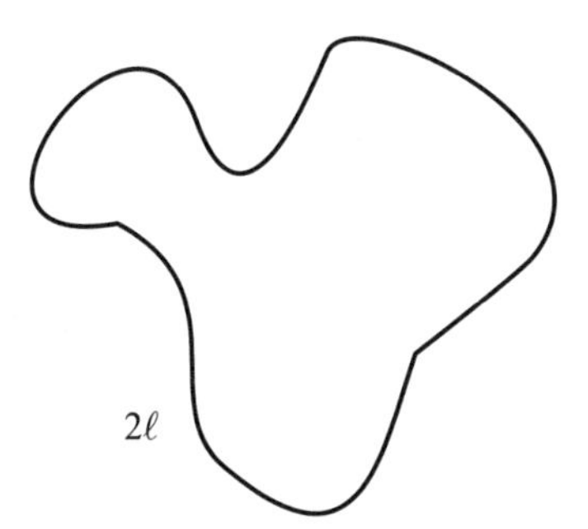

Figure 21.2. Area controlled by perimeter.

Thus our hypothesis is that

$$2 = \alpha^2 = \left(\frac{p}{q}\right)^2 = \left(\frac{p_1 \cdot p_2 \cdots p_k}{q_1 \cdot q_2 \cdots q_m}\right)^2 .$$

Clearing denominators, we see that

$$2 \cdot \left(q_1 \cdot q_2 \cdots q_m\right)^2 = \left(p_1 \cdot p_2 \cdots p_k\right)^2 .$$

We may rewrite this as

$$2 \cdot q_1^2 \cdot q_2^2 \cdots q_m^2 = p_1^2 \cdot p_2^2 \cdot p_k^2 .$$

But now we have a problem. Because every prime factor is repeated except the 2 on the left. There is no lone prime factor on the right to correspond to it. And that is a contradiction. So α cannot exist and 2 cannot have a rational square root. □

For You to Try Prove that it is impossible for two integers in sequence, both greater than 2, to be prime.

For You to Try Prove that if γ is a closed loop in the plane having length 2ℓ, then the area inside the loop cannot be more than ℓ^2. See Figure 21.2.

Further Reading

Bernard August and Thomas J. Osler, Divergence of series by rearrangement, *The College Mathematics Journal* 33(2002), 233–234.

J. L. Brenner and R. C. Lyndon, Proof of the fundamental theorem of algebra, *The American Mathematical Monthly* 88(1981), 253–256.

Enzo R. Gentile, Another proof of the irrationality of $\sqrt{2}$, *The College Mathematics Journal* 22(1991), 143.

Leslie Lamport, How to write a proof, *The American Mathematical Monthly* 102(1995), 600–608.

21.4 Direct Proof

A direct proof is one in which a sequence of logical steps leading ever closer to the desired conclusion is produced. There are no additional logical tricks, such as induction or proof by contradiction. The concept is best illustrated through examples.

21.4.1 Examples of Direct Proof

EXAMPLE 21.8. Prove that, if n is a positive integer, then the quantity $n^2 + 3n + 2$ is even.

Proof. Denote the quantity $n^2 + 3n + 2$ by K. Observe that

$$K = n^2 + 3n + 2 = (n + 1)(n + 2).$$

Thus K is the product of two successive integers: $n+1$ and $n+2$. One of those two integers must be even. So it is a multiple of 2. Therefore K itself is a multiple of 2. Hence K must be even. □

EXAMPLE 21.9. Prove that the sum of an even integer and an odd integer is odd.

Proof. An even integer e is divisible by 2, so it may be written in the form $e = 2m$, where m is an integer. An odd integer o has remainder 1 when divided by 2, so it may be written in the form $o = 2k + 1$, where k is an integer. The sum of these is

$$e + o = 2m + (2k + 1) = 2(m + k) + 1.$$

Thus we see that the sum of an even and an odd integer will have remainder 1 when it is divided by 2. As a result, the sum is odd. □

EXAMPLE 21.10. Prove that every even integer may be written as the sum of two odd integers.

Proof. Let the even integer be $K = 2m$, for m an integer. If m is odd, then we write

$$K = 2m = m + m$$

and we have written K as the sum of two odd integers. If, instead, m is even, then we write

$$K = 2m = (m - 1) + (m + 1).$$

Since m is even, then both $m - 1$ and $m + 1$ are odd. So again we have written K as the sum of two odd integers. □

EXAMPLE 21.11. Prove the Pythagorean theorem.

Remark. Of course we discussed the Pythagorean theorem in some detail in Chapter 1. But we have come a long distance since then, and learned quite a lot of mathematics. It is well to review those ideas now.

Proof. Examine Figure 21.3. It shows a square of side c inscribed inside a square of side $a + b$. Thus, on the one hand, the area of the larger square is $(a + b)^2$. On the other hand, the area of the larger square is the area of the smaller square plus the area of the four triangles. Thus we have

$$(a + b)^2 = c^2 + 4 \cdot \frac{ab}{2}.$$

Simplifying this equation gives the Pythagorean theorem. □

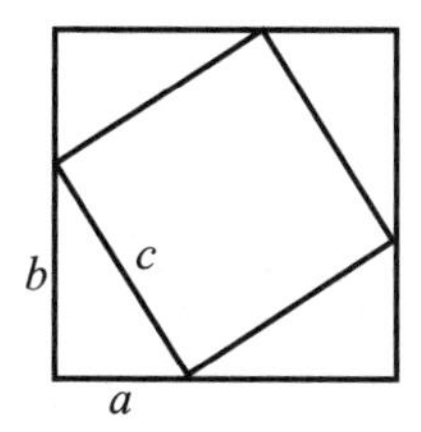

Figure 21.3. A square inscribed in a square.

For You to Try Prove that if n is an integer greater than 3, then $(n+1)^3 < n^4$.

For You to Try Prove that if X, Y, Z are points in the plane, then

$$\operatorname{dist}(X, Y) \leq \operatorname{dist}(X, Z) + \operatorname{dist}(Z, Y).$$

Further Reading

William Feller, A direct proof of Stirling's formula, *The American Mathematical Monthly* 74(1967), 1223–1225.

Arnold J. Insel, A direct proof of the integral formula for arctangent, *The College Mathematics Journal* 20(1989), 235–237.

S. Lefschetz, A direct proof of de Moivre's formula, *The American Mathematical Monthly* 23(1916), 366–368.

21.5 Other Methods of Proof

Induction, contradiction, and direct proof are the three most common proof techniques. Almost any proof can be shoehorned into one of these three paradigms. But there are other techniques that should be mentioned. One of these is enumeration, or counting. We illustrate this method with some examples.

21.5.1 Examples of Counting Arguments

EXAMPLE 21.12. Show that if there are 23 people in a room, then the odds are better than even that two of them have the same birthday.

Proof. The best strategy is to calculate the odds that *no two* of the people have the same birthday, and then to take complements.

Let us label the people $p_1, p_2, \ldots, p_{23}$. Then, assuming that none of the p_j have the same birthday, we see that p_1 can have a birthday on any of the 365 days in the year, p_2 can then have a birthday on any of the remaining 364 days, p_3 can have a birthday on any of the remaining 363 days, and so forth. Thus the number of different ways that 23 people can all have different birthdays is

$$365 \cdot 364 \cdot 363 \cdots 345 \cdot 344 \cdot 343.$$

On the other hand, the number of ways that birthdays could be distributed (with no restrictions) among 23 people is

$$\underbrace{365 \cdot 365 \cdot 365 \cdots 365}_{23 \text{ times}} = 365^{23}.$$

Thus, the probability that the 23 people all have different birthdays is

$$p = \frac{365 \cdot 364 \cdot 363 \cdots 343}{365^{23}}.$$

A quick calculation with a pocket calculator shows that $p \sim 0.4927 < .5$. Taking the complement, we see that the probability that at least two people will have the same birthday is $1 - p \sim 0.5073 > 0.5$. That is the desired result. □

EXAMPLE 21.13. Jill is dealt a poker hand of five cards from a standard deck of 52. What is the probability that she holds a straight flush?

Solution. A straight flush is five cards, all from the same suit, in sequence. Thus a straight flush could be

2, 3, 4, 5, 6 of spades
3, 4, 5, 6, 7 of hearts
10, J, Q, K, A of hearts

and so forth. Clearly, in any given suit, a straight could begin with 2, 3, 4, 5, 6, 7, 8, 9, or 10. So there are nine straight flushes in each suit and 36 straight flushes altogether. Since there are 2,598,960 possible poker hands altogether (see our discussions in Chapter 12), we see that the probability of being dealt a straight flush is

$$p = \frac{36}{2598960} \approx 0.00001385\,.$$

The last example is not quite a proof by contradiction and not quite a proof by exhaustion.

EXAMPLE 21.14. Let us show that there exist irrational numbers a and b such that a^b is rational.

Let $\alpha = \sqrt{2}$ and $\beta = \sqrt{2}$. If α^β is rational, then we are done, using $a = \alpha$ and $b = \beta$. If α^β is irrational, then observe that

$$\left(\alpha^\beta\right)^{\sqrt{2}} = \alpha^{[\beta \cdot \sqrt{2}]} = \alpha^2 = [\sqrt{2}]^2 = 2.$$

Thus, with $a = \alpha^\beta$ and $b = \sqrt{2}$ we have found two irrational numbers a, b such that $a^b = 2$ is rational.

□

Curiously, in this last example, we are unable to say which two irrational numbers do the job. But we have proved that two such numbers exist.

For You to Try Let T_k consist of all the point in the first quadrant of the plane (i.e., points (m, n) with $m \geq 0, n \geq 0$) having integer coordinates and satisfying $m + n \leq k$. Show that the number of points in T_k is $(k + 1)(k + 2)/2$.

For You to Try Show that the sum of the first k positive, even integers is $k(k+1)$.

Further Reading

Salomon Bochner, Mathematical reflections, *The American Mathematical Monthly* 81 (1974), 827–852.

Jerzy Czyz and William Self, The rationals are countable: Euclid's proof, *The College Mathematics Journal* 34 (2003), 367–369.

Fernando Q. Gouvêa, "A Marvelous Proof", *The American Mathematical Monthly* 101 (1994), 203–222.

Herve Lehning, Computer-aided or analytic proof?, *The College Mathematics Journal* 21 (1990), 228–239.

Exercises

1. Show that the number of disjoint discs of radius 2 that can be contained in the disc in the plane with center the origin and radius R cannot exceed $R^2/4$.
2. Show that
$$1^2 + 2^2 + 3^2 + \cdots + N^2 = \frac{2N^3 + 3N^2 + N}{6}.$$
3. Show that
$$1^3 + 2^3 + 3^3 + \cdots + N^3 = \frac{N^4 + 2N^3 + N^2}{4}.$$
4. Calculate the area inside a regular octagon.
5. Take five points in the plane which are not all colinear. Show that there is some line that passes through only two of them.
6. Show that, among all the rectangles in the plane with perimeter equal to 20, the square of side 5 has the greatest area.
7. Let C be a circle in the plane of radius 3. Let S be a square whose four corners lie on the circle. Calculate the area of S.
8. Let C be a circle in the plane of radius 3. Let T be an equilateral triangle whose three corners lie on the circle. Calculate the area of T.
9. Let P be a point on the unit circle centered at the origin in the plane with positive ordinate. Show that the triangle with vertices P, $(1,0)$, and $(-1,0)$ must be a right triangle.
10. Prove that the product of two odd natural numbers must be odd.
11. Prove that if n is an even natural number and if m is *any* natural number, then $n \cdot m$ must be even.
12. Prove that the sum of the first k odd natural numbers is k^2.
13. Prove that if n red letters and n blue letters are distributed among n mailboxes, then either some mailbox contains at least two red letters or some mailbox contains at least two blue letters or else some mailbox contains at least one red and one blue letter.

14. Prove that if m is a power of 3 and n is a power of 3 then $m + n$ is never a power of 3.

15. Prove that if the natural number n is a perfect square then $n + 1$ will never be a perfect square.

16. Prove that if the product of two integers is even then one of them must be even.

17. Prove that if the product of two integers is odd then both of them must be odd.

18. Prove that any integer can be written as the sum of at most two odd integers. Is the same true if "odd" is replaced by "even"?

19. **Project:** Give a proof by induction that the number of subsets of a set with k elements is 2^k. How many subsets are there with at least 3 three elements?

    How many functions are there from a set with k elements to a set with ℓ elements? In the case that $\ell = 2$, how does this result relate to the one in the last paragraph?

    Of course there are infinitely many functions from the integers $\mathbb{Z}$ to the integers $\mathbb{Z}$. Is the number countable or uncountable? See [KRA6].

20. **Project:** Prove that if $\{x_j\}$ is a Cauchy sequence in the real numbers $\mathbb{R}$, then there is a limit point x_0 such that $x_j \to x_0$ as $j \to \infty$. See [KRA1], [KRA6].

21. **Project:** It is a famous theorem that, if n is a positive integer greater than 2, then there will be a prime lying between n and $2n$. Gather evidence for this assertion. Discuss it in class. How does this jibe with the prime number theorem (Exercise 6 of Chapter 13)? See [STANPR] for more on this topic.

22

Alan Turing and Cryptography

22.0 Background on Alan Turing

Alan Mathison Turing was born in 1912 in London, England. He died tragically in 1954 in Wilmslow, Cheshire, England. Today Turing is considered to have been one of the great mathematical minds of the twentieth century. He did not invent cryptography (as we shall see, even Julius Caesar engaged in cryptography). But he ushered cryptography into the modern age. The current vigorous interaction of cryptography with computer science owes its genesis in significant part to the work of Turing. Turing also played a decisive role in many of the central ideas of modern logic. It is arguable that Turing had the key ideas for inventing the stored program computer (although it was John von Neumann (1903–1957), another twentieth-century mathematical genius, who together with Herman Goldstine (1913–2004) actually carried out the ideas).

Turing had a tragically short life, marked by disappointments and frustrations. Today he is considered to be one of the great geniuses of mathematical thought. His story is recounted in [TUR] and [HOD].

Turing had difficulty fitting in at the British "public schools" which he attended. [Note that a "public school" in Britain is what we in America would call a private school.] Young Turing was more interested in pursuing his own thoughts than in applying himself to the dreary school tasks that were designed for average students. At the Sherborne School, Turing had little patience for the tedious math techniques that the teachers taught. Yet he won almost every mathematics prize at the school. He was given poor marks in penmanship, and he struggled with English.

Turing had a passion for science beginning at a very young age. He later said that the book *Natural Wonders Every Child Should Know* had had a seminal influence on him. When he was still quite young, he read Einstein's papers on relativity and he read Arthur Eddington's account of quantum mechanics in the book *The Nature of the Physical World*.

In 1928, at the Sherborne School, Alan Turing became friends with Christopher Morcom. Now he had someone in whom he could confide, and with whom he could share scientific ideas and inquiries. Turing had never derived such intellectual companionship from either his classmates or his rather diffident schoolteachers. Sadly, Morcom died suddenly in 1930. This event had a shattering effect on the young Alan Turing. The loss of his companion led Turing to consider spiritual matters, and over time this led him to an interest in physics.

It may be mentioned that Turing developed early on an interest in sports. He was a very talented athlete—almost at the Olympic level—and he particularly excelled in running. He maintained an interest in sports throughout his life.

In 1931 Alan Turing entered King's College at Cambridge University. Turing earned a distinguished degree at King's in 1934, followed by a fellowship at King's. In 1936 he won the Smith Prize for his work in probability theory. In particular, Turing was one of the independent discoverers of the Central Limit Theorem.

In 1935 Turing took a course from Max Newman on the foundations of mathematics. Thus his scientific interests took an abrupt shift. The hot ideas of the time were Gödel's incompleteness theorem—which says that virtually any mathematical theory will have true statements in it that cannot be proved—and (what is closely related) David Hilbert's questions about decidability.

In 1936 Alan Turing published his seminal paper "On Computable Numbers, with an application to the Entscheidungsproblem." Here the *Entscheidungsproblem* is the fundamental question of how to decide—in a manner that can be executed by a machine—when a given mathematical question is provable. In this paper Turing first described his idea for what has now become known as the *Turing machine*. We now take a mathematical detour to talk about Turing machines.

Further Reading

I. Grattan-Guinness, On the development of logics between the two world wars, *The American Mathematical Monthly* 88(1981), 495–509.

Leon Harkleroad, How mathematicians know what computers can't do, *The College Mathematics Journal* 27(1996), 37–42.

S. L. Zabell, Alan Turing and the central limit theorem, *The American Mathematical Monthly* 102(1995), 483–494.

22.1 The Turing Machine

A *Turing machine* is a device for performing effectively computable operations. It consists of a machine through which a bi-infinite paper tape is fed. The tape is divided into an infinite sequence of congruent boxes (Figure 22.1). Each box has either a numeral 0 or a numeral 1 in it. The Turing machine has finitely many "states" $S_1, S_2, \ldots, S_n$. In any given state of the Turing machine, one of the boxes is being scanned.

After scanning the designated box, the Turing machine does one of three things:

(1) It either erases the numeral 1 that appears in the scanned box and replaces it with a 0, or it erases the numeral 0 that appears in the scanned box and replaces it with a 1, or it leaves the box unchanged.

(2) It moves the tape one box (or one unit) to the left or to the right.

(3) It goes from its current state S_j into a new state S_k.

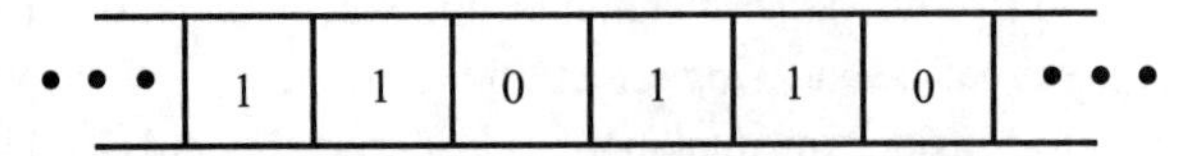

Figure 22.1. A deterministic Turing machine.

It turns out that every logical procedure, every algorithm, every mathematical proof, every computer program can be realized as a Turing machine. The Turing machine is a "universal logical device". The next section contains a simple instance of a Turing machine. In effect, Turing had designed a computer before technology had made it possible to actually build one.

22.1.1 An Example of a Turing Machine

Here is an example of a Turing machine for calculating $x + y$:

State	Old Value	New Value	Move (l. or r.)	New State	Explanation
0	1	1	R	0	pass over x
0	0	1	R	1	fill gap
1	1	1	R	1	pass over y
1	0	1	L	2	end of y
2	1	0	L	3	erase a 1
3	1	0	L	4	erase another 1
4	1	1	L	4	back up
4	0	0	R	5	halt

For You to Try If you look hard at the logic of this Turing machine, you will see that it thinks of x as a certain number of 1s, and it thinks of y as a certain number of 1s. It scans the x units, and writes a 1 to the right of these; then it scans y units, and writes a 1 to the right of these. The two blocks of 1s are joined into a single block (by erasing the space in between) and then the two extra 1s are erased. The result is $x + y$. Provide the details of this argument in simple English sentences.

Further Reading

James P. Jones, Recursive undecidability—an exposition, *The American Mathematical Monthly* 81(1974), 724–738.

Bill Marion, Turing machines and computational complexity, *The American Mathematical Monthly* 101(1994), 61–65.

Jan Mycielski, The meaning of the conjecture **P** $\neq$ **NP** for mathematical logic, *The American Mathematical Monthly* 90(1983), 129–130.

22.2 More on the Life of Alan Turing

The celebrated logician Alonzo Church published a paper closely related to Turing's at about the same time. As a result, Church and Turing ended up communicating and sharing ideas. Subsequently, in 1936, Turing went to Princeton for graduate study under Church's direction.

When Turing returned to Cambridge in 1938, he commenced work on actually building a computer. It was designed to be a rather crude, mechanical device, with a great many gears and wheels. In fact Turing had a very specific purpose in mind for his machine.

One of the great mathematical problems of the day (and it is still a hot open problem as of this writing) was to prove the *Riemann hypothesis*. The Riemann hypothesis, posed by Bernhard Riemann in 1859, concerns the location of the zeros of a certain complex function (the celebrated *Riemann zeta function*). An affirmative answer to the Riemann hypothesis would tell us a great deal about the distribution of prime numbers and have profound consequences for number theory and for cryptography.

According to Andrew Hodges (1949–), the Turing biographer,

> Apparently [Turing] had decided that the Riemann Hypothesis was probably false, if only because such great efforts have failed to prove it. Its falsity would mean that the zeta function did take the value zero at some point which was off the special line, in which case this point could be located by brute force, just by calculating enough values of the zeta function.

Turing did his own engineering work, hence he got involved in all the fine details of constructing this machine. He planned on eighty meshing gearwheels with weights attached at specific distances from their centers. The different moments of inertia would contribute different factors to the calculation, and the result would be the location of and an enumeration of the zeros of ζ (the zeta function).

Visits to Turing's apartment would find the guest greeted by heaps of gear wheels and axles and other junk strewn about the place. Although Turing got a good start cutting the gears and getting ready to assemble the machine, more pressing events (such as World War II) interrupted his efforts. His untimely death prevented the completion of the project.

When war broke out in 1939, Turing went to work for the Government Code and Cypher School at Bletchley Park. Turing played a seminal role in breaking German secret codes, and it has been said that his work saved more lives during the war than that of any other person. One of his great achievements during this time was the construction of the *Bombe* machine, a device for cracking all the encoded messages generated by the dreaded German *Enigma* machine. In fact Turing used ideas from abstract logic, together with some earlier contributions of Polish mathematicians, to design the *Bombe*. Turing's important contributions to the war effort were recognized with the award of an O.B.E. (Order of the British Empire) in 1945.

After the war Turing was invited by the National Physical Laboratory in London specifically to design a computer. He wrote a detailed proposal for the Automatic Computing Machine in 1946, and that document is in fact a discursive prospectus for a stored-program computer. The project that Turing proposed turned out to be too grandiose for practical implementation, and it was shelved.

Turing's interests turned to topics outside of mathematics, including neurology and physiology. But he maintained his passion for computers. In 1948 he accepted a position at the University of Manchester. There he became involved in a project, along with F. C. Williams and T. Kilburn, to construct a computing machine.

In 1951 Alan Turing was elected a Fellow of the Royal Society—the highest honor that can be bestowed upon a British scientist. This accolade was largely in recognition of his work on Turing machines.

Turing had a turbulent personal life. In 1952 he was arrested for violation of the British homosexuality statutes. He was convicted, and sentenced to take the drug oestrogen for one year. Turing subsequently re-dedicated himself to his scientific work, concentrating particularly on spinors and relatively theory. Unfortunately, because of his legal difficulties, Turing lost his security clearance and was labeled something of a "security risk". He had continued working with the cypher school at Bletchley, but his loss of clearance forced that collaboration to end. These events had a profound and saddening effect on Alan Turing.

Turing died in 1954 of potassium cyanide poisoning while conducting electrolysis experiments. The cyanide was found on a half-eaten apple. The police concluded that the death was a suicide, though people close to Turing argue that it was an accident.

Further Reading

Phillip A. Griffiths, Mathematics at the turn of the millennium, *The American Mathematical Monthly* 107 (2000), 1–14.

Rüdger Thiele, Hilbert's twenty-fourth problem, *The American Mathematical Monthly* 110 (2003), 1–24.

22.3 What is Cryptography?

We use Alan Turing's contributions as a touchstone for our study of cryptography. Cryptography is currently a very hot field, due in part to the availability of high speed digital computers to carry out decryption algorithms, in part to new and exciting connections between cryptography and number theory and logic, and in part to the need for practical coding methods both in industry and in government.

The discussion of cryptography that appears below is inspired by the lovely book [KOB]. We refer the reader to that source for additional ideas and further reading.

As we always do in mathematics, let us begin by introducing some terminology. Cryptography is the study of methods for sending text messages in disguised form in such a manner that only the intended recipient can remove the disguise and read the message. The original message that we wish to send is called the *plaintext* and the disguised message is called the *ciphertext*. We shall always assume that both our plaintext and our ciphertext are written in the standard roman alphabet (i.e., the letters A through Z) together perhaps with some additional symbols like "blank space (␣)", "question mark (?)", and so forth. The process of translating a plaintext message into a ciphertext message is called *encoding* or *enciphering* or *encrypting*. The process of translating an encoded message back to a plaintext message is called *deciphering* or sometimes *de-encrypting*.

For convenience, we usually break up both the plaintext message and the ciphertext message into blocks or units of characters. We call these pieces the *message units*, but we may think of them as "words" (however they are not necessarily English words). Sometimes we will declare in advance that all units are just single letters, or perhaps pairs of letters (these are called *digraphs*) or sometimes triples of letters (called *trigraphs*). Other times we will let the units be of varying sizes—just as the words in any body of text have varying sizes. An *enciphering transformation* is a function that assigns to each plaintext unit a ciphertext unit. The *deciphering transformation* is the inverse mapping that recovers the plaintext unit from the ciphertext unit. Any setup as we have just described is called a *cryptosystem*.

In general it is awkward to mathematically manipulate the letters of the alphabet. We have no notions of addition or multiplication on these letters. So it is convenient to associate to each letter a number. Then we can manipulate the numbers. For instance, it will be convenient to make the assignment

$$\begin{aligned} A &\leftrightarrow 0 \\ B &\leftrightarrow 1 \\ C &\leftrightarrow 2 \\ &\ldots \\ X &\leftrightarrow 23 \\ Y &\leftrightarrow 24 \\ Z &\leftrightarrow 25\,. \end{aligned}$$

Thus if we see the message

22 7 0 19 12 4 22 14 17 17 24

then we can immediately translate this to

WHATMEWORRY

or

WHAT ME WORRY

Notice that, in cryptography, we generally do not worry about capital and lowercase letters. Everything is uppercase. Quite often we will not have a symbol for "blank space", so a little extra thought must be given at the end (as in this last example) to extract the message from the decrypted sequence of letters.

One device of which we will make frequent and consistent use is *modular arithmetic*. Recall that if n and k are integers, then $n \bmod k$ is that unique integer n' between 0 and $k-1$ inclusive such that $n - n'$ is divisible by k. For example,

$$13 \bmod 5 = 3\,,$$

$$-23 \bmod 7 = 5\,,$$

$$82 \bmod 14 = 12\,,$$

$$10 \bmod 3 = 1\,.$$

How do we calculate these values? Look at the first of these. To determine $13 \bmod 5$, we *divide* 5 into 13: Of course 5 goes into 13 with quotient 2 and remainder 3. It is the *remainder* that we seek. Thus

$$13 \bmod 5 = 3\,.$$

It is similar with the other examples. To determine $82 \bmod 14$, divide 14 into 82. It goes 5 times with remainder 12. Hence

$$82 \bmod 14 = 12\,.$$

It is convenient that modular arithmetic respects the arithmetic operations. For example,

$$8 \times 7 = 56 \qquad \text{and} \qquad 56 \bmod 6 = 2\,.$$

But

$$8 \bmod 6 = 2 \quad \text{and} \quad 7 \bmod 6 = 1 \quad \text{and} \quad 2 \times 1 = 2\,.$$

So it does not matter whether we pass to mod 6 *before* multiplying or *after multiplying*. Either way we obtain the same result 2. Similar properties hold for addition and subtraction. One must be a bit more cautious with division, as we shall see below.

We supply some further examples:

$$[3 \bmod 5] \times [8 \bmod 5] = 24 \bmod 5 = 4\,;$$

$$[7 \bmod 9] + [5 \bmod 9] = 12 \bmod 9 = 3\,;$$

$$[4 \bmod 11] - [9 \bmod 11] = -5 \bmod 11 = 6\,.$$

Now we begin to learn some cryptography by way of examples.

EXAMPLE 22.1. We use the ordinary 26-letter Roman alphabet A–Z, with the numbers 0-25 assigned to the letters as indicated above. Let $S = \{0, 1, 2, \ldots, 25\}$. We will consider units consisting of single letters. Thus our cryptosystem will consist of a function $f : S \to S$ which assigns to each unit of plaintext a new unit of ciphertext. In particular, let us consider the specific example

$$f(P) = \begin{cases} P + 5 & \text{if} \quad P < 21\,, \\ P - 21 & \text{if} \quad P \geq 21\,. \end{cases}$$

Put in other words,

$$f(P) = P + 5 \bmod 26\,. \tag{$*$}$$

Next let us use this cryptosystem to encode the message

GOAWAY

or

GO AWAY .

The first step is that we transliterate the letters into numbers (because, as noted earlier, numbers are easier to manipulate). Thus GOAWAY becomes 6 14 0 22 0 24 .

Now we apply the "shift encryption" ($*$) to this sequence of numbers. Notice that

$$f(6) = 6 + 5 \bmod 26 = 11 \bmod 26 = 11\,,$$

$$f(14) = 14 + 5 \bmod 26 = 19\,,$$

$$f(0) = 0 + 5 \bmod 26 = 5\,,$$

$$f(22) = 22 + 5 \bmod 26 = 1\,,$$

$$f(0) = 0 + 5 \bmod 26 = 5\,,$$

$$f(24) = 24 + 5 \bmod 26 = 3\,.$$

Thus our ciphertext is 11 19 5 1 5 3 . In practice, we may convert this ciphertext back to roman letters using our standard correspondence ($A \leftrightarrow 0$, $B \leftrightarrow 1$, etc.). The result is LTFBFD . Thus the encryption of "GO AWAY" is "LTFBFD". Notice that we have no coding for a blank space, so we ignore it.

This is a very simple example of a cryptosystem. It is said that Julius Caesar used this system with 26 letters and a shift of 3. We call this encryption system a "shift transformation".

Now let us use this same cryptosystem to encode the word "BRAVO". First, we translate our plaintext word to numbers:

$$1\ 17\ 0\ 21\ 14\,.$$

Now we add 5 mod 26 to each numerical entry. The result is

$$6\ 22\ 5\ 0\ 19\,.$$

Notice that the fourth entry is 0 because

$$21 + 5 \bmod 26 = 26 \bmod 26 = 0 \bmod 26\,.$$

Thus if we wanted to send the message "BRAVO" in encrypted form, we would send 6 22 5 0 19. We can translate the encrypted message to roman letters as "GWFAT".

Conversely, we *decrypt* a message by subtracting 5 mod 26. Suppose, for instance, that you receive the encrypted message

YMFSPDTZ

Of course we first transliterate the message (using our usual correspondence) to a sequence of numbers:

$$24\ 12\ 5\ 18\ 15\ 3\ 19\ 25\,.$$

We decrypt by applying the function $f^{-1}(Q) = Q - 5 \bmod 26$. The result is

$$19\ 7\ 0\ 13\ 10\ 24\ 14\ 20\,.$$

This easily translates to

THANKYOU

or

THANK YOU .

In a typical, real-life circumstance, you receive an encrypted message and *you do not know the method of encryption*. It is your job to figure out how to decode the message. We call this process *breaking the code*, and the science of codebreaking is called *cryptoanalysis*.

EXAMPLE 22.2. If the codebreaker happens to know that the message he/she has received is encrypted using a shift transformation, then there is a reasonable method to proceed. Imagine that you receive the message

CQNKNJCUNBOXANENA

Looks like nonsense. But the cryptographer has reason to believe that this message has been encoded using a shift transformation on single letters of the 26-letter alphabet. It remains to find the numerical value of the shift.

We use a method called *frequency analysis*. The idea of this technique is that it is known that "E" is the most frequently occurring letter in the English language. Thus we may suppose that the most frequently occuring character in the ciphertext is the encryption of "E" (*not* "E" itself). In fact we see that the character "N" occurs five times in the ciphertext, and that is certainly the most frequently occurring letter. If we hypothesize that "N" is the encryption of "E", then we see that "4" has been translated to "13" in the encryption. Thus the encryption key is $P \mapsto P + 9 \mod 26$. And therefore the decryption scheme is $P \mapsto P - 9 \mod 26$. If this putative decryption scheme gives a sensible message, then it is likely the correct choice (as any other decryption scheme will likely give nonsense). Let us try this scheme and see what result it gives. We have

CQNKNJCUNBOXANENA

has numerical realization

2 16 13 10 13 9 2 20 13 1 14 23 0 13 4 13 0 .

Under our decryption scheme, this translates to

19 7 4 1 4 0 19 11 4 18 5 14 17 4 21 4 17

which has textual realization

THEBEATLESFOREVER.

In other words, the secret message is

THE BEATLES FOREVER.

The trouble with the shift transformation is that it is just too simple-minded. It is too easy to break. There are variants that make it slightly more sophisticated. For example, suppose that the East Coast and the West Coast branches of National Widget Corporation cook up a system for sending secret messages back and forth. They will use a shift transformation, but *in each week of the year they will use a different shift*. This adds a level of complexity to the process. But the fact remains that, using a frequency analysis, the code can likely be broken in any given week.

Further Reading

Lester S. Hill, Cryptography in an algebraic alphabet, *The American Mathematical Monthly* 36(1929), 306–312.

Jack Levine, Some elementary cryptanalysis of algebraic cryptography, *The American Mathematical Monthly* 68(1961), 411–418.

Dennis Luciano and Gordon Prichett, Cryptology: From Caesar ciphers to public-key cryptosystems, *The College Mathematics Journal* 18(1987), 2–17.

22.4 Encryption by Way of Affine Transformations

We can add a genuine level of sophistication to the encryption process by adding some new mathematics. Instead of considering a simple shift of the form $P \mapsto P + b$ for some fixed integer b, we instead consider an *affine transformation* of the form $P \mapsto aP + b$. Now we are both multiplying (or dilating) the element P by an integer a and then translating it by b.

22.4.1 Division in Modular Arithmetic

There is a subtlety in the application of the affine transformation method that we must consider before we can look at an example. If the encryption scheme is $P \mapsto Q \equiv aP + b$, then the decryption scheme must be the inverse function. In other words, we solve for P in terms of Q. This just involves elementary algebra, and we find that

$$P = [1/a](Q - b) \bmod 26\,.$$

We see that decryption, in the context of an affine transformation, involves division in arithmetic modulo 26. This is a new idea, and we should look at a couple of simple examples before we proceed with our cryptographic considerations.

We want to consider division modulo 26. Thus if a and b are whole numbers, then we want to calculate b/a and we want the answer to be another whole number modulo 26. This is possible only because we are cancelling multiples of 26, and it will only work when a has no common prime factors with 26.

First let us calculate $4/7 \bmod 26$. What does this mean? We are dividing the whole number 4 by the whole number 7, and this looks like a fraction. But things are a bit different in modular arithmetic. We seek a number k such that

$$\frac{4}{7} \bmod 26 = k$$

or

$$4 = 7 \cdot k \bmod 26$$

or

$$4 - 7 \cdot k \quad \text{is divisible by } 26\,.$$

We simply try different values for k, and we find with $k = 8$ that

$$4 - 7 \cdot 8 = 4 - 56 = -52 \quad \text{is indeed divisible by } 26\,.$$

In conclusion,

$$\frac{4}{7} \bmod 26 = 8\,.$$

We see the somewhat surprising conclusion that the fraction $4/7$ can be realized as a whole number in arithmetic modulo 26.

Next let us try to calculate $1/4 \bmod 26$. This is doomed to fail, because 4 and 26 have the prime factor 2 in common. We seek an integer k such that

$$1 = 4 \cdot k \bmod 26\,,$$

or in other words

$$1 - 4k \quad \text{is a multiple of } 26\,.$$

But of course $4k$ will always be even so $1 - 4k$ will always be odd—*it cannot be a multiple of the even number 26*. This division problem cannot be solved.

For You to Try We conclude this brief discussion with the example $2/9 \bmod 26$. We invite the reader to discover that the answer is $6 \bmod 26$.

There is in fact a mathematical device for performing division in modular arithmetic. It is the classical Euclidean algorithm. This simple idea is one of the most powerful in all of number theory. It says this: If n and d are integers, then d divides into n some whole number q times with some remainder r, and $0 \le r < d$. In other words,

$$n = d \cdot q + r\,.$$

You have been using this idea all your life when you calculate a long division problem (not using a calculator, of course). We shall see in the next example that the Euclidean algorithm is a device for organizing information so that we can directly perform long division in modular arithmetic.

EXAMPLE 22.3. Let us calculate $1/20$ in arithmetic mod 57. We apply the Euclidean algorithm to 57 and 20. Thus we begin with

$$57 = 2 \cdot 20 + 17\,.$$

We continue by repeatedly applying the Euclidean algorithm to divide the divisor by the remainder:

$$\begin{aligned} 20 &= 1 \cdot 17 + 3 \\ 17 &= 5 \cdot 3 + 2 \\ 3 &= 1 \cdot 2 + 1 \end{aligned}$$

Now, as previously indicated, we utilize this Euclidean algorithm information to organize our calculations. Begin with the last line to write

$$\begin{aligned} 1 &= 3 - 1 \cdot 2 \\ &= 3 - 1 \cdot (17 - 5 \cdot 3) \\ &= [20 - 17] - 1 \cdot ([57 - 2 \cdot 20] - 5 \cdot [20 - 17]) \\ &= 20 \cdot 8 + 17 \cdot (-6) - 57 \\ &= 20 \cdot 8 + (57 - 2 \cdot 20) \cdot (-6) - 57 \\ &= 20 \cdot 20 - 7 \cdot 57\,. \end{aligned}$$

This calculation tells us that $1 = 20 \cdot 20 \bmod 57$. In other words, $1/20 = 20 \bmod 57$.

For You to Try We offer the reader the exercise of calculating $1/25 \bmod 64$ using the Euclidean algorithm.

22.4.2 Instances of the Affine Transformation Encryption

EXAMPLE 22.4. Let us encrypt the message "GO AWAY" using the affine transformation $P \mapsto 5P + 6 \bmod 26$. As usual,

GO AWAY has numerical realization 6 14 0 22 0 24 .

Under the affine transformation, we obtain the new numerical realization

$$10 \;\; 24 \;\; 6 \;\; 12 \;\; 6 \;\; 22 \;\; .$$

In roman letters, the message becomes the ciphertext

KYGMGW .

In order to decrypt the message, we must use the *inverse* affine transformation. If $R = 5P + 6 \bmod 26$, then $P = [1/5](R - 6) \bmod 26$. Using modular arithmetic, we see that 10 corresponds to

$$[1/5](10 - 6) = [1/5] \cdot 4 = 6 \bmod 26$$

(because $5 \cdot 6 \bmod 26 = 30 \bmod 26 = 4 \bmod 26$). Likewise 24 corresponds to

$$[1/5](24 - 6) = [1/5] \cdot 18 = 14 \bmod 26$$

(because $5 \cdot 14 \bmod 26 = 70 \bmod 26 = 18 \bmod 26$). We calculate the rest of the correspondences:

$$[1/5](6 - 6) = [1/5] \cdot 0 = 0 \bmod 26$$

(because $5 \cdot 0 \bmod 26 = 0 \bmod 26$). Next,

$$[1/5](12 - 6) = [1/5] \cdot 6 = 22 \bmod 26$$

(because $5 \cdot 22 \bmod 26 = 110 \bmod 26 = 6 \bmod 26$). Again,

$$[1/5](6 - 6) = [1/5] \cdot 0 = 0 \bmod 26 \, .$$

And, finally,

$$[1/5](22 - 6) = [1/5] \cdot 16 = 24 \bmod 26$$

(because $5 \cdot 24 \bmod 26 = 16 \bmod 26$).

In sum, we have applied our decryption algorithm to recover the message

$$6 \;\; 14 \;\; 0 \;\; 22 \;\; 0 \;\; 24 \;\; .$$

This transliterates to

GOAWAY

or

GO AWAY .

In a real-life situation—if we were attempting to decrypt a message—we would not know in advance which affine transformation was used for the encoding. We now give an example to illustrate how to deal with such an eventuality.

EXAMPLE 22.5. We continue to work with the 26-letter Roman alphabet. We receive a block of ciphertext and wish to decode it. We notice that the most frequently occurring character in the ciphertext is "M" and the second most frequently occurring character in the ciphertext is "R". It is well known that, in ordinary English, the most commonly occurring letter is "E" and the second most commonly occurring letter is "T". So it is natural to hypothesize that we are dealing with an affine transformation that assigns "E" to "M" and "T" to "R".

This means that we seek an affine transformation $f(P) = aP + b$ such that $f(4) = 12 \bmod 26$ and $f(19) = 17 \bmod 26$. All arithmetic is, as usual, modulo 26. We are led then to the equations

$$12 = a \cdot 4 + b \bmod 26\,,$$
$$17 = a \cdot 19 + b \bmod 26\,.$$

We subtract these two equations to eliminate b and obtain

$$-5 = a \cdot (-15) \bmod 26$$

or

$$a = [-5/(-15)] \bmod 26\,.$$

The solution is $a = 9$. Substituting this value into the first equation gives $b = -24 = 2 \bmod 26$.

Thus our affine encoding transformation is (we hope) $f(P) = 9P + 2$. It is also easy to determine that the inverse (or decoding) transformation is $f^{-1}(Q) = [Q - 2]/9$.

For You to Try Use the affine decryption scheme in the last example to decode the message "ZMDEMRILMRRMZ".

Next we present an example in which an expanded alphabet is used.

EXAMPLE 22.6. Consider the standard Roman alphabet of 26 characters along with the additional characters "blank space" (denoted ␣), "question mark" (?), "period (.)", and "exclamation point (!)". So now we have 30 characters, and arithmetic will be module 30. As usual, we assign a positive integer to each of our characters. Thus we have

$$A \leftrightarrow 0$$
$$B \leftrightarrow 1$$
$$C \leftrightarrow 2$$
$$\cdots$$
$$X \leftrightarrow 23$$
$$Y \leftrightarrow 24$$
$$Z \leftrightarrow 25$$
$$\sqcup \leftrightarrow 26$$
$$? \leftrightarrow 27$$
$$. \leftrightarrow 28$$
$$! \leftrightarrow 29$$

Because there are now 30 different characters, we also use 30 different numerical codes—the numbers from 0 to 29.

Imagine that we receive a block of ciphertext, and that we wish to decode it. We notice that the most commonly used characters in the ciphertext are "D" and "!". It is known that the most commonly used characters in ordinary English are "␣" and "E".[1] If we assume that

[1] We formerly said that "E" was the most commonly used letter. But that was before we added the blank space ␣ to our alphabet.

the ciphertext was encrypted with an affine transformation, then we seek an affine mapping $f(P) = aP + B$ such that $f(\sqcup) = D$ and $f(E) =!$. Thus we are led to $f(26) = 3$ and $f(4) = 29$ and then to the system of equations

$$3 = a \cdot 26 + b \bmod 30\,,$$
$$29 = a \cdot 4 + b \bmod 30\,.$$

As before, we subtract the equations to eliminate b. The result is

$$-26 = 22a \bmod 30\,.$$

This equation is equivalent (dividing by 2) to

$$-13 = 11a \bmod 30\,.$$

Since 11 and 30 have no factors in common, we may easily find the unique solution $a = 7$. Substituting this value in the second equation gives $b = 1$. We conclude that our affine transformation is $f(P) = 7P + 1$.

If the ciphertext we have received is

21 7 29 3 14 29 12 14 7 14 19 18 29 24

then we can apply $f^{-1}(Q) = [Q - 1]/7$ to obtain the plaintext message

20 18 4 26 19 4 23 19 26 18 19 24 11 4 29 .

This transliterates to

USE TEXT STYLE!

A nice feature of this example is that the spaces and the punctuation are built into our system of characters. Hence the translated message is quite clear, and requires no further massaging.

Further Reading

Fred Krakowski, Affine transformations and mirror-symmetry, *The American Mathematical Monthly* 70(1963), 189–190.

Susan Landau, Polynomials in the nation's service: Using algebra to design the advanced encryption standard, *The American Mathematical Monthly* 111(2004), 89–117.

Gerald J. Porter, Linear algebra and affine planar transformations, *The College Mathematics Journal* 24(1993), 47–51.

22.5 Digraph Transformations

Just to give an indication of how cryptographers think, we shall now consider digraphs. Instead of thinking of our message units as single characters, we will now have units that are *pairs* of characters. Put in other words, the plaintext message is broken up into two-character segments or words. [It should be stressed that these will not, in general, correspond to English words. Certainly words from the English language are generally longer than two letters. Here, when we say "word", we simply mean a unit of information.]

In case the plaintext message has an odd number of characters, then of course we cannot break it up evenly into units of two characters. In this instance we add a "dummy" character like "X" to the end of the message so that an even number of characters will result. Any English message will still be readable if an "X" is tacked on the end.

Let K be the number of elements in our alphabet (in earlier examples, we have seen alphabets with 26 characters and also alphabets with 30 characters). Suppose now that MN is a digraph (i.e., an ordered pair of characters from our alphabet). Let x be the numerical equivalent of M and let y be the numerical equivalent of N. Then we assign to the digraph MN the number $x \cdot K + y$. Roughly speaking, we are now working in base-K arithmetic.

EXAMPLE 22.7. Let us work in the familiar Roman alphabet of 26 characters. A common digraph in English is "th". Notice that the numerical equivalent of "T" is 19 and the numerical equivalent of "H" is 7. According to our scheme, we assign to this digraph the single number $19 \cdot 26 + 7 = 501$.

It is not difficult to see that each positive integer corresponds to a unique digraph. Consider the number 358. Then 26 divides into 358 a total of 13 times with a remainder of 20. We conclude that 358 corresponds to the digraph with numerical equivalents 13 20. This is the digraph "NU".

It is straightforward to see that the greatest integer that can arise in this labeling scheme for digraphs is for the digraph $\Omega\Omega$, where Ω is the last character in our alphabet. If the first character is assigned to 0 (as we have done in the past) then the last character is assigned to $K - 1$ (where K is the number of characters in the alphabet). The numerical label is then $(K-1) \cdot K + (K-1) = K \cdot K - 1$. So it is safe to say that $K^2 - 1$ is an upper bound for numerical labels in our digraph system.

We conclude, then, that an enciphering transformation is a function that consists of a rearrangement of the integers $\{0, 1, 2, \ldots, K^2 - 1\}$. One of the simplest such transformations is an *affine transformation* on $\{0, 1, 2, \ldots, K^2 - 1\}$. We think of this set of integers as $\mathbb{Z}$ modulo K^2. So the encryption has the form $f(P) = aP + b \bmod K^2$. As usual, the integer a must have no prime factors in common with K^2 (and hence no prime factors in common with K).

EXAMPLE 22.8. We work as usual with the 26-letter Roman alphabet. There are then 26×26 digraphs, and these are enumerated by means of the integers $0, 1, 2, \ldots, 26^2 - 1$. In other words, we work in arithmetic modulo 676, where of course $676 = 26^2$. As an instance, the digraph "ME" has letters "M" corresponding to "12" and "E" corresponding to "4". Thus we assign the digraph number $12 \cdot 26 + 4 = 316 \bmod 676$.

If our affine enciphering transformation is $f(P) = 97 \cdot P + 230$, then the digraph "ME" is encrypted as $97 \cdot 316 + 230 = 462 \bmod 676$.

If instead we consider the digraph "EM" then we assign the integer $4 \cdot 26 + 12 = 116$. And now the encryption is $97 \cdot 116 + 230 = 666 \bmod 676$.

EXAMPLE 22.9. Suppose that we want to break a digraphic encryption system that uses an affine transformation. So we need to determine a and b. This will require two pieces of information.

Let us attempt a frequency analysis. From statistical studies, it is known that some of the most common digraphs are "TH", "HE", and "EA". The most common ones that include the "blank space" character are "E␣ ", "S␣", and "␣T". If we examine a good-sized block of ciphertext and notice the most commonly occurring digraphs, then we might suppose that those are the encryptions of "TH"or "HE" or "EA". Consider for example the ciphertext (based on the 27-character alphabet consisting of the usual 26 letters of the Roman alphabet plus the blank space, and numbered 0 through 26)

XIHZYIQHRCZJSDXIDCYIQHPS .

We notice that the digraphs "XI", "YI", and "QH" each occur twice in the message. We might suppose that one of these is the encryption of "TH", one is the encryption of "HE", and one is the encryption of "EA" (although, as indicated above, there are other possibilities). Let us attempt to directly solve for the affine transformation that will decript our ciphertext. The affine transformation will have the form $f^{-1}(Q) = a'Q + b'$ and our job is to find a' and b'.

To be specific, let us guess that

TH encrypts as YI

HE encrypts as XI .

This means that we have the numerical correspondences

$$520 \leftrightarrow 656$$

and

$$193 \leftrightarrow 629 .$$

So we have the algebraic equations

$$520 = a' \cdot 656 + b' \bmod 729 ,$$
$$193 = a' \cdot 629 + b' \bmod 729 .$$

Subracting the equations as usual (to eliminate b'), we see that

$$327 = a' \cdot 27 \bmod 729 .$$

Unfortunately this equation does not have a unique solution, because 27 and 729 have prime factors in common (such as 3).

We make another guess. Let us suppose that

TH encrypts as QH

HE encrypts as YI .

This means that we have the numerical correspondences

$$520 \leftrightarrow 439$$

and

$$193 \leftrightarrow 656 .$$

So we have the algebraic equations

$$520 = a' \cdot 439 + b' \bmod 729$$
$$193 = a' \cdot 656 + b' \bmod 729 .$$

Subracting the equations as usual (to eliminate b'), we see that

$$327 = a' \cdot 217 \bmod 729 .$$

Now 217 and 729 have no prime factors in common, so we may solve for a' uniquely. The answer is $a' = 408$. Substituting into our first equation gives $b' = 13$. So our decryption algorithm is

$$f^{-1}(Q) = 408Q + 13 . \qquad (*)$$

We apply this rule to the ciphertext

XIHZYIQHRCZJSDXIDCYIQHPS.

For example, the digraph "XI" has numerical equivalent 629. It translates, with decryption rule (∗), to 37. This in turn corresponds to the plaintext digraph "BK". We can already tell we are in trouble, because there is no word in the English language that contains the two letters "BK" in sequence.

It is our job then to try all the other possible correspondences of encrypted digraphs"XI", "YI", and "QH" to the plaintext digraphs. We shall not work them all out here. It turns out that the one that does the trick is

XI is the encryption of TH

and

QH is the encryption of EA .

Let us try it and see that it succesfully decrypts our secret message.

The proposed correspondences have numerical interpretation

$$629 \leftrightarrow 520$$

and

$$439 \leftrightarrow 108 .$$

This leads to the equations

$$520 = a' \cdot 629 + b' \bmod 729 ,$$
$$108 = a' \cdot 439 + b' \bmod 729 .$$

Subtracting as usual, we obtain

$$412 = a' \cdot 190 \bmod 729 .$$

Since 190 and 729 have no prime factors in common, we can certainly divide by 190 and solve for a'. We find that $a' = 547$. Substituting into the second equation gives $b' = 545$. In conclusion, the decrypting transformation is $f^{-1}(Q) = 547Q + 545 \bmod 729$.

Now we can systematically apply this affine transformation to the digraphs in the ciphertext and recover the original message. Let us begin:

$$XI \to 629 \stackrel{f^{-1}}{\longrightarrow} 520 \to TH ,$$

$$HZ \to 214 \stackrel{f^{-1}}{\longrightarrow} 234 \to IS .$$

The calculations continue, and the end result is the original plaintext message

THIS HEART OR THAT HEADX

As you can see, an "X" is affixed to the end to force the message to have an even number of characters (counting blank spaces) so that the digraph method will work.

One important point that the last example illustrates is that cryptography will always entail a certain amount of (organized) guesswork.

Further Reading

Steve M. Cohen and Paul R. Coe, Card shuffling in discrete mathematics, *The College Mathematics Journal* 26(1995), 224–227.

Fred S. Roberts and Thomas A. Brown, Signed digraphs and the energy crisis, *The American Mathematical Monthly* 82(1975), 577–594.

William Staton and Clifton Wingard, Independent sets and the golden ratio, *The College Mathematics Journal* 26(1995), 292–296.

Exercises

1. Use the shift encryption system given by $P \mapsto P - 3$ to encrypt the message

BYE BYE, BIRDIE .

2. Use the shift *decription* scheme $P \mapsto P - 12$ to decrypt the code EAXAZSNMNK .

3. Use a frequency analysis on the ciphertext ZRRGZRURER to determine the shift encryption scheme. Then decrypt the message.

4. Use the affine encryption system given by $P \mapsto 3P + 11$ to encrypt the message

HELLO MY HONEY.

5. Use the affine *decryption* scheme $P \mapsto [P - 3]/7$ to decrypt the code RDQYPHZY-DQYP .

6. Use a frequency analysis on the ciphertext VQNXZAVQDURLX to determine the affine encryption scheme. Then decrypt the message. Discuss this problem in class.

7. Break the message

THIS WAS NOT THE END

up into two-character digraphs. Now tranlate each digraph into a pair of numbers, and then encrypt each digraph according to the rule $P \mapsto 13P + 29$. Now translate back to a new encrypted word expressed with roman characters.

8. Use the digraph technique and the affine transformation $P \mapsto 11P - 5$ to encrypt the message

NO GOOD WILL COME OF IT.

Your answer should be a string of roman characters.

9. It is known that the ciphertext PZCAILRNSXVC was obtained from a certain message with the digraph method using the affine transformation $P \mapsto 9P + 3$. Find the original message.

10. A certain message is broken up into digraphs and converted to a sequence of numerical expressions in the standard fashion described in the text. Then it is encrypted with an affine transformation. The resulting ciphertext is GGANFTNXCQNDSKQC. Use a frequency analysis to discover the affine transformation and then decipher the text to a standard English sentence. Discuss this problem in class.

11. Consider the message

NOW IS THE TIME FOR FUN.

Transliterate this to a list of numerals, one character at a time, in the usual way. Now apply the encryption algorithm

$$P \mapsto 5P^2 + P \bmod 26\,.$$

What ciphertext results?

12. Project: The ciphertext

AUACCEE

results from applying the encryption scheme

$$P \longmapsto 3P^2 - P + 2$$

to a certain 7-letter text. The trouble with this encryption scheme is that it is not one-to-one. It encrypts more than one letter in the same way. For example, both G and T get encrypted as A. In spite of this liability, determine what the original message was. Discuss the problem in class. Refer to [KOB] and [WAS] for further details.

13. Project: Produce a digraph encryption scheme that is not one-to-one. Use it to encrypt the message

One for the money.

What ciphertext results? What would be another plain English message that encrypts to the same ciphertext? If you had to decrypt the message, how would you distinguish these two originals? See [KOB] and [WAS].

Bibliography

[AAB] A. Aaboe, *Episodes from the Early History of Mathematics*, Mathematical Association of America, Washington, D.C., 1964.

[ALD] A. A. al'Daffa, *The Muslim Contribution to Mathematics*, London, 1978.

[APP] P. Appell, *Henri Poincaré*, Plon-Nourrit et cie, Paris, 1925.

[BAR] J. Barnes, *The Presocratic Philosophers*, Routledge and Kegan Paul, London, 1979.

[BEL] B. Belhoste, *Augustin-Louis Cauchy. A Biography*, Springer-Verlag, New York, 1991.

[BEM] P. H. Berg and J. L. McGregor, *Elementary Partial Differential Equations*, Holden-Day, San Francisco, 1964.

[BIE] G. Biedenkapp, *Sophie Germain, ein Weiblicher Denker*, Jena, 1910.

[BIB] G. D. Birkhoff and R. Beatley, *Basic Geometry*, Chelsea, New York, 1940.

[BLK] B. Blank and S. G. Krantz, *Calculus*, Key College Press, 2006.

[BOM] C. B. Boyer and U. Merzbach, *History of Mathematics*, John Wiley & Sons, New York, 1988.

[BSM] J. W. Brewer and M. K. Smith, *Emmy Noether: A Tribute to Her Life and Work*, M. Dekker, New York, 1981.

[BRE] D. Brewster, *Memoirs of the Life, Writings, and Discoveries of Sir Isaac Newton*, 1855, 2 volumes, T. Constable and Co., Edinburgh, reprinted 1965.

[BUH] W. K. Bühler, *Gauss: A Biographical Study*, Springer-Verlag, New York, 1981.

[CAS] R. Casse, *Projective Geometry : An Introduction*, Oxford University Press, Oxford, 2006.

[CGK] M. Cohen, E. D. Gaughan, A. Knoebel, D. S. Kurtz, and D. Pengelley, *Student Research Projects in Calculus*, Mathematical Association of America, Washington, D.C., 1992.

[DED] R. Dedekind, *Essays on the Theory of Numbers*, Dover Publications, New York, 1963.

[DOC] M. Do Carmo, *Differential Geometry of Curves and Surfaces*, Prentice-Hall, Englewood Cliffs, NJ, 1976.

[DAU] J. W. Dauben, Joseph Warren, *Georg Cantor: His Mathematics and Philosophy of the Infinite*, Princeton University Press, Princeton, 1990.

[DEV] C. J. de Vogel, *Pythagoras and Early Pythagoreanism*, Van Gorcum, Assen, 1966.

[DIC] A. Dick, *Emmy Noether*, Birkhäuser, Basel, 1970.

[DIJ] E. J. Dijksterhuis, *Archimedes*, Princeton University Press, Princeton, NJ, 1987.

[DUA] F. J. Duarte, *Bibliografia: Euclides, Arquimedes, Newton*, Caracas, 1967.

[DUN] W. Dunham, *The Calculus Gallery: Masterpieces from Newton to Lebesgue*, Princeton University Press, Princeton, NJ, 2008.

[DUN] G. W. Dunnington, *Carl Friedrich Gauss: Titan of Science*, Mathematical Association of America, Washington, D.C., 2004.

[DZI] M. Dzielska, *Hypatia of Alexandria*, Harvard University Press, Cambridge, MA, 1995.

[STANPR] Paul Erdős, *The St. Andrews Mathematics History Project*, `www-history.mcs.st-andrews.ac.uk/Biographies/Erdos.html`.

[EUC] Euclid, *The Elements*, 2nd ed., Cambridge University Press, Cambridge, 1926.

[EXO] G. Exoo, A Euclidean Ramsey problem, *Disc. Comput. Geom.* 29(2003), 223–227.

[FAD] E. Fadell and A. G. Fadell, *Calculus*, Van Nostrand Reinhold Co., New York, 1970.

[FEH] H Fehr, *Intermédiare des Mathématiciens* 9(1902), 74.

[FIE] M. Fierz, *Girolamo Cardano, 1501–1576 : Physician, Natural Philosopher, Mathematician, Astrologer, and Interpreter of Dreams*, Birkhäuser, Boston, 1983.

[FRA] W. Franz, *Euklid aus der Sicht der Mathematischen und der Naturwissenschaftlichen Welt der Gegenwart. Rede beim Antritt des Rektorats*, Klostermann, Frankfurt, 1965.

[FRE] M. Freedman, The topology of four-dimensional manifolds, *J. Diff. Geom.* 17(1982), 357–453.

[FUE] R. Fueter, *Leonhard Euler*, Birkhäuser, Basel, 1979.

[FUL] W. Fulton, *Algebraic Curves; An Introduction to Algebraic Geometry*, Notes written with the collaboration of Richard Weiss, W. A. Benjamin, Reading, MA, 1969.

[GIL] C. C. Gillespie, *Dictionary of Scientific Biography*, Scribners, New York, 1981.

[GLE] J. Gleick, *Isaac Newton*, Pantheon Press, New York, 2003.

[GOL] M. Golasiński, On generalized "ham sandwich" theorems, *Archivum Mathematicum* 42(2006), 25–30.

[GRA] J. V. Grabiner, *The Origins of Cauchy's Rigorous Calculus*, Dover Publications, Mineola, NY, 2005.

[GRAH] R. L. Graham, *Rudiments of Ramsey Theory*, Conference Board of the Mathematical Sciences, American Mathematical Society, Providence, RI, 1981.

[WIKIGN] Graham's number, *Wikipedia* `en.wikipedia.org/wiki/Grahams_Number`.

[GOS] W. C. Graustein and W. F. Osgood, *Plane and Solid Analytic Geometry*, Macmillan, New York, 1953.

[TAG] B. Green and T. Tao, The primes contain arbitrarily long arithmetic progressions, *Annals of Math.* 167(2008), 481–547.

[GRE] M. J. Greenberg, *Euclidean and non-Euclidean Geometries*, 4th ed., W. H. Freeman, New York, 2008.

[GRH] M. J. Greenberg and J. R. Harper, *Algebraic Topology*, Benjamin/Cummings Publishers, Reading, MA, 1981.

[GRK] R. E. Greene and S. G. Krantz, *Function Theory of One Complex Variable*, 3rd ed., American Mathematical Society, Providence, RI, 2006.

[HAL] E. S. Haldane, *Descartes: His Life and Times*, American Scholar Publications, New York, 1966.

[HAW] G. H. Hardy and E. M. Wright, *An Introduction to the Theory of Numbers*, Oxford: the Clarendon Press, New York, 2000.

[HAR] W. Hart, *Plane Trigonometry*, D. C. Heath, Boston, 1933.

[HEA] T. L. Heath, *A History of Greek Mathematics* (2 Vols.), The Clarendon Press, Oxford, 1921.

[HEM] E. M. Hemmerling, *College Plane Geometry*, John Wiley & Sons, New York, 1958.

[HER] I. Herstein, *Abstract Algebra*, 3rd ed., J. Wiley and Sons, New York, 1999.

[HOD] A. Hodges, *Alan Turing : The Enigma*, Simon and Schuster, New York, 1983.

[HRJ] K. Hrbacek and T. Jech, *Introduction to Set Theory*, 3rd ed., Marcel Dekker, New York, 1999.

[KAP1] R. Kaplan, *The Nothing That Is: A Natural History of Zero*, Oxford University Press, Oxford, UK, 2000.

[KAP2] ——, *The Art of the Infinite: The Pleasures of Mathematics*, Oxford University Press, Oxford, UK, 2003.

[KAT] V. Katz, *A History of Mathematics: An Introduction*, Addison-Wesley, Reading, MA, 1998.

[KEI] H. J. Keisler, *Elementary Calculus; An Approach Using Infinitesimals*, Bogden & Quigley, Tarrytown-on-Hudson, NY, 1971.

[KEN] K. Kendig, *Sink or Float*, Mathematical Association of America, Washington, D.C., 2008.

[KEP] St. Andrews Web page on Johannes Kepler `www-groups.dcs.st-and.ac.uk/~history/Biographies/Kepler.html`.

[KNU] D. Knuth, *Surreal Numbers: How Two Ex-Students Turned on to Pure Mathematics and Found Total Happiness*, Addison-Wesley, Reading, MA, 1974.

[KON] S. Kobayashi and Nomizu, *Foundations of Differential Geometry*, Interscience Publishers, New York, 1963.

[KOBA] A. H. Koblitz, *A Convergence of Lives: Sofia Kovalevskaia: Scientist, Writer, Revolutionary*, Rutgers University Press, New Brunswick, NJ, 1993.

[KOB] N. Koblitz, *A Course in Number Theory and Cryptography*, Springer-Verlag, New York, 1987.

[KRA1] S. G. Krantz, *Real Analysis and Foundations*, 2nd ed., CRC Press, Boca Raton, Florida, 2005.

[KRA2] ——, *Calculus Demystified*, McGraw-Hill, New York, 2002.

[KRA3] ——, *Complex Analysis: The Geometric Viewpoint*, 2nd ed., Mathematical Association of America, Washington, D.C., 2004.

[KRA4] ——, *A Guide to Complex Variables*, Mathematical Association of America, Washington, D.C., 2008.

[KRA5] ——, *Techniques of Problem Solving*, American Mathematical Society, Providence, RI, 1996.

[KRA6] ——, *The Elements of Advanced Mathematics*, 2nd ed., CRC Press, Boca Raton, FL, 2002.

[KRP] S. G. Krantz and H. R. Parks, *The Implicit Function Theorem*, Birkhäuser, Boston, 2002.

[KUM] E. E. Kummer, *Peter Gustav Lejeune Dirichlet*, in L. Kronecker and L. Fuchs, *G. Lejeune Dirichlet's Werke*, Chelsea Publishing, Bronx, NY, 1969.

[LAE] Diogenes Laertius, *The Lives, Teachings, and Sayings of Famous Philosophers*, George Bell & Sons, London, 1895.

[LAU] D. Laugwitz, *Bernhard Riemann 1826–1866*, Birkhäuser, Boston, 1999.

[MAH] M. S. Mahoney, *The Mathematical Career of Pierre de Fermat (1601–1665)*, Princeton University Press, Princeton, 1994.

[MAO] E. Maor, *The Pythagorean Theorem : A 4,000-Year History*, Princeton University Press, Princeton, NJ, 2007.

[MOS] D. C. Marshall, E. Odell, and M. Starbird, *Number Theory through Inquiry*, Mathematical Association of America, Washington, D.C., 2007.

[MTW] C. W. Misner, K. S. Thorne, and J. A. Wheeler, *Gravity*, Freeman, San Francisco, 1973.

[MOR] M. E. Mortenson, *Geometric Transformations*, Industrial Press, New York, 1995.

[NAS] S. Nasar and D. Gruber, Manifold destiny, *The New Yorker*, August 21, 2006.

[NIV] I. Niven, *Numbers: Rational and Irrational*, Random House, New York, 1961.

[OME] D. J. O'Meara, *Pythagoras Revived : Mathematics and Philosophy in Late Antiquity*, Cover image, reviews, etc. Clarendon Press ; New York : Oxford University Press, 1989.

[ONE] B. O'Neill, *Elementary Differential Geometry*, Academic Press, New York, 1966.

[ORE1] O. Ore, *Niels Henrik Abel, Mathematician Extraordinary*, Chelsea Publishing, New York, 1974.

[ORE2] ——, Biography of Dirichlet in *Dictionary of Scientific Biography*, C. C. Gilliespie, ed., Scribner, New York, 1970–1990.

[PER1] G. Perelman, The entropy formula for the Ricci flow and its geometric applications, `arXiv:math.DG/0211159v1`.

[PER2] ——, Ricci flow with surgery on three-manifolds, `arXiv:math.DG/0303109v1`.

[PER3] ——, Finite extinction time for the solutions to the Ricci flow on certain three-manifolds, `arXiv:math.DG/0307245v1`.

[POL] G. Pólya, *How to Solve It*, Princeton University Press, Princeton, 1988.

[POR] Porphyry, *Vita Pythagorae*, Amstelodani, 1707.

[PRA] V. V. Prasolov, *Polynomials*, translated from the Russian by Dimitry Leites, Springer, Berlin, 2004.

[PUR] W. Purkert and H. J. Ilgauds, *Georg Cantor 1845–1918*, Birkhäuser, Basel, 1987.

[RAM] F. P. Ramsey, On a problem of formal logic, *Proc. London Math. Soc.* 30(1930), 264-286.

[RAS] R. Rashed, *The Development of Arabic Mathematics: Between Arithmetic and Algebra*, London, 1994.

[RIG] L. Toti Rigatelli, *Evariste Galois (1811–1832)*, Birkhäuser, Boston, 1996.

[ROS] F. Rosen (trs.), *Muhammad ibn Musa Al-Khwarizmi : Algebra*, London, 1831.

[ROZ] B. A. Rozenfel'd and A. P. Yushkevich, *Omar Khayyam* (Russian), Akademija Nauk SSSR Izdat. Nauka, Moscow, 1965.

[RUD] W. Rudin, *Principles of Mathematical Analysis*, 3rd ed., McGraw-Hill, New York, 1976.

[RUS] B. Russell, *A History of Western Philosophy*, Simon and Schuster, New York, 1972.

[STE] H. Selin, ed., *Mathematics Across Cultures*, Kluwer Academic Publishers, Dordrecht, 2000.

[SIK] G. F. Simmons and S. G. Krantz, *Differential Equations: Theory, Technique, and Practice*, McGraw-Hill, New York, 2006.

[SMA1] S. Smale, Generalized Poincaré's conjecture in dimensions greater than four, *Ann. of Math.* 74(1961), 391–406.

[SMA2] ——, Mathematical problems for the next century, *Mathematics: Frontiers and Perspectives*, American Mathematical Society, Providence, RI, 2000, 271–294.

[SMI] D. E. Smith, *History of Mathematics*, Dover, New York, 1951.

[COSM] D. A. Smith and J. H. Conway, *On Quaternions and Octonions: Their Geometry, Arithmetic, and Symmetry*, AK Peters, Natick, MA, 2003.

[SOK] Z. K. Sokolovskaya, The 'pretelescopic' period of the history of astronomical instruments. al-Khwarizmi in the development of precision instruments in the Near and Middle East (Russian), in *The great medieval scientist al-Khwarizmi* (Tashkent, 1985), 165-178.

[STE] S. Stein, *Archimedes : What Did He Do Besides Cry Eureka?*, Mathematical Association of America, Washington, D.C., 1999.

[STT] A. H. Stone and J. W. Tukey, Generalized "sandwich" theorems, *Duke Math. J.* 9(1942), 356–359.

[STR] K. Stromberg, *An Introduction to Classical Real Analysis*, Wadsworth, Belmont, 1981.

[SUP] P. Suppes, *Axiomatic Set Theory*, Van Nostrand, Princeton, 1972.

[TAB] S. Tabachnikov, *Geometry and Billiards*, American Mathematical Society, Providence, RI, 2005.

[THI] R. Thiele, *Leonhard Euler*, Teubner, Leipzig, 1982.

[TOU] E. Toulouse, *Henri Poincaré*, Paris, 1910.

[TUR] S. Turing, *Alan M. Turing*, W. Heffer, Cambridge, 1959.

[WIKIPR] Twin Primes, *Wikipedia*, `en.wikipedia.org/wiki/Twin_prime`.

[VAN] B. L. van der Waerden, *Modern Algebra*, in part based on lectures by E. Artin and E. Noether, translated by Fred Blum and John R. Schulenberger, Ungar, New York, 1970.

[VRO] J. R. Vrooman, *René Descartes: A Biography*, Putnam & Sons, New York, 1970.

[WAS] L. Washington, *Elliptic Curves : Number Theory and Cryptography*, Chapman & Hall/CRC, Boca Raton, FL, 2008.

[WEN] G. A. Wentworth, *Wentworth's Plane Geometry*, rev. by George Wentworth and David Eugene Smith, Ginn and Company, Boston, 1910.

[WES] R. S. Westfall, *The Life of Isaac Newton*, Cambridge University Press, Cambridge, 1993.

[WID] D. V. Widder, *The Heat Equation*, Academic Press, New York, 1975.

[WIL] A. Wiles, Modular elliptic curves and Fermat's last theorem, *Annals of Mathematics* 141(1995), 443–551.

[YEO] A. Yeo, *Trig or Treat : An Encyclopedia of Trigonometric Identity Proofs (TIPs), Intellectually Challenging Games*, World Scientific Publishing, Singapore, 2007.

Index

About the Author

Steven G. Krantz was born in San Francisco, California in 1951. He received the B.A. degree from the University of California at Santa Cruz in 1971 and the Ph.D. from Princeton University in 1974.

Krantz has taught at UCLA, Penn State, Princeton University, and Washington University in St. Louis. He served as Chair of the latter department for five years.

Krantz has published more than 50 books and more than 150 scholarly papers. He is the recipient of the Chauvenet Prize and the Beckenbach Book Award of the MAA. He has received the UCLA Alumni Foundation Distinguished Teaching Award and the Kemper Award. He has directed 17 Ph.D. theses and 9 Masters theses.

VERMONT STATE COLLEGES
0 0003 0849559 7